Advances in Nutritional Manipulation of Rumen Fermentation

Advances in Nutritional Manipulation of Rumen Fermentation

Guest Editor

Lizhi Wang

MDPI

Basel • Beijing • Wuhan • Barcelona • Belgrade • Novi Sad • Cluj • Manchester

Guest Editor
Lizhi Wang
Animal Nutrition Institute
Sichuan Agricultural
University
Chengdu
China

Editorial Office
MDPI AG
Grosspeteranlage 5
4052 Basel, Switzerland

This is a reprint of the Special Issue, published open access by the journal *Animals* (ISSN 2076-2615), freely accessible at: www.mdpi.com/journal/animals/special_issues/S707M7BR9X.

For citation purposes, cite each article independently as indicated on the article page online and using the guide below:

Lastname, A.A.; Lastname, B.B. Article Title. *Journal Name* **Year**, *Volume Number*, Page Range.

ISBN 978-3-7258-3914-8 (Hbk)
ISBN 978-3-7258-3913-1 (PDF)
https://doi.org/10.3390/books978-3-7258-3913-1

Contents

About the Editor

Lizhi Wang

Lizhi Wang (professor and doctoral supervisor) is affiliated with the Institute of Animal Nutrition at Sichuan Agricultural University, China, where he serves as the Director of the Ruminant Nutrition Research Laboratory. He is recognized as an Academic and Technical Leader in Sichuan Province, China, and a Key Expert of the Sichuan Cattle Innovation Team. His primary research focus is on Ruminant Nutrition and Feed Science. In recent years, his work has primarily centered on regulating the structure and composition of the gastrointestinal microbiota in ruminants through nutritional strategies, aiming to enhance their disease resistance, maintain their health status, promote the efficient utilization of nutrients, and reduce methane emissions. To date, he has led a project that received the Second Prize of the Sichuan Provincial Science and Technology Progress Award (winning once) and co-led projects that received the Second Prize of the Sichuan Provincial Science and Technology Progress Award (winning thrice), in addition to completing over 10 provincial- and ministerial-level scientific research projects. He has published more than 80 research papers (including 45 SCI-indexed papers), obtained 6 Chinese national patents, participated in the compilation of the Sichuan Province Dairy Cattle Feeding Standards, and authored 10 monographs on ruminant livestock production.

animals

Editorial

Advances in Nutritional Manipulation of Rumen Fermentation

Lizhi Wang

Institute of Animal Nutrition, Sichuan Agricultural University, Chengdu 611130, China; wanglizhi08@aliyun.com

1. Introduction

Rumen, a complex and dynamic ecosystem, plays a pivotal role in the digestion and nutrient utilization of ruminant animals. The complex microbial communities and fermentation processes in rumen have a profound impact on feed utilization efficiency, animal performance, and environmental sustainability. In this regard, this Special Issue, 'Advances in Nutritional Manipulation of Rumen Fermentation' (*Animals*, 2025), brings together the latest research in this field and provides an important scientific basis and practical guidance for optimizing ruminant production.

Rumen fermentation is a critical process in ruminant nutrition, as it determines the efficiency of feed utilization and the production of volatile fatty acids (VFAs), which are essential for energy metabolism. However, aberrant rumen fermentation can lead to various health issues, including rumen acidosis and rumen bloat, which can have severe consequences for animal welfare and productivity. Nutritional manipulation offers a promising means of addressing these challenges by regulating rumen fermentation, improving feed efficiency, and reducing the incidence of rumen disorders.

2. Methodological Breakthroughs

This Special Issue presents examples of how advanced approaches have deepened our understanding of rumen fermentation. The following two key approaches stand out.

2.1. In Vitro Fermentation Models

This Special Issue highlights the utility of in vitro systems for hypothesis testing without ethical or logistical constraints. Lukbun et al. (2024) utilized a dual-flow continuous culture system to simulate rumen conditions, demonstrating that cassava-based diets supplemented with Enterococcus gallinarum KKU-BC15 reduced cyanide toxicity by 98% while increasing propionate concentration by 8.97% [1]. This model not only accelerates the screening of feed additives, but also mitigates risks associated with in vivo trials, such as acute acidosis or cyanide poisoning. Such systems are pivotal for translating lab discoveries into practical farm solutions.

2.2. Application of Meta-Analysis Methods

Meta-analysis is now widely used in various fields. In the present Special Issue, the study by Yanza et al. (2024) presents the first systematic meta-analysis of the effects of different sources of saponin extracts in ruminants, revealing their potential role in ruminant nutrition and environmental management [2]. Their database collection of 26 articles, containing 66 in vivo studies, was analyzed using mixed models with SAS software, and their results show that saponin extracts can significantly affect production performance, rumen fermentation, nitrogen utilization, and blood metabolites in ruminants, but that the effects of saponin extracts varied among different sources and levels.

Received: 4 March 2025
Accepted: 11 March 2025
Published: 14 March 2025

Citation: Wang, L. Advances in Nutritional Manipulation of Rumen Fermentation. *Animals* **2025**, *15*, 833. https://doi.org/10.3390/ani15060833

3. Addressing Persistent Challenges in Ruminant Nutrition

3.1. Methane Mitigation

Rumen fermentation accounts for ~30% of anthropogenic methane emissions, a critical target for climate action. This Special Issue contains articles that explore the reduction in methane emissions using additives. Li et al. (2024) investigated the methane mitigation effects and the mechanisms of *Sargassum mcclurei* using in vitro rumen fermentation [3]. The study found that freeze-dried *S. mcclurei* at a 2% supplementation level significantly reduced methane emissions and enhanced crude protein degradability. This research highlights the potential use of *S. mcclurei* as a feed additive to reduce methane emissions and improve rumen fermentation.

3.2. Protein and Energy Levels

Optimizing the ratio of energy to protein in diets can effectively improve feed costs and increase feed utilization efficiency. Jo et al. (2024) explored the effects of different incubation temperatures on ruminal fermentation and rumen microorganisms, aiming to determine the appropriate protein and energy levels for the enhancement of microbial protein synthesis [4]. The study found that higher incubation temperatures increased NH3-N and total volatile fatty acids (TVFAs), but decreased liquid-associated bacterial (LAB) amounts. The findings suggest that dietary adjustments can significantly impact rumen fermentation and microbial activity, providing insights into metabolic adaptation under different ruminal temperatures.

3.3. Development and Utilization of Feed Resources

The rational development and utilization of feed resources is an effective measure to effectively reduce feed costs and improve economic efficiency. The study by Lukbun et al. (2024) investigated the effects that adding cyanide-utilizing bacteria (CUB) to cassava diets had on rumen fermentation, gas production, and cyanide concentrations [1]. The results show that supplementing cassava diets with CUB, particularly at high cyanide levels, significantly improved cyanide degradation, gas production, and in vitro digestibility. This study highlights the potential for CUB to mitigate cyanide toxicity and enhance feed quality, which is particularly relevant for regions where cassava is a primary feed source.

3.4. Innovations and Applications of Nutritional Strategies

The research in the Special Issue also explores the practical effects of a variety of nutritional regulation strategies. For example, the addition of specific feed additives (e.g., saponins, active dry yeast, and plant extracts) significantly improve rumen fermentation patterns, reduce methane emissions, and lower the risk of rumen acidosis and bloat. Liu et al. (2024) assessed the impact of active dry yeast (ADY) on nutrient digestibility and rumen fermentation using both in vitro and in vivo experiments [5]. Their results indicate that ADY supplementation improves nutrient digestibility and rumen fermentation, with the specific effects varying depending on the type of ADY used. This study underscores the potential for ADY to be used as a feed additive to enhance lamb growth and optimize rumen fermentation.

4. Future Research Directions

The studies in this Special Issue lay the foundation for future research in the field of rumen fermentation. For instance, the potential for cyanide-utilizing bacteria (CUB) to mitigate cyanide toxicity in cassava diets warrants further investigation to explore their applications in different ruminant species and feeding conditions. Similarly, the effects of different yeast strains and supplementation levels on rumen fermentation and animal performance need to be explored in more depth to optimize their use in practical feeding strategies.

The findings in regard to methane mitigation using *Sargassum mcclurei* open up new avenues for research into the use of seaweed and other natural products as feed additives to reduce greenhouse gas emissions from ruminants. Additionally, attempts to explore the use of dietary adjustments to enhance microbial protein synthesis and improve feed efficiency under different environmental conditions provides a basis for developing tailored nutritional strategies for ruminants.

5. Conclusions

In summary, this Special Issue, 'Advances in Nutritional Manipulation of Rumen Fermentation', presents a comprehensive collection of research articles that explore the latest advancements in the field of ruminant nutrition. The studies highlight the potential for various nutritional strategies to optimize rumen fermentation, enhance feed efficiency, and address common rumen disorders. The findings provide valuable insights for researchers, farmers, and industry stakeholders, offering practical solutions to improve animal productivity and sustainability.

By showcasing methodological breakthroughs and addressing persistent challenges, this Special Issue contributes significantly to the current academic discourse on rumen fermentation. The research presented here not only advances our understanding of the complex interactions between nutrition and rumen microbiota, but also provides a roadmap for future research and applications in the field of ruminant nutrition.

Conflicts of Interest: The author declares no conflicts of interest.

References

1. Lukbun, S.; Suntara, C.; Prachumchai, R.; Khota, W.; Cherdthong, A. Exploring the In Vitro Effects of Cassava Diets and Enterococcus Strains on Rumen Fermentation, Gas Production, and Cyanide Concentrations. *Animals* **2024**, *14*, 3269. [CrossRef] [PubMed]
2. Yanza, Y.R.; Irawan, A.; Jayanegara, A.; Ramadhani, F.; Respati, A.N.; Fitri, A.; Hidayat, C.; Niderkorn, V.; Cieslak, A.; Szumacher-Strabel, M.; et al. Saponin Extracts Utilization as Dietary Additive in Ruminant Nutrition: A Meta-Analysis of In Vivo Studies. *Animals* **2024**, *14*, 1231. [CrossRef] [PubMed]
3. Li, S.; Sun, Y.; Guo, T.; Liu, W.; Tong, X.; Zhang, Z.; Sun, J.; Yang, Y.; Yang, S.; Li, D.; et al. *Sargassum mcclurei* Mitigating Methane Emissions and Affecting Rumen Microbial Community in In Vitro Rumen Fermentation. *Animals* **2024**, *14*, 2057. [CrossRef] [PubMed]
4. Jo, Y.H.; Kim, W.S.; Kim, Y.R.; Ju, M.S.; Nejad, J.G.; Lee, H.G. Impacts of Protein and Energy Levels on Rumen Fermentation and Microbial Activity Under Different Incubation Temperatures. *Animals* **2024**, *14*, 3093. [CrossRef] [PubMed]
5. Liu, H.; Li, F.; Ma, Z.; Ma, M.; Ungerfeld, E.; Zhang, Z.; Weng, X.; Liu, B.; Deng, X.; Guo, L. Effects of Active Dry Yeast Supplementation in In Vitro and In Vivo Nutrient Digestibility, Rumen Fermentation, and Bacterial Community. *Animals* **2024**, *14*, 2916. [CrossRef] [PubMed]

animals

MDPI

Article

Effect of Adding Yeast Cultures to High-Grain Conditions on Production Performance, Rumen Fermentation Profile, Microbial Abundance, and Immunity in Goats

Pei Qi and Lizhi Wang *

Institute of Animal Nutrition, Sichuan Agricultural University, Chengdu 611130, China; qipei@stu.sicau.edu.cn
* Correspondence: wanglizhi08@aliyun.com

Simple Summary: In China, high-grain diets for ruminants due to forage scarcity cause metabolic disorders. Yeast culture promotes rumen health by stabilizing microbial composition and enhancing nutrition and immunity. Therefore, the objective of this experiment was mainly to investigate the effect of the addition of two different species of yeast culture on goat production performance, rumen fermentation profile, microbial balance, and immunity under high-grain conditions. The results indicated that the incorporation of yeast culture into a high-grain diet significantly enhanced the production performance of goats, augmented their immunity, and stabilized the rumen environment. However, different types of yeast cultures acted by different mechanisms. The results provide a reference for the rational application of yeast culture in production.

Abstract: It is a common practice among farmers to utilize high-grain diets with the intention of promoting ruminant growth. However, this approach bears the risk of inducing rumen disorders and nutrient metabolism diseases. Yeast culture (YC) showed advantages in ruminant applications. The objective of this study was to evaluate the effects of adding two different types of YC to high-grain conditions on production performance, rumen fermentation profile, microbial abundance, and immunity in goats. A total of 30 male goats with similar body condition were randomly distributed into 3 dietary treatments with 10 replicates per treatment as follows: basic diet group (CON); basic diet + 0.5% yeast culture 1 (YC1) group; basic diet + 0.5% yeast culture 2 (YC2) group. The trial lasted for 36 days. The results demonstrated that dietary YC supplementation led to an increase in the average daily gain and a reduction in feed intake and weight gain ratio in goats. It increased the apparent digestibility of crude protein, NDF, and ADF ($p < 0.05$). The serum concentrations of interleukin (IL)-1β, IL-6, and Tumor Necrosis Factor-α in the control group were significantly higher than those of the YC groups ($p < 0.05$). The serum concentrations of Immunoglobulin (Ig)A and IgG in the control group were significantly lower than those in the YC groups ($p < 0.05$). The rumen concentration of microbial protein (MCP) in the control group was significantly lower than that in the YC groups ($p < 0.05$). There was a negative correlation between the concentration of IL-10 and *Bacteroidota*, *Spirochaetota*, and *Succinivibrio*, while there was a positive correlation between concentrations of IL-10 and Firmicutes. Nevertheless, discrepancies were observed in the impact of the two different types of YC on the physiological and biochemical indicators of the animals. The concentration of triglyceride in the YC1 group was significantly higher than that of the CON and YC2 groups, while the concentration of urea in the YC2 group was significantly higher than that of the CON and YC1 groups ($p < 0.05$). At the phylum level, the addition of YC2 to the diet significantly increased the relative abundance of Bacteroidota and Fibrobacterota and significantly decreased Firmicutes compared to the control. At the genus level, the addition of YC1 to the HGD significantly reduced the relative abundance of *Rikenellaceae_RC9_gut_group*, while the addition of YC2 to the HGD significantly increased the relative abundance of *Prevotellace-ae_UCG-001*, *Fibrobacter*, and *Prevotellaceae_UCG-003* ($p < 0.05$). The addition of YC significantly improved growth performance, increased nutrient digestibility, beneficially manipulated ruminal fermentation and microbial diversity, and improved immune function. The choice of yeast cultures can be customized according to specific production conditions.

Citation: Qi, P.; Wang, L. Effect of Adding Yeast Cultures to High-Grain Conditions on Production Performance, Rumen Fermentation Profile, Microbial Abundance, and Immunity in Goats. *Animals* **2024**, *14*, 1799. https://doi.org/10.3390/ani14121799

Academic Editor: Giuseppe Pulina

Received: 4 May 2024
Revised: 7 June 2024
Accepted: 11 June 2024
Published: 16 June 2024

Keywords: yeast culture; high-grain diet; immunity; ruminal fermentation; microbial abundance

1. Introduction

In China, due to the severe shortage of high-quality forage, it has become increasingly common to feed ruminants high-grain diets (HGD) to ensure high growth performance [1]. However, the concentration of harmful substances such as lipopolysaccharide (LPS) and histamine (His) in the rumen increases significantly when goats are fed HGD for prolonged periods. In addition, an imbalance in the rumen bacterial community may occur, leading to a significant reduction in the diversity of the rumen microflora [2–4]. The rumen is a unique digestive organ of ruminants, which contains many anaerobic microorganisms, including fungi, bacteria, and protozoa [5]. The growth of ruminants is dependent on the symbiotic relationship between the host and the rumen microorganisms, which provide the host with energy, protein, vitamins, and other essential nutrients [6]. The microbial community composition of the digestive tract in ruminants is influenced by various factors, including species, age, environment, and diet [7–11]. Of these factors, diet has the greatest impact on rumen microorganisms [12,13]. The composition of rumen microorganisms not only affects the digestion and absorption of the host's diet but also has a significant impact on the host's health and immune function. Therefore, maintaining rumen environment homeostasis, improving animal immunity, and reducing inflammatory reactions have become hot topics under the condition of HGD feeding. Yeast culture (YC), as a prebiotic micro-ecological product, consists of a small number of residual yeast cells, yeast metabolites, yeast cell wall fragments, and part of the culture medium [14]. YC contains a wide variety of biologically active substances, including proteins, small peptides, oligosaccharides such as β-glucan and mannan, vitamins, minerals, enzymes, and numerous 'unknown growth factors', which have beneficial nutritional and health effects on animals [15]. Probiotics in yeast cultures can regulate the structure of the rumen microbial community, inhibit the growth of lactic acid-producing bacteria such as Lactobacillus, reduce the accumulation of lactic acid, maintain the acid-base balance in the rumen, and reduce the adverse effects on animals [16,17]. The active ingredients in yeast cultures have been demonstrated to stimulate the animal's immune system, promote the production and secretion of antibodies, and enhance the animal's resistance to disease [18]. Although numerous studies have been conducted on the role of YC in ruminants, there are limited studies on the effects of yeast culture added into HGD [19–22]. Meanwhile, the effects of different types and processes of YC may vary significantly. Therefore, we hypothesized that YC may influence rumen fermentation patterns by affecting the abundance of major functional flora, therefore mitigating the adverse effects of prolonged feeding of high-grain diets. This experiment was conducted to evaluate the effects of YC on the production performance, ruminal fermentation characteristics, rumen bacterial populations, and immunity in goats fed with HGD.

2. Materials and Methods

The research protocol for this study was approved by the Animal Policy and Welfare Committee of the Agricultural Research Organization of Sichuan Province, Chengdu, China (approval code: SCAUAC201408-3), and it adheres to the guidelines of the Animal Care and Ethical Committee of Sichuan Agricultural University.

2.1. Experimental, Animals, Design, and Diet

The experiment took place at Sichuan Agricultural University's Animal Nutrition Institute's experimental farm in Ya'an, China. A total of 30 healthy fattening goats (6 months of age, male, vaccinated and dewormed, uncastrated, BW = 25.63 ± 0.36 kg) with similar body condition were randomly distributed into 3 dietary treatments (10 replicates/treatment) as follows: Basic diet group (CON); basic diet + 0.5% yeast culture 1 (YC1) group; basic diet + 0.5% yeast culture 2 (YC2) group. The animals were both vaccinated and dewormed. The YC1 was supplied by Bio-Nutrition International Company (Madison, MI, USA), and the YC2 was supplied by Vitech Ultra Bioscience Corporation (Orange, CA, USA). The main components of YC1 include mannan, dextran, proteins, peptides, amino acids, organic acids, vitamins, mineral salts, nucleotides, active peptides, esters, and alcohols. The main components of YC2 include mannan-oligosaccharides, beta-glucan, surfactants, proteins, peptides, amino acids, organic acids, vitamins, mineral salts, and nucleotides. Following the guidelines of the Chinese Feeding Standard for Meat-producing Sheep and Goats [23], the experimental diet was designed, as detailed in Table 1. Goats received a Total Mixed Ration (TMR). Each goat was kept separately in metabolic crates that included feeders and water containers. The experimental diet was provided twice daily at approximately 9:00 a.m. and 5:00 p.m. for ad libitum intake. During the pre-feeding period, the quantity of feed and the quantity of leftovers were recorded daily for each test sheep. The quantity of the following day's feed was then adjusted according to the quantity of leftovers from the previous day, with the objective of ensuring that approximately 5% of the feed remained in the trough daily. Fresh water was continuously available. After an acclimation period of 10 days, the formal trial lasted 36 days, consisting of a 30-day feeding phase and a 6-day digestion phase.

Table 1. Ingredients and chemical composition of experimental diets (%, dry matter basis).

Item	Group		
	CON	**YC1**	**YC2**
Ingredients			
Corn	50.70	50.20	50.20
Soybean meal	10.00	10.00	10.00
DDGS	6.00	6.00	6.00
Alfalfa hay	20.00	20.00	20.00
Corn husk	10.00	10.00	10.00
$CaCO_3$	0.80	0.80	0.80
NaCl	0.40	0.40	0.40
Premix	0.55	0.55	0.55
$NaHCO_3$	1.00	1.00	1.00
MgO	0.50	0.50	0.50
Antioxidant	0.05	0.05	0.05
Yeast culture	0.00	0.50	0.50
Chemical composition			
Dry matter	89.05	89.05	89.05
Crude protein	14.20	14.20	14.21
Crude ash	6.44	6.44	6.44
Crude fat	4.31	4.31	4.31
Metabolizable energy, M Cal/kg	2.57	2.57	2.57
Neutral detergent fiber	48.05	48.05	48.05
Acid detergent fiber	12.12	12.12	12.12
Calcium	0.56	0.56	0.56
Phosphorus	0.38	0.38	0.38
Concentrate:roughage	80:20	80:20	80:20

The premix provided the following per kg of diet: vitamin A 2200 IU; vitamin D 250 IU; vitamin E 20 IU; Fe 40 mg; Cu 10 mg; Zn 30 mg; Mn 40 mg; I 0.8 mg; Se 0.2 mg; Co 0.11 mg.CON = control group; YC1 = yeast culture1 group; YC2 = yeast culture 2 group.

2.2. Product Performance

During the treatment period, the initial individual and final body weight (BW) at the beginning of the feeding experiment were recorded. A daily record of the amount of feed consumed was kept, and the average daily feed intake (ADFI), average daily gain (ADG), and feed-to-gain ratio (F/G) were calculated.

2.3. Sample Collection of Blood and Ruminal Fluid

Collection of feed samples: Appropriate samples were collected from the diets to be tested using the tetrad method, and the feed samples collected in each period were mixed and stored using a sealing bag for nutrient determination.

Collection of fecal samples: During the digestion test period, all fecal matter from each sheep was collected and weighed at approximately 09:00 and 17:00 to record the weight. Every day, the feces were mixed and weighed to take 10% of them. One part of them was added to sulfuric acid for nitrogen fixation. The method of addition was to add 10% sulfuric acid (10 mL per 100 g of fresh feces) to avoid the loss of ammonia nitrogen in the feces. The samples were preserved at −20 °C for the determination of fecal nitrogen. The other part was used for air-dried sample preparation and determination of nutrient composition. At the conclusion of the experiment, the freshly collected manure samples were mixed with sulfuric acid and then stored at −20 °C until the determination of manure nitrogen. The remaining manure samples from each day were combined to create air-dried samples for the analysis of dry matter (DM) and other nutrient contents.

Collection of rumen content and blood samples. Rumen content and blood samples were collected at 9:00 a.m. on the 30th day of the experiment, prior to the morning feed. The blood samples were collected from the jugular vein of the goat. The samples were maintained on ice until all had been collected, after which they were processed immediately in the laboratory. Centrifugation at 4000 rpm for 15 min was performed to obtain serum, after which the samples were frozen at −20 °C for the testing index. The rumen content was collected via oral sampling, with the oral collector being inserted into the oral cavity and then into the rumen. The initial rumen content sample was discarded to prevent contamination from reticulum fluid, salivary fluid, or bacteria on the animal's body surface. A subsequent sample of 300 mL of rumen content was then collected. The rumen content was divided into two portions, one of which was squeezed through four layers of gauze. The rumen fluid was collected, and the pH was tested immediately [24].

2.4. Laboratory Analyses

The samples were dried at 55 °C and then analyzed for dry matter content at 105 °C (DM, method 934.01), crude protein (CP, method 984.13), and ether extract (EE, method 973.1) in accordance with the protocols established by the Association of Official Analytical Chemists [25]. The levels of neutral detergent fiber (NDF) and acid detergent fiber (ADF) were quantified using the methodology described by Van Soest et al. [26]. The apparent nutrient digestibility was calculated using the following formula [27,28]:

$$\text{Apparent digestibility of nutrient (\%)} = (\text{ingested nutrient} - \text{excreted nutrient in feces})/(\text{ingested nutrient}) \times 100$$

Metaphosphoric acid (0.25 mL) was added to 1 mL rumen fluid and centrifuged at 15,000× *g* for 15 min, and gas chromatography (GC-MS, Agilent Technologies, Palo Alto, CA, USA) was used to detect acetic, propionic and butyric acid concentrations in the rumen fluid. Serum Immunoglobulin (Ig)G, IgA, IgM, Tumor Necrosis Factor-α(TNF-α), Interleukin (IL)-1β, IL-6, IL-10, and lipopolysaccharide (LPS) were detected by double antibody sandwich ELISA (ELISA kit: Meimian Biotechnology, Yancheng, Jiangsu, China).

The concentrations of glutamic pyruvic transaminase (ALT), glutamic oxaloacetic transaminase (AST), urea creatinine (CREA), alkaline phosphatase (ALP), lactate dehydrogenase (LDH), creatine kinase (CK), total protein (TP), albumin (ALB), total cholesterol (TC), triglyceride (TG), low-density lipoprotein cholesterol (LDL-C), high-density lipoprotein cholesterol (HDL-C) in serum were detected by automatic biochemical analyzer (Hitachi 3100, Hitachi Limited, Tokyo, Japan) using standard kit (Zhongsheng Beiqin Biotechnology Co., Ltd., Beijing, China).

2.5. DNA Extraction and Amplification

To extract DNA, 1 mL of rumen fluid was centrifuged at 12,000× *g* for 10 min at 4 °C. The DNA extraction was conducted using the TIANamp Bacteria DNA Kit (Tianjin, China) in accordance with the methodology proposed by Guo [29] and in accordance with the instructions provided by the manufacturer. The purity and concentration of the extracted DNA were assessed via gel electrophoresis. The quality of the extracted bacterial DNA was assessed via agarose electrophoresis and a NanoDrop 8000 spectrophotometer (Thermo Fisher Scientific, Melbourne, Australia). The high-quality DNA was then amplified with bacteria-specific primers targeting the V4 hypervariable region of the 16S rRNA gene. The forward primer was 5′-GTGCCAGCMGCCGCGGTAA-30′ (515F), and the reverse primer was 5′-GGACTACVSGGGTATCTAAT-3′ (806R) [30]. Each sample was assigned a unique 5-8-base error-correcting barcode on the 515F primer for multiplex sequencing. The amplicons were subsequently submitted to Novogene Technology Company (Beijing, China) for sequencing on the MiSeq Illumina Sequencing Platform in accordance with the protocols outlined by Caporaso [31].

2.6. Bioinformatics and Statistical Analysis

The reads acquired from Novogene were analyzed using the QIIME pipeline software (version 1.8.0) [32] in accordance with the previously described method. Sequences containing uncertain nucleotides, unmatched barcodes, or three consecutive nucleotides with Q values below 20 were discarded. The Uchime algorithm in QIIME was then employed to eliminate chimeric sequences using USEARCH V7.0 [33]. To reduce the impact of sequencing noise, a pre-clustering methodology was employed [34]. The sequences were then clustered into operational taxonomic units (OTUs) using the Uclust method at 97% similarity, with a representative sequence selected for each OUT [35]. The representative sequences were then aligned against the Greengenes database (http://greengenes.lbl.gov, accessed on 6 December 2023) and assigned taxonomy via the RDP Classifier [36]. Six alpha-diversity indices of the bacterial communities (Chao1, Dominance, goods_coverage, observed_features, pielou_e, and Shannon) were calculated. Beta diversity was visualized using principal coordinate analysis (PCoA), which is based on an unweighted UniFrac distance matrix [37]. The relative abundance of bacterial phyla and genera was depicted using histograms created with OriginPro software (version 9.0). Furthermore, a heatmap generated by the R software (version 3.4.2) was employed to illustrate the genera shared by all samples. Finally, R software was used to analyze the significant species differences between groups.

2.7. Correlation between Rumen Microbiota and Blood Parameters Variables

Nonparametric Spearman rank correlation coefficient analysis implemented in SPSS statistical software (Ver. 27.0 for Windows; SPSS Inc., Chicago, IL, USA) was used to analyze the relationship between blood parameters and the relative abundance of rumen bacteria in rumen fluid. The correlation matrix was displayed as a heatmap using the corrplot package in R (Corrplot: visualization of a correlation matrix, R package version 0.2-0, 2010) [38].

2.8. Statistical Analysis

All data are presented as the mean ± SE. Each index was analyzed with 10 replicates. Statistical analyses were carried out using the SPSS statistical software (Ver. 27.0 for Windows; SPSS Inc., Chicago, IL, USA). The Shapiro–Wilk test and Levene's test were performed to test the data for normality and homoscedasticity, respectively. The differences in the relative bacterial abundance among the three groups were analyzed by a nonparametric test, and other parameters were analyzed by one-way analysis of variance (ANOVA) followed by Duncan's multiple comparisons to determine significant differences among the treatments. Differences were considered significant at $p < 0.05$.

3. Results

3.1. Product Performance

The results of the growth performance of goats are shown in Table 2. Dietary YC supplementation increased the goat average daily gain ($p < 0.001$) and average daily feed intake ($p = 0.009$) and reduced the goat F/G ($p = 0.005$).

Table 2. Growth performance of different groups in goats.

	Group			p-Value
	CON	YC1	YC2	
Initial body weight, kg	25.75 ± 0.42	25.42 ± 0.40	25.72 ± 0.96	0.828
BW at 30 days, kg	30.99 ± 0.23	31.95 ± 0.39	32.20 ± 0.86	0.297
ADG, g/d	174.67 ± 6.82 ^{b}	213.33 ± 4.77 ^{a}	216.67 ± 9.33 ^{a}	<0.001
ADFI, kg/d	1.11 ± 0.02 ^{b}	1.15 ± 0.05 ^{a}	1.17 ± 0.06 ^{a}	0.009
F/G	6.35 ± 0.26 ^{a}	5.39 ± 0.21 ^{b}	5.41 ± 0.16 ^{b}	0.005

CON = control group; YC1 = yeast culture1 group; YC2 = yeast culture 2 group. All values were expressed as means ± standard error. ^{a,b} Means with different superscripts within the same column differ significantly ($p < 0.05$). BW = Body weight. ADG = Average daily gain. ADFI = Average daily feed intake. F/G = Feed intake and weight gain ratio.

3.2. The Digestibility of Nutrients

The digestibility of nutrients is shown in Table 3. The digestibility of DM, ADF, and NDF in the YC1 and YC2 groups was significantly higher than that in the CON group ($p < 0.05$). There was no significant difference between YC1 and YC2 groups. The digestibility of CP in the YC2 group was significantly higher than that in the CON group, and there was no significant difference between the YC1 and the other groups. The digestibility of EE in the YC1 group was significantly higher than that in the YC2 and CON groups ($p = 0.009$), and there was no significant difference between the YC2 and CON groups.

Table 3. Apparent total tract digestibility of nutrients (%) of different groups in goats.

Item	Group			p-Value
	CON	YC1	YC2	
Dry matter	68.07 ± 1.20 ^{b}	73.45 ± 1.47 ^{a}	72.54 ± 1.55 ^{a}	0.002
Crude protein	70.34 ± 2.02 ^{b}	72.78 ± 1.22 ^{ab}	74.22 ± 1.60 ^{a}	0.009
Acid detergent fiber	44.79 ± 1.91 ^{b}	52.20 ± 0.84 ^{a}	52.43 ± 1.63 ^{a}	0.012
Neutral detergent fiber	69.47 ± 1.76 ^{b}	74.78 ± 1.16 ^{a}	73.48 ± 0.60 ^{a}	0.016
Ether extract	71.43 ± 2.27 ^{b}	78.38 ± 1.42 ^{a}	72.58 ± 2.26 ^{b}	0.009

CON = control group; YC1 = yeast culture1 group; YC2 = yeast culture 2 group. All values were expressed as means ± standard error. ^{a,b} Means with different superscripts within the same column differ significantly ($p < 0.05$).

3.3. Blood Parameters

The blood parameters of the JV are shown in Table 4. The concentration of urea in the YC2 group was significantly higher than that of the CON and YC1 groups ($p = 0.005$). The concentration of triglyceride in the YC1 group was significantly higher than that of the CON and YC2 groups ($p = 0.044$). The concentration of high-density lipoprotein cholesterol in the YC2 group was significantly higher than that of the YC1 group ($p < 0.05$). The concentrations of IL-1β, IL-6, and TNF-α in the CON group were significantly higher than those of the YC1 and YC2 groups ($p < 0.001$). The concentration of IL-10 in the YC1 group was significantly higher than that observed in the control and YC2 groups ($p < 0.001$). However, no significant difference was observed in the concentration of IL-10 between the control and YC2 groups ($p > 0.05$). The concentrations of IgA and IgG in the CON group were significantly lower than those in the YC1 and YC2 groups ($p < 0.05$). The concentration of IgM in the CON group was significantly lower than that in the YC2 group ($p < 0.05$).

However, no significant difference was observed in the concentrations of ALT, AST, CREA, ALP, LDH, CK, TP, ALB, TC, LDL-C, LPS, HIS, MAP, and HSP-70 among groups ($p > 0.05$).

Table 4. Jugular vein blood parameters of different groups in goats.

Items	Group			p-Value
	CON	YC1	YC2	
Glutamic pyruvic transaminase, ALT, U/L	21.99 ± 1.63	19.93 ± 1.26	20.85 ± 1.63	0.634
Glutamic oxaloacetic transaminase, AST, U/L	66.47 ± 6.26	66.99 ± 6.24	68.43 ± 5.44	0.972
Urea, mmol/L	5.50 ± 0.40 ^{b}	6.11 ± 0.45 ^{b}	7.61 ± 0.29 ^{a}	0.005
Creatinine, CREA, umol/L	55.34 ± 5.34	60.64 ± 5.65	65.98 ± 3.10	0.314
Alkaline phosphatase, ALP, U/L	307.3 ± 87.41	512.10 ± 145.27	348.2 ± 107.02	0.428
Lactate dehydrogenase, LDH, U/L	278.05 ± 18.89	291.41 ± 22.47	296.88 ± 25.35	0.830
Creatine kinase, CK, U/L	109.89 ± 11.22	110.07 ± 14.77	110.67 ± 11.24	0.998
Total protein, TP, g/L	55.34 ± 3.72	54.13 ± 2.65	59.81 ± 2.63	0.393
Albumin, ALB, g/L	27.43 ± 1.72	26.12 ± 1.21	30.35 ± 1.52	0.143
Total cholesterol, TC, mmol/L	1.54 ± 0.12	1.45 ± 0.12	1.68 ± 0.14	0.466
Triglyceride, TG, mmol/L	0.25 ± 0.03 ^{b}	0..43 ± 0.07 ^{a}	0.28 ± 0.04 ^{b}	0.044
Low-density lipoprotein cholesterol, LDL-C, mmol/L	0.54 ± 0.05	0.57 ± 0.05	0.57 ± 0.05	0.904
High-density lipoprotein cholesterol, HDL-C, mmol/L	1.51 ± 0.10 ^{ab}	1.33 ± 0.11 ^{b}	1.72 ± 0.13 ^{a}	0.065
Lipopolysaccharide, LPS, ng/L	249.37 ± 5.89	241.71 ± 7.59	263.88 ± 11.30	0.197
Histamine, HIS, ng/mL	9.59 ± 0.26	9.94 ± 0.21	10.26 ± 0.38	0.286
Interleukin-1β, IL-1β, pg/mL	78.60 ± 10.46 ^{a}	36.05 ± 4.17 ^{b}	34.07 ± 15.96 ^{b}	<0.001
Interleukin-6, IL-6, pg/mL	92.84 ± 8.95 ^{a}	39.99 ± 4.53 ^{b}	37.69 ± 3.94 ^{b}	<0.001
Interleukin-10, IL-10, pg/mL	20.53 ± 1.94 ^{b}	41.66 ± 1.48 ^{a}	25.70 ± 2.87 ^{b}	<0.001
Tumor Necrosis Factor-α, TNF-α, pg/mL	157.76 ± 12.31 ^{a}	91.84 ± 5.81 ^{b}	97.19 ± 5.21 ^{b}	<0.001
Immunoglobulin M, IgM, μg/mL	1733.44 ± 73.71 ^{b}	1926.57 ± 87.40 ^{ab}	2052.36 ± 80.37 ^{a}	0.031
Immunoglobulin A, IgA, μg/mL	152.45 ± 21.18 ^{b}	206.60 ± 8.11 ^{a}	199.92 ± 11.17 ^{a}	0.028
Immunoglobulin G, IgG, mg/mL	5.24 ± 0.43 ^{b}	7.01 ± 0.23 ^{a}	7.52 ± 0.26 ^{a}	<0.001
Major acute phase protein, MAP, mg/L	112.33 ± 9.79	107.59 ± 5.50	97.78 ± 6.79	0.395
Heat shock protein 70, HP-70, pg/mL	375.41 ± 32.46	317.08 ± 24.80	337.57 ± 22.65	0.315

CON = control group; YC1 = yeast culture1 group; YC2 = yeast culture 2 group. All values were expressed as means ± standard error. ^{a,b} Means with different superscripts within the same column differ significantly ($p < 0.05$).

3.4. Rumen Fermentation Characteristics

The results of rumen fluid characteristics are presented in Table 5. The concentration of MCP in the CON group was significantly lower than that in the YC1 and YC2 groups ($p = 0.039$). There was no significant difference in rumen pH, the concentration of NH_3-N, and rumen volatile fatty acid among three different groups ($p > 0.05$).

Table 5. Rumen fermentation characteristics of different groups in goats.

Item	Group			p-Value
	CON	YC1	YC2	
pH	7.05 ± 0.08	7.08 ± 0.06	7.06 ± 0.07	0.655
TVFA, mmol/l	40.36 ± 5.17	39.63 ± 5.49	36.86 ± 3.41	0.871
Acetic acid, mmol/L	25.70 ± 3.54	25.18 ± 23.04	23.04 ± 2.36	0.820
Propionic acid, mmol/L	9.80 ± 1.88	11.25 ± 2.14	10.01 ± 1.08	0.824
Butyric acid, mmol/L	2.88 ± 0.54	1.60 ± 0.26	2.03 ± 0.50	0.136
Isobutyric acid, mmol/L	0.66 ± 0.10	0.55 ± 0.08	0.60 ± 0.07	0.715
Isovaleric acid, mmol/L	0.89 ± 0.11	0.72 ± 0.12	0.81 ± 0.06	0.542
Pentanoic acid, mmol/L	0.49 ± 0.07	0.36 ± 0.08	0.38 ± 0.04	0.306
Acetic acid/Propionic acid	2.94 ± 0.27	2.83 ± 0.60	2.34 ± 0.20	0.565
Microbial protein, μg/L	16.64 ± 2.24 ^{b}	28.13 ± 1.90 ^{a}	28.60 ± 5.37 ^{a}	0.039
NH_3-N, mg/dl	70.90 ± 13.10	67.09 ± 6.92	60.23 ± 1.66	0.656

CON = control group; YC1 = yeast culture1 group; YC2 = yeast culture 2 group. All values were expressed as means ± standard error. ^{a,b} Means with different superscripts within the same column differ significantly ($p < 0.05$).

3.5. Statistics of 16SrRNA Sequencing Results of Rumen Microorganisms

3.5.1. OTU Analysis

According to the OTUs results and research requirements of noise reduction, the common and unique OTUs among different groups were analyzed. The CON group shared 472 and 418 OTUs, with YC1 and YC2 groups, and the number of shared OTUs among the three groups was 995 (Figure 1).

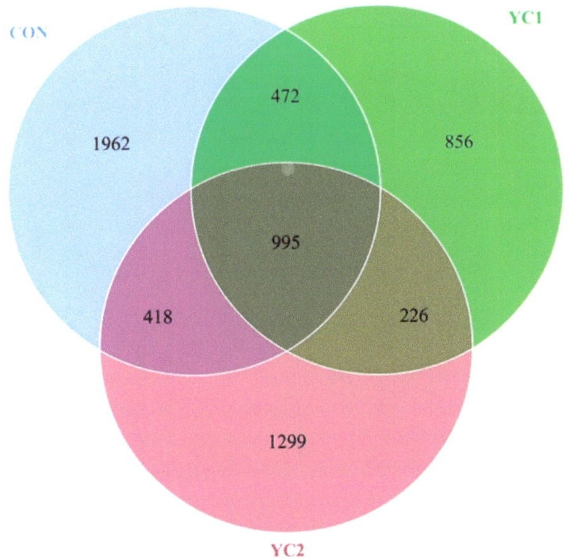

Figure 1. The Venn plot of OTU.

3.5.2. Alpha-Diversity Analysis

The alpha-diversity indexes estimation of the 16S rRNA gene libraries of the goat rumen that emerged from the sequencing analysis were presented in Table 6. In the current study, the results showed that the chao1 index and observed_otus index in the YC1 group were the lowest and significantly lower than that in the CON and YC2 groups ($p = 0.002$). The pielou_e index, Shannon, and Simpson indexes in the YC1 group were significantly lower than those in the CON group, and there was no significant difference between the YC2 and CON groups ($p > 0.05$).

Table 6. Differences in Alpha-diversity indices.

Item	Group			p-Value
	CON	YC1	YC2	
Chao1	802.11 ± 57.60 [a]	522.66 ± 44.54 [b]	719.69 ± 47.48 [a]	0.002
Dominance	0.04 ± 0.00 [b]	0.10 ± 0.02 [a]	0.06 ± 0.01 [ab]	0.051
goods_coverage	0.99 ± 0.00	0.99 ± 0.00	0.99 ± 0.00	0.087
observed_otus	761.6 ± 52.94 [a]	500.40 ± 43.23 [b]	677.11 ± 43.42 [a]	0.002
pielou_e	0.71 ± 0.03 [a]	0.61 ± 0.03 [b]	0.68 ± 0.03 [ab]	0.071
Shannon	6.81 ± 0.31 [a]	5.49 ± 0.34 [b]	6.39 ± 0.31 [ab]	0.020
Simpson	0.96 ± 0.01 [a]	0.90 ± 0.02 [b]	0.94 ± 0.01 [ab]	0.051

CON = control group; YC1 = yeast culture1 group; YC2 = yeast culture 2 group. All values were expressed as means ± standard error. [a,b] Means with different superscripts within the same column differ significantly ($p < 0.05$).

3.5.3. β Diversity Analysis

In the β diversity analysis, weighted UniFrac was selected to measure the difference coefficient between two samples, and PCoA was used to analyze the similarity of each sample. If the sample distance is closer, it indicates that the species composition structure is more similar. As shown in Figure 2, the sample dispersion of the CON group was the highest, while that of the YC2 group was the most concentrated.

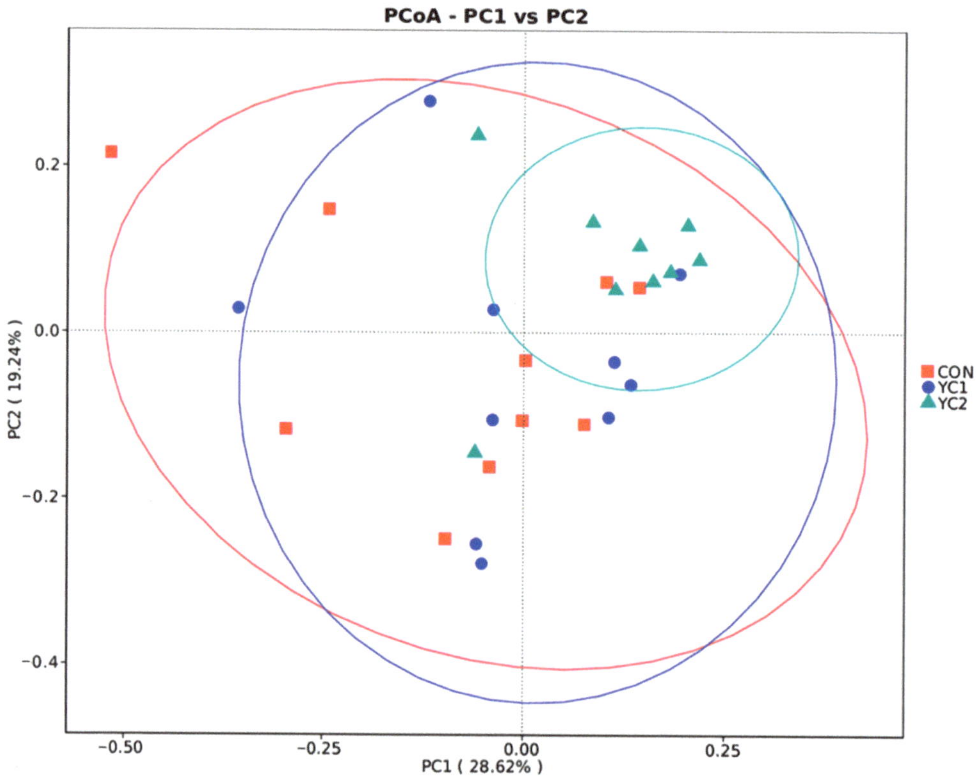

Figure 2. PCoA map of rumen microorganisms.

3.5.4. Microbiota Composition in Rumen

At the phylum level, 23 taxa were identified in the rumen. At the genus level, a total of 237 genera were detected in the rumen. Data of the top 10 microorganism populations were analyzed (Figure 3). It was found that a greater relative abundance of Bacteroidota ($p < 0.05$) and Fibrobacterota ($p < 0.05$) in YC2 was higher than in CON (Table 7). Then, the top 10 microorganism populations were analyzed at the genus level. It was found that the relative abundance of *Prevotellaceae_UCG-001* ($p = 0.006$), *Fibrobacter* ($p < 0.001$), and *Prevotellaceae_UCG-003* ($p = 0.004$) in YC2 was higher than in CON and YC1 (Table 7). In contrast, *Prevotellaceae_UCG-010* abundance was lower ($p = 0.018$) in YC1 and YC2 goats compared with the control. In addition, the result showed that *Rikenellaceae_RC9_gut_group* abundance was lower ($p = 0.012$) in YC1 goats compared with control and YC2.

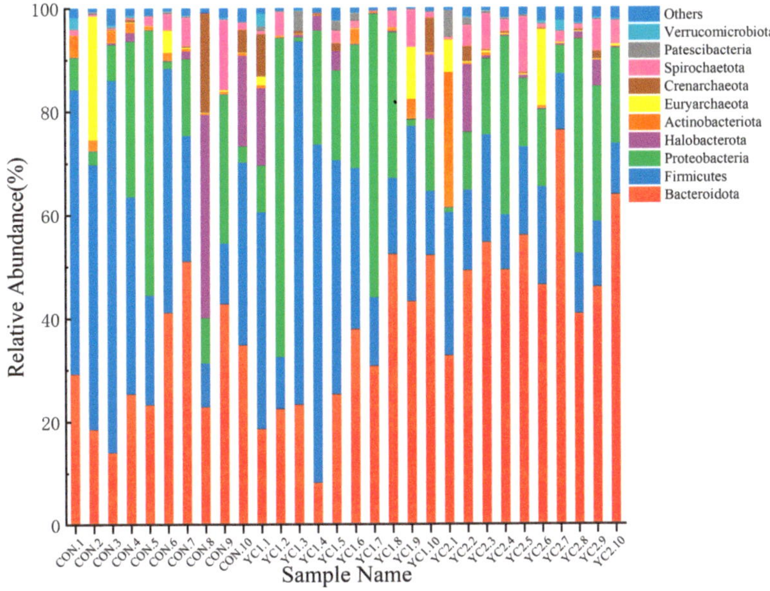

(**A**)

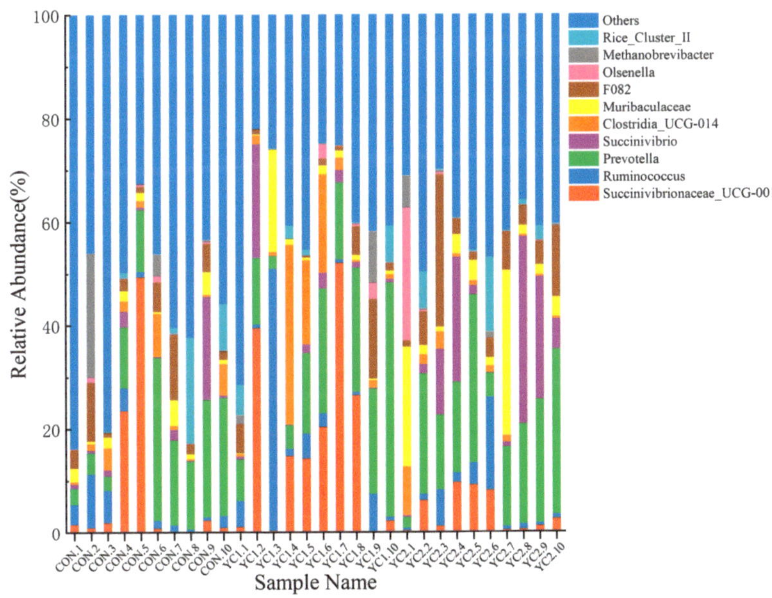

(**B**)

Figure 3. The relative abundance (%) of microbiota in the rumen at the phylum level (**A**) and genus level (**B**). Data represent the relative abundance of the top 10 of the community among the three groups.

Table 7. The differential microorganisms in the rumen at phylum and genus levels among groups.

Item	Group			*p*-Value
	CON	YC1	YC2	
Phylum				
Bacteroidota	30.27 ± 3.74 [b]	31.38 ± 4.64 [b]	52.03 ± 4.25 [a]	0.002
Firmicutes	36.44 ± 6.44 [a]	33.78 ± 6.94 [a]	15.11 ± 1.97 [b]	0.033
Fibrobacterota	0.24 ± 0.06 [b]	0.10 ± 0.03 [b]	0.91 ± 0.21 [a]	<0.001
genus				
Rikenellaceae_RC9_gut_group	3.13 ± 0.69 [a]	0.71 ± 0.23 [b]	4.17 ± 0.1.18 [a]	0.012
Prevotellaceae_UCG-001	0.96 ± 0.20 [b]	0.81 ± 0.19 [b]	2.19 ± 0.45 [a]	0.006
Prevotellaceae_UCG-003	0.41 ± 0.20 [b]	0.30 ± 0.07 [b]	1.26 ± 0.28 [a]	0.004
Fibrobacter	0.24 ± 0.06 [b]	0.10 ± 0.03 [b]	0.91 ± 0.21 [a]	<0.001
Prevotellaceae_UCG-010	0.45 ± 0.11 [a]	0.13 ± 0.05 [b]	0.22 ± 0.04 [b]	0.018

CON = control group; YC1 = yeast culture1 group; YC2 = yeast culture 2 group. All values were expressed as means ± standard error. [a,b] Means with different superscripts within the same column differ significantly ($p < 0.05$). Data represent the relative abundance at greater than 0.1% of the community among the three groups.

3.6. Correlation between Rumen Microbiota and Blood Parameters

The relationship between ruminal microbiota relative abundance (representing at least 0.1% of the bacterial community in at least one sample (in phyla and genus level)) and serum index was analyzed. The results showed (Figure 4) that the concentration of IgA was negatively correlated with Bacteroidota (R = −0.407, $p < 0.05$), *Prevotella* (R = −0.407, $p < 0.05$), *Succinivibrio* (R = −0.374, $p < 0.01$), and Muribaculaceae (R = −0.376, $p < 0.05$), while it was positively correlated with *Clostridia_UCG-014* (R = 0.421, $p < 0.05$). The concentration of IgG was negatively correlated with Bacteroidota (R = −0.406, $p < 0.05$), Verrucomicrobiota (R = −0.558, $p < 0.01$), *Succinivibrio* (R = −0.497, $p < 0.01$), and Muribaculaceae (R = −0.521, $p < 0.01$). The concentration of IgM was negatively correlated with Spirochaetota (R = −0.411, $p < 0.05$) and *Succinivibrio* (R = −0.430, $p < 0.05$). The concentration of IL-6 was negatively correlated with Actinobacteriota (R = −0.479, $p < 0.01$), while it was positively correlated with Bacteroidota (R = 0.458, $p < 0.05$) and Muribaculaceae (R = 0.373, $p < 0.05$). IL-10 was negatively correlated with the relative abundance of Bacteroidota (R = −0.526, $p < 0.01$), Spirochaetota (R = −0.430, $p < 0.05$), and *Succinivibrio* (R = −0.437, $p < 0.05$), while it was positively correlated with Firmicutes (R = 0.484, $p < 0.01$) and Actinobacteriota (R = 0.530, $p < 0.01$). The concentration of IL-1β was negatively correlated with Firmicutes (R = −0.456, $p < 0.05$), Actinobacteriota (R = −0.538, $p < 0.01$), and *ClostridiaUCG-014* (R = −0.478, $p < 0.01$), while it was positively correlated with Muribaculaceae (R = 0.453, $p < 0.05$). The concentration of TNF-α was positively correlated with Bacteroidota (R = 0.389, $p < 0.05$). The concentration of MAP was negatively correlated with Firmicutes (R = −0.392, $p < 0.05$). The concentration of HP-70 was negatively correlated with Firmicutes (R = −0.479, $p < 0.01$), Patescibacteria (R = −0.446, $p < 0.05$), and *Clostridia_UCG-014* (R = −0.517, $p < 0.01$), while it was positively correlated with Spirochaetota (R = 0.434, $p < 0.05$), Verrucomicrobiota (R = 0.377, $p < 0.05$). The concentration of LPS was negatively correlated with *Succinivibrionaceae_UCG-001* (R = −0.378, $p < 0.05$), while it was positively correlated with Euryarchaeota (R = 0.568, $p < 0.01$), *F082* (R = 0.450, $p < 0.05$), *Methanobrevibacter* (R = 0.568, $p < 0.01$). The concentration of His was positively correlated with *Prevotella* (R = 0.546, $p < 0.01$).

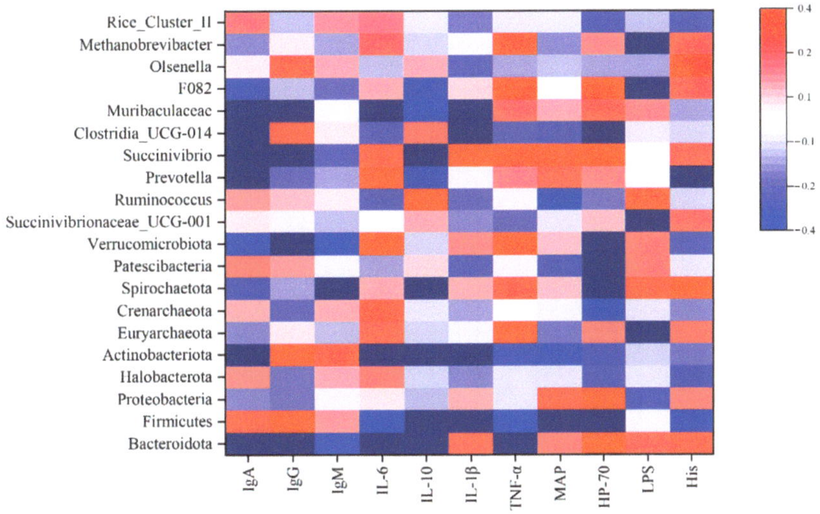

Figure 4. Spearman-parametric rank correlation matrix between blood parameters and microbiota relative abundance (representing at least 0.1% of the bacterial community in at least one sample). The blue color represents a negative correlation, the red color represents a positive correlation, and the white color represents no correlation. Spearman correlations between bacterial and biological parameters at corresponding were analyzed.

4. Discussion

The present results indicate that the inclusion of YC in high-grain diets led to a significant increase in the average daily weight gain of goats. Improvements in ADG have also been reported previously in goats supplemented with YC [39]. Additionally, it decreased F/G and increased the apparent digestibility of crude protein, NDF, and ADF. The relative abundance of rumen microorganisms in goats was changed. Therefore, it could be speculated that YC may enhance goat performance in two ways. First, it may be related to the effect of YC on the rumen internal environment. Second, the active components in YC may improve the body's immunity, enhancing the goat's production performance [40,41].

The pH is a reliable indicator of the anaerobic fermentation status of the rumen in ruminants. It is crucial to maintain the pH within the normal range to ensure normal rumen fermentation. Previous studies in cattle have shown that feeding high-grain diets significantly reduces rumen pH and induces rumen acidosis, and YC can decrease the occurrence of rumen acidosis by stabilizing the pH of the rumen through decreasing lactic acid-producing bacteria and increasing lactic acid-utilizing bacteria or rumen protozoa [42–46]. In this trial, it was expected that YC supplementation would prevent rumen acidosis in goats. Rumen pH values in all treatment groups in this trial were higher than 7. Previous studies have also found that rumen pH does not fall below 6 when goats are fed high-grain diets [47]. This phenomenon may be related to differences in rumen organization between cattle and goats. Acetic acid, propionic acid, and butyric acid account for more than 95% of the total volatile fatty acids produced by rumen fermentation [48]. The addition of YC did not significantly affect the content of total volatile fatty acids, acetic acid, propionic acid, and butyric acid in this experiment. The addition of YC to mid-lactation cows also did not have a significant effect on total volatile fatty acids [19,49]. This experiment showed that the addition of YC to the diet tended to decrease the ratio of acetic acid to propionic acid, which may reflect the type of rumen fermentation. Furthermore, the addition of YC to high-grain diets significantly increased MCP content. The results indicated that YC groups promoted rumen nitrogen metabolism, improved ammonia and nitrogen utilization, increased rumen microbial protein production, and facilitated protein deposition compared to the control group. Positive changes in the rumen environment can affect bacterial populations and their fermentation products, leading to improved productivity and nutrient utilization in ruminants [50,51].

Maintaining a micro-ecological balance in the gastrointestinal tract is crucial for the overall health of the organism. The dominant microflora plays a significant role in achieving this balance. An imbalance in the dominant gastrointestinal flora can lead to a dysfunctional microecosystem [52,53]. This study found that the YC groups had a positive effect on bacterial populations, mainly by increasing the relative abundance of Bacteroidota and Fibrobacterota. Supplementation of YC increased the number and activity of fiber-digesting bacteria, resulting in improved fiber digestibility. The relative abundance of *Prevotellaceae_UGG_001* and *Prevotellaceae_UGG_003* among rumen bacteria significantly affected fiber digestibility. Although the relative abundance of these two bacteria did not increase significantly in the YC1 group during this experiment, the fiber digestibility of the YC1 group still showed a significant increase. This may be because fiber degradation is not only caused by the action of one bacterial group alone but by the action of multiple bacterial groups working together [54]. To better understand the effect of YC on the immune performance of goats, the correlation between rumen microbiota and immune index was analyzed in this study. The results showed that there was a negative correlation between the concentration of IL-10 and Bacteroidota, Spirochaetota, and *Succinivibrio*, while there was a positive correlation between the concentration of IL-10 and Firmicutes. The rumen microbiota of ruminants is diverse, with some bacteria considered beneficial, such as *Prevotellaceae_UCG-001* and *Clostridia_UCG-014*. Conversely, other bacteria, such as Streptococcus, may produce harmful substances that cause health problems [55]. YC may improve immunity mainly by altering the composition of rumen microorganisms [56].

Ruminants produce ammonia in their rumen, which is then converted to urea in the liver. The concentration of urea in the blood is highly correlated with the nitrogen cycle and rumen protein degradation and can reflect protein metabolism in the animal [57]. This study found that the urea concentration in the YC2 group was significantly higher than in the CON and YC1 groups. The additional NH_3-N was used to synthesize bacterial proteins, increasing the efficiency of feed protein utilization by ruminants. Serum TG reflects lipid levels, and its content reflects lipid absorption and metabolism. An increase in serum TG concentration may be due to increased lipase activity and dietary fat utilization by yeast preparations [58]. The serum TG concentration in the YC1 group was significantly higher than that in the control and YC2 groups. YC has the potential to improve the production performance of ruminants by promoting lipid metabolism.

The nutritional modulation of yeast cultures was studied to improve the immunity of ruminants and achieve high production performance. Immunoglobulins are proteins produced by plasma cells that have antibody activity and exert antibacterial and antiviral effects. The IgG, IgM, and IgA are usually tested to reflect the body's immune status [59]. This experiment's results showed that the serum IgA and IgG concentrations of the YC1 and YC2 groups were significantly higher than those of the control group. Additionally, the serum IgM concentration of the YC2 group was significantly higher than that of the control group. The addition of yeast cultures to HGD can significantly enhance the immunity of goats by reducing the release of pro-inflammatory factors that impair the integrity of the gastrointestinal tract and the intestinal barrier function [60]. In periparturient cows, the addition of YC to the diet resulted in reduced serum levels of IL-6 and IL-8 and down-regulated the expression level of IL-6 mRNA in the uterus. This study found that the serum levels of IL-1β, IL-6, and TNF-α significantly decreased in the YC1 and YC2 groups. Additionally, the serum level of IL-10 significantly increased in the YC1 group. These results suggest that yeast culture may have anti-inflammatory effects and can reduce the level of inflammatory factors in serum. Wang et al. also found that yeast products have antioxidant, anti-inflammatory, and immunomodulatory effects in vitro [61]. Yeast cell wall polysaccharides, such as β-glucan, mannan, and B-complex vitamins, are the primary nutrient factors that produce these effects. It has been shown that biotin and B-complex vitamins alleviate rumen acidosis in ruminants and are effective in the prevention and treatment of hoof disease in dairy cows [62,63]. The immunomodulatory effects of yeast cultures may be more pronounced when animals are under pathogenic challenges and environmental stress.

Since the composition and mechanism of action of yeast culture products from different companies vary due to differences in production processes and yeast strains, their effects on animal physiological and biochemical indicators are also not consistent. The concentration of triglyceride in the YC1 group was significantly higher than that of the CON and YC2 groups, while the concentration of urea in the YC2 group was significantly higher than that of the CON and YC1 groups. The relative abundance of rumen microorganisms was also differently affected by different types of yeast cultures. At the phylum level, the addition of YC2 to the diet significantly increased the relative abundance of Bacteroidota and Fibrobacterota and significantly decreased Firmicutes compared to the control. At the genus level, the addition of YC1 to the HGD significantly reduced the relative abundance of *Rikenellaceae_RC9_gut_group*, while the addition of YC2 to the HGD significantly increased the relative abundance of *Prevotellaceae_UCG-001*, Fibrobacter, and *Prevotellaceae_UCG-003*. These three bacteria were highly correlated with fiber degradation, suggesting that YC2 may increase the digestibility of cellulose in goats.

5. Conclusions

The results observed in this study suggested that supplementation of the two commercial YC in goats' high-grain diet improved growth performance and improved immunity. The use of yeast cultures also positively influenced rumen fermentation and microbial diversity, particularly by increasing the relative abundance of beneficial bacteria such as Bacteroidota and Fibrobacterota. In the meantime, the microbiota indirectly affects immune function by regulating metabolites. Furthermore, the different types of yeast cultures exerted distinct effects on physiological and biochemical parameters. YC1 significantly increased serum triglyceride levels, while YC2 elevated urea concentrations, reflecting differences in how these cultures modulate lipid and protein metabolism. However, different types of yeast cultures acted by different mechanisms. The selection of a specific yeast culture should be tailored to the production goals and the desired physiological outcomes. Further research is necessary to investigate the long-term effects and underlying mechanisms of different yeast culture formulations in various ruminant species.

Author Contributions: P.Q. and L.W.: conceptualization, methodology; P.Q. and L.W.: writing—original draft preparation, reviewing, and editing; P.Q. and L.W.: investigation; P.Q. and L.W.: supervision. All authors have read and agreed to the published version of the manuscript.

Funding: This work was financially supported by the Science and Technology Department of Sichuan Province (grant number 2021YFH0154).

Institutional Review Board Statement: The research procedure used in the current study was approved by the Animal Policy and Welfare Committee of the Agricultural Research Organization of Sichuan Province, China (approval code: SCAUAC201408-3), and agrees with the rules of the Animal Care and Ethical Committee of the Sichuan Agricultural University.

Informed Consent Statement: Not applicable.

Data Availability Statement: Original data are available from the corresponding author upon reasonable request.

Acknowledgments: This work was funded by the projects of the international science and technology innovation cooperation program of Sichuan province: study on the mechanism of rumen flatulence induced by high concentrate diet in ruminants (2021YFH0154). We would like to thank all staff members of Ruminant Nutrition Laboratory, Sichuan Agricultural University. The authors gratefully acknowledge the technical support of Zehao Tan and Yidan Wang.

Conflicts of Interest: The authors declare no conflicts of interest.

References

1. Cavallini, D.; Mammi, L.M.E.; Palmonari, A.; García-González, R.; Chapman, J.D.; McLean, D.J.; Formigoni, A. Effect of an Immunomodulatory Feed Additive in Mitigating the Stress Responses in Lactating Dairy Cows to a High Concentrate Diet Challenge. *Animals* **2022**, *12*, 2129. [CrossRef] [PubMed]
2. Khafipour, E.; Krause, D.O.; Plaizier, J.C. A grain-based subacute ruminal acidosis challenge causes translocation of lipopolysaccharide and triggers inflammation. *J. Dairy Sci.* **2009**, *92*, 1060–1070. [CrossRef] [PubMed]
3. Tian, J. Effects of High Concentrate Diet on Rumen Fermentation and Epithelial Inflammation Related Gene Expression in Lactating Goats. Master's Thesis, Nanjing Agricultural University, Nanjing, China, 2018.
4. Lu, Z.; Shen, H.; Shen, Z. High-Concentrate Diet-Induced Change of Cellular Metabolism Leads to Decreases of Immunity and Imbalance of Cellular Activities in Rumen Epithelium. *Cell Physiol. Biochem.* **2018**, *45*, 2145–2157. [CrossRef] [PubMed]
5. Boyd, R.D.; Bauman, D.E.; Fox, D.G.; Scanes, C.G. Impact of metabolism modifiers on protein accretion and protein and energy requirements of livestock. *J. Anim. Sci.* **1991**, *69*, 56–75. [CrossRef]
6. Reuben, R.C.; Elghandour, M.M.M.Y.; Alqaisi, O.; Cone, J.W.; Márquez, O.; Salem, A.Z.M. Influence of microbial probiotics on ruminant health and nutrition: Sources, mode of action and implications. *J. Sci. Food Agric.* **2022**, *102*, 1319–1340. [CrossRef] [PubMed]
7. Wen, K.; Zhao, M.M.; Liu, L.; Khogali, M.K.; Geng, T.Y.; Wang, H.R.; Gong, D.Q. Thiamine modulates intestinal morphological structure and microbiota under subacute ruminal acidosis induced by a high-concentrate diet in Saanen goats. *Animal* **2021**, *15*, 100370. [CrossRef] [PubMed]
8. Li, F.; Shah, A.M.; Wang, Z.; Peng, Q.; Hu, R.; Zou, H.; Tan, C.; Zhang, X.; Liao, Y.; Wang, Y.; et al. Effects of Land Transport Stress on Variations in Ruminal Microbe Diversity and Immune Functions in Different Breeds of Cattle. *Animals* **2019**, *9*, 599. [CrossRef] [PubMed]
9. Li, L.P.; Peng, K.L.; Xue, M.Y.; Zhu, S.L.; Liu, J.X.; Sun, H.Z. An Age Effect of Rumen Microbiome in Dairy Buffaloes Revealed by Metagenomics. *Microorganisms* **2022**, *10*, 1491. [CrossRef] [PubMed]
10. Han, X.; Liu, H.; Hu, L.; Xu, S.; Ai, D.; Fu, H.; Chen, Y.; Zhao, J. Characteristics and sex differences of rumen microflora of yaks around the lake. *Acta Anim. Nutr.* **2020**, *32*, 234–243.
11. Wu, D.; Vinitchaikul, P.; Deng, M.; Zhang, G.; Sun, L.; Wang, H.; Gou, X.; Mao, H.; Yang, S. Exploration of the effects of altitude change on bacteria and fungi in the rumen of yak (*Bos grunniens*). *Arch. Microbiol.* **2021**, *203*, 835–846. [CrossRef]
12. Fliegerova, K.O.; Podmirseg, S.M.; Vinzelj, J.; Grilli, D.J.; Kvasnová, S.; Schierová, D.; Sechovcová, H.; Mrázek, J.; Siddi, G.; Arenas, G.N.; et al. The Effect of a High-Grain Diet on the Rumen Microbiome of Goats with a Special Focus on Anaerobic Fungi. *Microorganisms* **2021**, *9*, 157. [CrossRef] [PubMed]
13. Granja-Salcedo, Y.T.; Ribeiro Junior, C.S.; de Jesus, R.B.; Gomez-Insuasti, A.S.; Rivera, A.R.; Messana, J.D.; Canesin, R.C.; Berchielli, T.T. Effect of different levels of concentrate on ruminal microorganisms and rumen fermentation in Nellore steers. *Arch. Anim. Nutr.* **2016**, *70*, 17–32. [CrossRef] [PubMed]
14. Pang, Y.; Zhang, H.; Wen, H.; Wan, H.; Wu, H.; Chen, Y.; Li, S.; Zhang, L.; Sun, X.; Li, B.; et al. Yeast Probiotic and Yeast Products in Enhancing Livestock Feeds Utilization and Performance: An Overview. *J. Fungi* **2022**, *8*, 1191. [CrossRef] [PubMed] [PubMed Central]
15. Perricone, V.; Sandrini, S.; Irshad, N.; Savoini, G.; Comi, M.; Agazzi, A. Yeast-Derived Products: The Role of Hydrolyzed Yeast and Yeast Culture in Poultry Nutrition-A Review. *Animals* **2022**, *12*, 1426. [CrossRef] [PubMed] [PubMed Central]
16. Marden, J.P.; Julien, C.; Monteils, V.; Auclair, E.; Moncoulon, R.; Bayourthe, C. How does live yeast differ from sodium bicarbonate to stabilize ruminal pH in high-yielding dairy cows? *J. Dairy Sci.* **2008**, *91*, 3528–3535. [CrossRef] [PubMed]
17. Newbold, C.J.; Wallace, R.J.; Chen, X.B.; McIntosh, F.M. Different strains of Saccharomyces cerevisiae differ in their effects on ruminal bacterial numbers in vitro and in sheep. *J. Anim. Sci.* **1995**, *73*, 1811–1818. [CrossRef] [PubMed]
18. Lesmeister, K.E.; Heinrichs, A.J.; Gabler, M.T. Effects of supplemental yeast (Saccharomyces cerevisiae) culture on rumen development, growth characteristics, and blood parameters in neonatal dairy calves. *J. Dairy Sci.* **2004**, *87*, 1832–1839. [CrossRef] [PubMed]
19. Song, B.; Wu, T.; You, P.; Wang, H.; Burke, J.L.; Kang, K.; Yu, W.; Wang, M.; Li, B.; He, Y.; et al. Dietary Supplementation of Yeast Culture Into Pelleted Total Mixed Rations Improves the Growth Performance of Fattening Lambs. *Front. Vet. Sci.* **2021**, *8*, 657816. [CrossRef] [PubMed]
20. Garcia Diaz, T.; Ferriani Branco, A.; Jacovaci, F.A.; Cabreira Jobim, C.; Bolson, D.C.; Pratti Daniel, J.L. Inclusion of live yeast and mannan-oligosaccharides in high grain-based diets for sheep: Ruminal parameters, inflammatory response and rumen morphology. *PLoS ONE* **2018**, *13*, e0193313. [CrossRef]
21. Emu, Q.; Guan, H.; Zhu, J.; Zhang, L.; Fan, J.; Ji, Y.; Lin, Y.; Li, C.; Dan, X.; Aguo, Y.; et al. Grazing and Supplementation of Dietary Yeast Probiotics Shape the Gut Microbiota and Improve the Immunity of Black Fattening Goats (*Capra hircus*). *Front. Microbiol.* **2021**, *12*, 666837. [CrossRef]
22. Nocek, J.E.; Holt, M.G.; Oppy, J. Effects of supplementation with yeast culture and enzymatically hydrolyzed yeast on performance of early lactation dairy cattle. *J. Dairy Sci.* **2011**, *94*, 4046–4056. [CrossRef] [PubMed]
23. *NY/T816-2004*; Feed Standard of Meat-Producing Sheep and Goats. Industry Standards-Agriculture: Beijing, China, 2004.
24. Wang, Y.; Li, Q.; Wang, L.; Liu, Y.; Yan, T. Effects of a High-Concentrate Diet on the Blood Parameters and Liver Transcriptome of Goats. *Animals* **2023**, *13*, 1559. [CrossRef] [PubMed]

25. AOAC. *Official Methods of Analysis of the Association of Official's Analytical Chemists*; Association of Official Analytical Chemists: Arlington, VA, USA, 2003.
26. Van Soest, P.V.; Robertson, J.B.; Lewis, B.A. Methods for dietary fiber, neutral detergent fiber, and nonstarch polysaccharides in relation to animal nutrition. *J. Dairy Sci.* **1991**, *74*, 3583–3597. [CrossRef] [PubMed]
27. Adeola, O. Digestion and balance techniques in pigs. In *Swine Nutrition*, 2nd ed.; Lewis, A.J., Sothern, L.L., Eds.; CRC Press: Boca Raton, FL, USA, 2000; 906p.
28. Jang, K.B.; Kim, S.W. Evaluation of standardized ileal digestibility of amino acids in fermented soybean meal for nursery pigs using direct and difference procedures. *Anim. Biosci.* **2023**, *36*, 275–283. [CrossRef] [PubMed]
29. Guo, W.; Li, Y.; Wang, L.; Wang, J.; Xu, Q.; Yan, T.; Xue, B. Evaluation of composition and individual variability of rumen microbiota in yaks by 16S rRNA high-throughput sequencing technology. *Anaerobe* **2015**, *34*, 74–79. [CrossRef] [PubMed]
30. Caporaso, J.G.; Lauber, C.L.; Walters, W.A.; Berg-Lyons, D.; Lozupone, C.A.; Turnbaugh, P.J.; Fierer, N.; Knight, R. Global patterns of 16S rRNA diversity at a depth of millions of sequences per sample. *Proc. Natl. Acad. Sci. USA* **2011**, *108* (Suppl. S1), 4516–4522. [CrossRef]
31. Caporaso, J.G.; Lauber, C.L.; Walters, W.A.; Berg-Lyons, D.; Huntley, J.; Fierer, N.; Owens, S.M.; Betley, J.; Fraser, L.; Bauer, M.; et al. Ultra-high-throughput microbial community analysis on the Illumina HiSeq and MiSeq platforms. *ISME J.* **2012**, *6*, 1621–1624. [CrossRef]
32. Caporaso, J.G.; Kuczynski, J.; Stombaugh, J.; Bittinger, K.; Bushman, F.D.; Costello, E.K.; Fierer, N.; Peña, A.G.; Goodrich, J.K.; Gordon, J.I.; et al. QIIME allows analysis of high-throughput community sequencing data. *Nat. Methods* **2010**, *7*, 335–336. [CrossRef] [PubMed]
33. Edgar, R.C.; Haas, B.J.; Clemente, J.C.; Quince, C.; Knight, R. UCHIME improves sensitivity and speed of chimera detection. *Bioinformatics* **2011**, *27*, 2194–2200. [CrossRef]
34. Schloss, P.D.; Westcott, S.L.; Ryabin, T.; Hall, J.R.; Hartmann, M.; Hollister, E.B.; Lesniewski, R.A.; Oakley, B.B.; Parks, D.H.; Robinson, C.J.; et al. Introducing mothur: Open-source, platform-independent, community-supported software for describing and comparing microbial communities. *Appl. Environ. Microbiol.* **2009**, *75*, 7537–7541. [CrossRef]
35. Huse, S.M.; Welch, D.M.; Morrison, H.G.; Sogin, M.L. Ironing out the wrinkles in the rare biosphere through improved OTU clustering. *Environ. Microbiol.* **2010**, *12*, 1889–1898. [CrossRef] [PubMed]
36. Edgar, R.C. Search and clustering orders of magnitude faster than BLAST. *Bioinformatics* **2010**, *26*, 2460–2461. [CrossRef] [PubMed]
37. Cole, J.R.; Wang, Q.; Cardenas, E.; Fish, J.; Chai, B.; Farris, R.J.; Kulam-Syed-Mohideen, A.S.; McGarrell, D.M.; Marsh, T.; Garrity, G.M.; et al. The Ribosomal Database Project: Improved alignments and new tools for rRNA analysis. *Nucleic Acids Res.* **2009**, *37*, D141–D145. [CrossRef] [PubMed]
38. Kembel, S.W.; Cowan, P.D.; Helmus, M.R.; Cornwell, W.K.; Morlon, H.; Ackerly, D.D.; Blomberg, S.P.; Webb, C.O. Picante: R tools for integrating phylogenies and ecology. *Bioinformatics* **2010**, *26*, 1463–1464. [CrossRef]
39. Lozupone, C.; Lladser, M.E.; Knights, D.; Stombaugh, J.; Knight, R. UniFrac: An effective distance metric for microbial community comparison. *ISME J.* **2011**, *5*, 169–172. [CrossRef] [PubMed]
40. Desnoyers, M.; Giger-Reverdin, S.; Bertin, G.; Duvaux-Ponter, C.; Sauvant, D. Meta-analysis of the influence of Saccharomyces cerevisiae supplementation on ruminal parameters and milk production of ruminants. *J. Dairy Sci.* **2009**, *92*, 1620–1632. [CrossRef] [PubMed]
41. Poppy, G.D.; Rabiee, A.R.; Lean, I.J.; Sanchez, W.K.; Dorton, K.L.; Morley, P.S. A meta-analysis of the effects of feeding yeast culture produced by anaerobic fermentation of Saccharomyces cerevisiae on milk production of lactating dairy cows. *J. Dairy Sci.* **2012**, *95*, 6027–6041. [CrossRef] [PubMed]
42. Ogbuewu, I.P.; Mbajiorgu, C.A. Meta-analysis of Saccharomyces cerevisiae on enhancement of growth performance, rumen fermentation and haemato-biochemical characteristics of growing goats. *Heliyon* **2023**, *9*, e14178. [CrossRef]
43. Dai, H.; Ma, N.; Chang, G.; Aabdin, Z.U.; Shen, X. Long-term high-concentrate diet feeding induces apoptosis of rumen epithelial cells and inflammation of rumen epithelium in dairy cows. *Anim. Biotechnol.* **2022**, *33*, 289–296. [CrossRef]
44. Ma, N.; Abaker, J.A.; Wei, G.; Chen, H.; Shen, X.; Chang, G. A high-concentrate diet induces an inflammatory response and oxidative stress and depresses milk fat synthesis in the mammary gland of dairy cows. *J. Dairy Sci.* **2022**, *105*, 5493–5505. [CrossRef]
45. Jiang, Y.; Dai, P.; Dai, Q.; Ma, J.; Wang, Z.; Hu, R.; Zou, H.; Peng, Q.; Wang, L.; Xue, B. Effects of the higher concentrate ratio on the production performance, ruminal fermentation, and morphological structure in male cattle-yaks. *Vet. Med. Sci.* **2022**, *8*, 771–780. [CrossRef] [PubMed]
46. Ogata, T.; Makino, H.; Ishizuka, N.; Iwamoto, E.; Masaki, T.; Ikuta, K.; Kim, Y.H.; Sato, S. Long-term high-grain diet altered the ruminal pH, fermentation, and composition and functions of the rumen bacterial community, leading to enhanced lactic acid production in Japanese Black beef cattle during fattening. *PLoS ONE* **2019**, *14*, e0225448. [CrossRef] [PubMed]
47. Lu, Z.; Kong, L.; Ren, S.; Aschenbach, J.R.; Shen, H. Acid tolerance of lactate-utilizing bacteria of the order Bacteroidales contributes to prevention of ruminal acidosis in goats adapted to a high-concentrate diet. *Anim. Nutr.* **2023**, *14*, 130–140. [CrossRef] [PubMed]
48. Halfen, J.; Carpinelli, N.; Del Pino, F.A.B.; Chapman, J.D.; Sharman, E.D.; Anderson, J.L.; Osorio, J.S. Effects of yeast culture supplementation on lactation performance and rumen fermentation profile and microbial abundance in mid-lactation Holstein dairy cows. *J. Dairy Sci.* **2021**, *104*, 11580–11592. [CrossRef] [PubMed]

49. Carpinelli, N.A.; Halfen, J.; Trevisi, E.; Chapman, J.D.; Sharman, E.D.; Anderson, J.L.; Osorio, J.S. Effects of peripartal yeast culture supplementation on lactation performance, blood biomarkers, rumen fermentation, and rumen bacteria species in dairy cows. *J. Dairy Sci.* **2021**, *104*, 10727–10743. [CrossRef] [PubMed]
50. Sivinski, S.E.; Meier, K.E.; Mamedova, L.K.; Saylor, B.A.; Shaffer, J.E.; Sauls-Hiesterman, J.A.; Yoon, I.; Bradford, B.J. Effect of Saccharomyces cerevisiae fermentation product on oxidative status, inflammation, and immune response in transition dairy cattle. *J. Dairy Sci.* **2022**, *105*, 8850–8865. [CrossRef] [PubMed]
51. Dias, A.L.G.; Freitas, J.A.; Micai, B.; Azevedo, R.A.; Greco, L.F.; Santos, J.E.P. Effect of supplemental yeast culture and dietary starch content on rumen fermentation and digestion in dairy cows. *J. Dairy Sci.* **2018**, *101*, 201–221. [CrossRef] [PubMed]
52. Ma, L.; Xu, S.; Liu, H.; Xu, T.; Hu, L.; Zhao, N.; Han, X.; Zhang, X. Yak rumen microbial diversity at different forage growth stages of an alpine meadow on the Qinghai-Tibet Plateau. *PeerJ* **2019**, *7*, e7645. [CrossRef] [PubMed]
53. Petersen, C.; Round, J.L. Defining dysbiosis and its influence on host immunity and disease. *Cell Microbiol.* **2014**, *16*, 1024–1033. [CrossRef]
54. Mei, S.; He, G.; Chen, Z.; Zhang, R.; Liao, Y.; Zhu, M.; Xu, D.; Shen, Y.; Zhou, B.; Wang, K.; et al. Probiotic-Fermented Distillers Grain Alters the Rumen Microbiome, Metabolome, and Enzyme Activity, Enhancing the Immune Status of Finishing Cattle. *Animals* **2023**, *13*, 3774. [CrossRef]
55. Wang, Y.; Nan, X.; Zhao, Y.; Jiang, L.; Wang, H.; Zhang, F.; Hua, D.; Liu, J.; Yao, J.; Yang, L.; et al. Dietary Supplementation of Inulin Ameliorates Subclinical Mastitis via Regulation of Rumen Microbial Community and Metabolites in Dairy Cows. *Microbiol. Spectr.* **2021**, *9*, e0010521. [CrossRef] [PubMed]
56. Jiang, Q.; Sherlock, D.N.; Elolimy, A.A.; Vailati-Riboni, M.; Yoon, I.; Loor, J.J. Impact of a Saccharomyces cerevisiae fermentation product during an intestinal barrier challenge in lactating Holstein cows on ileal microbiota and markers of tissue structure and immunity. *J. Anim. Sci.* **2023**, *101*, skad309. [CrossRef] [PubMed]
57. Li, D. *Effects of Compound Bacteria Culture and Yeast β-Glucan on Antioxidant Function and Inflammatory Factors in Mutton Sheep*; Inner Mongolia Agricultural University: Hohhot, China, 2021.
58. Kowalik, B.; Skomiał, J.; Pająk, J.J.; Taciak, M.; Majewska, M.; Bełżecki, G. Population of ciliates, rumen fermentation indicators and biochemical parameters of blood serum in heifers fed diets supplemented with yeast (Saccharomyces cerevisiae) preparation. *Anim. Sci. Pap. Rep.* **2012**, *30*, 329–338.
59. Zhang, J. Study on the Immune Enhancement Effect of Yeast Cell Wall Extract on Newcastle Disease Vaccine. Master's Thesis, Southwest University, Chongqing, China, 2022. [CrossRef]
60. Oshima, T.; Koseki, J.; Chen, X.; Matsumoto, T.; Miwa, H. Acid modulates the squamous epithelial barrier function by modulating the localization of claudins in the superficial layers. *Lab. Invest.* **2012**, *92*, 22–31. [CrossRef] [PubMed]
61. Wang, H.; Su, M.; Wang, C.; Li, D.; Li, Q.; Liu, Z.; Qi, X.; Wu, Y.; Zhao, Y.; Li, T.; et al. Yeast culture repairs rumen epithelial injury by regulating microbial communities and metabolites in sheep. *Front. Microbiol.* **2023**, *14*, 1305772. [CrossRef] [PubMed]
62. Pan, X.; Xue, F.; Nan, X.; Tang, Z.; Wang, K.; Beckers, Y.; Jiang, L.; Xiong, B. Illumina Sequencing Approach to Characterize Thiamine Metabolism Related Bacteria and the Impacts of Thiamine Supplementation on Ruminal Microbiota in Dairy Cows Fed High-Grain Diets. *Front. Microbiol.* **2017**, *8*, 1818. [CrossRef]
63. Bampidis, V.A.; Lymberopoulos, A.G.; Christodoulou, V.; Belibasaki, S. Impacts of supplemental dietary biotin on lameness in sheep. *Anim. Feed. Sci. Technol.* **2006**, *134*, 162–169. [CrossRef]

animals

Article

Saponin Extracts Utilization as Dietary Additive in Ruminant Nutrition: A Meta-Analysis of In Vivo Studies

Yulianri Rizki Yanza [1,*,†,‡], Agung Irawan [2,*,†,‡], Anuraga Jayanegara [3,‡], Fitri Ramadhani [4], Adib Norma Respati [5], Ainissya Fitri [6,‡], Cecep Hidayat [7,‡], Vincent Niderkorn [8], Adam Cieslak [9], Malgorzata Szumacher-Strabel [9], Rahmat Hidayat [1] and Ujang Hidayat Tanuwiria [1]

1 Department of Animal Nutrition and Feed Technology, Faculty of Animal Husbandry, Universitas Padjadjaran, Jatinangor, Sumedang 45363, West Java, Indonesia; r.hidayat@unpad.ac.id (R.H.); ujang.hidayat@unpad.ac.id (U.H.T.)
2 Vocational School, Universitas Sebelas Maret, Surakarta 57126, Central Java, Indonesia
3 Department of Animal Nutrition and Feed Technology, Faculty of Animal Science, IPB University, Bogor 16680, West Java, Indonesia; anuragaja@apps.ipb.ac.id
4 Department of Biology Education, Islamic University of Riau, Pekanbaru 28284, Riau, Indonesia; fitrirmdhani01@gmail.com
5 Department of Animal Science, Politeknik Negeri Jember, Jember 68101, Jawa Timur, Indonesia; adib@polije.ac.id
6 Research Center for Applied Zoology, National Research and Innovation Agency (BRIN), Jl. Raya Jakarta-Bogor Km 46, Cibinong, Bogor 16911, West Java, Indonesia; aini002@brin.go.id
7 Research Center for Animal Husbandry, Research Organization for Agriculture and Food, National Research and Innovation Agency (BRIN), Jl. Raya Jakarta-Bogor Km 46, Cibinong, Bogor 16911, West Java, Indonesia; hidayat_c2p@yahoo.com
8 INRAE, VetAgro Sup, UMRH, Université Clermont Auvergne, 63122 Saint-Genès-Champanelle, France; vincent.niderkorn@inrae.fr
9 Department of Animal Nutrition, Faculty of Veterinary Medicine and Animal Science, Poznan University of Life Sciences, Wolynska 33, 60637 Poznan, Poland; adam.cieslak@up.poznan.pl (A.C.); malgorzata.szumacher@up.poznan.pl (M.S.-S.)
* Correspondence: y.r.yanza23@unpad.ac.id (Y.R.Y.); a.irawan@staff.uns.ac.id (A.I.); Tel.: +62-822-6081-1535 (Y.R.Y.)
† These authors contribute to the paper equally.
‡ Current address: Animal Feed Nutrition Modeling (AFENUE) Research Group, Department of Animal Nutrition and Feed Technology, Faculty of Animal Science, IPB University, Bogor 16680, West Java, Indonesia.

Citation: Yanza, Y.R.; Irawan, A.; Jayanegara, A.; Ramadhani, F.; Respati, A.N.; Fitri, A.; Hidayat, C.; Niderkorn, V.; Cieslak, A.; Szumacher-Strabel, M.; et al. Saponin Extracts Utilization as Dietary Additive in Ruminant Nutrition: A Meta-Analysis of In Vivo Studies. *Animals* **2024**, *14*, 1231. https://doi.org/10.3390/ani14081231

Academic Editor: Lizhi Wang

Received: 23 January 2024
Revised: 31 March 2024
Accepted: 11 April 2024
Published: 19 April 2024

Simple Summary: The present meta-analysis was conducted to investigate the use of saponin extracts as dietary supplements for ruminants, based on in vivo studies. This study aimed to highlight the benefits of saponin extracts and their impact on ruminant health and production. A total of 26 articles, comprising 66 studies, were included in the database. The meta-analysis was designed to elucidate the effects of saponin extracts from different sources on production performance, milk yield, digestibility, rumen fermentation, nitrogen utilization, and blood metabolites. The results revealed that increased saponin supplementation linearly decreased milk production and altered rumen fermentation profiles without affecting digestibility rates, particularly influencing volatile fatty acid production and protozoa population. However, the efficacy and safety of different levels of saponin extracts vary, and further research is required to optimize its use for enhancing ruminant productivity, mitigating environmental impacts, and exploring the specific effects of saponin extracts on ruminant health. It can be concluded that the utilization of saponin extracts in ruminant diets is complex, and a comprehensive understanding of the optimal application of various saponin extracts across different ruminant physiological conditions is necessary.

Abstract: The present meta-analysis aimed to determine the underlying effects of different saponins extracted from different sources on the production performance, milk yield, digestibility, rumen fermentation, blood metabolites, and nitrogen utilization of ruminants. A total of 26 papers comprising 66 in vivo studies (148 data points of dietary treatments) were evaluated in the present study. The databases were statistically analyzed using the mixed model procedure of SAS, where experiments considered random effects and tannin-related factors were treated as fixed effects. Statistical procedures were then continued in comparing different sources of saponin extract through Mixed Model analysis, where experiments were also random factors and sources of saponin extract were fixed factors. The evidence revealed in the present meta-analysis that saponin supplementation of up to 40 g/kg DM appears to have no detrimental impact on feed intake across ruminant types, suggesting that it does not significantly affect diet palatability. However, the results indicated that there are species-specific responses to saponin supplementation, particularly in relation to palatability and nutrient absorption efficiency, with larger ruminants being better able to tolerate the bitterness induced by saponin extracts. Furthermore, the study found that saponin extracts can influence nutrient digestibility and rumen fermentation dynamics, with different effects observed in large and small ruminants. While some saponin extracts can enhance average daily weight gain and milk yield, others can have adverse effects, highlighting the importance of considering both saponin sources and animal physiological condition when developing nutritional strategies. Additionally, optimization of ruminant production by utilizing saponin extracts is necessary to avoid negative health implications, such as increased blood creatinine levels. Different saponin extracts utilization in ruminant nutrition and environmental management, have a distinct understanding associated to their various bioactive properties. However, among the saponin sources, saponin extracted from *Quilaja saponaria* is more likely to improve large ruminant production performance while maintaining ruminant health and metabolism, but negatively affect small ruminants. Further research is needed to unravel the intricate effects of different saponin sources on ruminant health and productivity, emphasizing the importance of tailored dietary strategies that consider the unique physiological and metabolic characteristics of the target livestock.

Keywords: saponin extract; ruminant; methane; performance; milk; N utilization; meta-analysis

1. Introduction

Ruminant nutrition has evolved its interest in utilizing natural plant compounds to sustainably produce good quality products of ruminant origin, such as meat and milk, effectively and efficiently, as well as maintaining animal health. One of the generally known natural plant compounds is saponins, which have recently emerged as a significant concern because of their beneficial mode of action in ruminant production [1]. Saponins are glycosides characterized by sugar and non-sugar bonds (such as aglycone or sapogenin) that have a soap-like mode of action. They are naturally found in a wide variety of plants, such as legumes and medicinal herb plants, at various concentrations because of their anti-nutritional factors that interfere with digestive metabolism [2]. In recent years, saponins have been introduced as effective natural compound agents to modulate the rumen fermentation and digestibility in ruminant [3].

Ruminant feed supplemented with saponins has demonstrated various beneficial effects, including promoting health and immune metabolism, modulating ruminal fermentation and digestibility, and mitigating methane production, which contributes to reducing the environmental impact of GHG emissions [4,5]. Extensive research has revealed that saponins possess potent antimicrobial properties that directly decrease the population of microorganisms, such as bacteria, protozoa, and methanogens, which are linked to a reduction in enteric methane production [6–8]. However, evidence regarding the beneficial effects of saponin extracts as health-promoting agents and their performance in ruminants is inconclusive [9]. For instance, a study revealed that saponins extracted from *Yucca schidigera* fed to dairy cattle at about 25–50 g/d had lower ruminal VFA concentration [10], but showed no effects on VFA concentration when saponin from the whole part of *Terminalia chebula* Retz. was introduced to goats [11]. Moreover, several studies have confirmed that saponins whether in a whole part sources of saponin or extracted form influence ruminal N-ammonia concentrations [12,13] and have detrimental effects on excreted N in feces and urine [13,14].

Inconsistencies regarding the effects of saponin supplementation were also found in performance parameter analyses, such as average daily gain, milk yield, N-utilization and ruminal digestibility [14–16]. Therefore, there is a need for extended evaluation of saponin utilization in ruminants, especially in extracted forms originating from various plant sources. Different plants containing saponin compound extracts have different efficacies in modulating ruminal fermentation in various ruminant species as well as their influence on animal performance and ruminal metabolism. Preeminently, saponin extract utilization in ruminant nutrition attracts interest, which can be related to the type of animals, dosage levels, and plant sources. Hence, the present me-ta-analysis aimed to determine the effects of saponin extract supplementation at various levels and sources (types) on production performance, milk yield, digestibility, ruminal fermentation, blood metabolites, and nitrogen utilization of ruminants.

2. Materials and Methods

2.1. Database Development

The present study database was constructed from various studies which reported the nutritional utilization of saponins in extracted form. The data included in the constructed database were based on published articles written in English describing in vivo experiments. All included in vivo data were obtained from journals indexed in Google scholar, Crossref, Scopus and Web of Science such as graphically described in Figure 1.

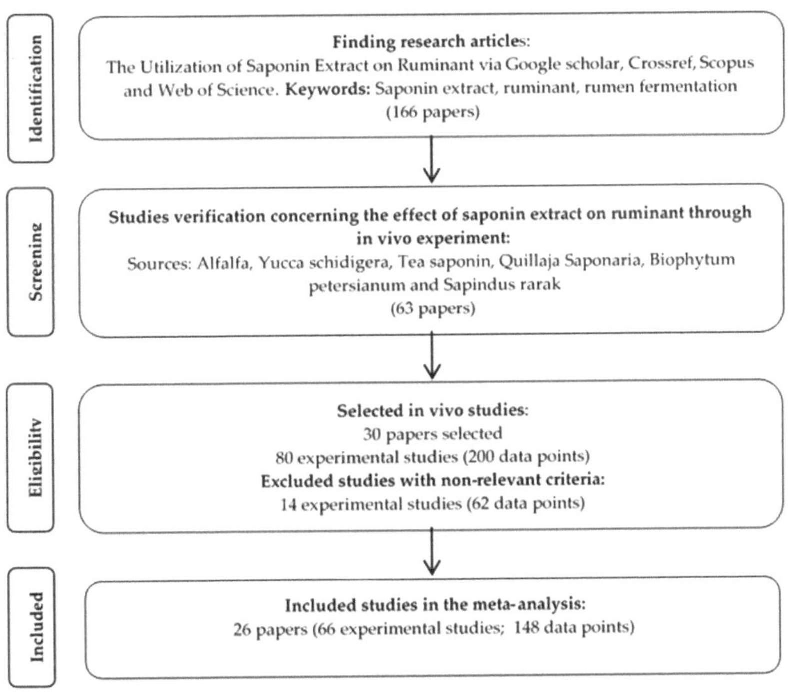

Figure 1. Diagram flow for selection of the studies on the influence of saponin extract.

The final database was set up from 26 experimental studies consisting of 66 dietary treatments (148 data points; Table 1). The criteria for incorporating articles into the database were as follows: (a) detailed description of in vivo experiments conducted on ruminants; (b) inclusion of extracted saponins into basal feeds within the studies; (c) consists information of observed variables such as dietary intake, average daily gain (ADG), milk yield and nutrient profile, total digestibility, ruminal fermentation characteristics, nitrogen utilization, and blood parameters; and (d) articles are written in English. In the present study, the source of saponins was determined from the extracts of alfalfa or *Medicago sativa* (MS), *Quillaja Saponaria* (QS), tea saponin or *Camelia sinensis* (CS), *Yucca schidigera* (YS), *Agave americana* (AA), *Biophytum petersianum* (BP), and *Sapindus rarak* (SR). Most studies have used a mixed basal diet composed of corn silage, wheat straw, canola meal, barley concentrate, elephant grass, wheat pollard, maize silage, ensiled brewer's grains, ensiled beet pulp, meadow hay, rapeseed meal, and hay pelleted concentrate. Type of ruminants that were included in the database by the manner of saponin extract supplementation were steers, dairy cows and buffalo, categorized as type of large ruminants, while sheep, lamb, and goats, were categorized as type of small ruminants.

The saponin extracts supplementation units were standardized and presented as g/kg dry matter (DM) of feed. Moreover, the units of each measured value in the database for each parameter were standardized. Hence, all measured values in an observed parameter expressed in other than the common units were converted to get similar units. For instance, dry matter intake (DMI), organic matter intake (OMI), and neutral detergent fibre intake (NDFI) that are presented as g/d or kg/d were converted and expressed as g/kg metabolic body weight unit (g/kg $BW^{0.75}$). The average daily gain (ADG) expressed as grams per day (g/d) and kilograms per day (kg/d) were converted into g/kg $BW^{0.75}$ unit, while milk yield units reported as g/d and kg/d were converted and expressed as g/kg $BW^{0.75}$ and g/kg DMI. Such unit conversion relative to metabolic body weight is necessary to reduce the variability of presented data by considering the type, and weight of trialed animals across studies. Similarly, N utilization units expressed as g/d and kg/d were also converted into g/kg $BW^{0.75}$ unit. Moreover, digestibility rate, milk nutrient composition, milk nitrogen utilization, and VFA proportion units were standardized and presented as g/100 g or percentages. Additionally, ruminal fermentation profiles expressed in total VFA concentration, ruminal ammonia concentration, and blood plasma parameters were converted and presented in mmol/L, μmol/L, g/dL or mg/dL. Meanwhile overall data of protozoa population expressed in 10^x/mL were converted into log_{10}/mL unit.

2.2. Statistical Analysis

The statistical analysis was performed using a similar approach to the previous meta-analysis using SAS version 9.4 [17,18]. The raw dataset was assessed for outliers using the PROC REG of SAS and a minimum sample size of at least three studies was included in the analysis. To examine the effects of saponin extracts, we initially performed a meta-regression analysis by considering the inclusion levels of saponin extracts in the diets, type of animals, sources of saponins, and other relevant covariates that might have affected the observed parameters. Then, a categorical meta-analysis was performed followed Yanza et al. [19] and Respati et al. [18] statistical methods with some modification to explore the specific effects of saponin sources and their interactions with other factors in the dataset. For the meta-regression analysis, multiple models based on linear mixed models (LMM) were tested using the following model:

$$\Delta\Upsilon_{ij} = \beta_0 + \beta_1 X_{ij} + \beta_2 {X_{ij}}^2 + (\beta_1 \times \beta_3 \ldots n) X_{ij} \times S_i + \varepsilon_{ij}, \quad \text{([full model])}$$

$$\Delta\Upsilon_{ij} = \beta_0 + \beta_1 X_{ij} + \beta_2 {X_{ij}}^2 + (\beta_1 \times \beta_3 \ldots n - 1) X_{ij} \times S_i + \varepsilon_{ij}, \quad \text{([reduced model])}$$

where $\Delta\Upsilon_{ij}$ = estimated outcome of the dependent variable based on *j* observation in *i* experiment, β_0 = estimated intercept (fixed effect), β_1 = linear model coefficient of continuous predictor (fixed effect), β_2 = quadratic term coefficient of continuous predictor (fixed effect), X_{ij} = saponin extracts' levels of *j* observation in *i* experiment, the matrix of the continuous predictor variable, $\beta_3 \ldots \beta_n$ = coefficient of the categorical variables, S_i = the random effect of studies, and ε_i = the residual error at ~$N(0,\sigma^2)$.

The study employed a rigorous methodology to find accurate model. For instance, the CLASS statement was utilized to analyze varying concentrations of saponin extract supplementation and study variables without quantitative data, while the RANDOM statement was declared based on different experiments across studies. The models used weighing factors to represent the characteristic variability of each study, with the weights divided by the mean of all weights relatively to the number of observed variables of each level, as suggested by St-Pierre [20]. This methodology ensured that the results can maintain the expressions of dispersion in the original scale of the measurements and expressed robust, accurate, and reliable results.

A backward elimination procedure was followed to obtain the best-fitted model using Akaike's information criterion (AIC), root-means square errors (RMSE), and between-models F-test of presented results. The linear model was retained if quadratic effect and other interaction effects were not statistically significant ($p < 0.05$) or tend to significant ($0.10 < p < 0.05$). Then, the categorical meta-analysis was performed refered to Yanza et al. [21] protocol to examine the effects of the sources of saponins using the following statistical model:

$$Y_{ij} = \mu + \beta_a + (\beta_a \times \beta_b)x_{ij} + s\beta_{ij} + S_i + e_{ij}$$

where Y_{ij} = the estimated means of response variable Y of j observation in i study, μ = overall mean, β_a = fixed effect of categorical variables, β_b = fixed effect of the co-variates, $\beta_a \times \beta_b$ = interaction terms between categorical variables and covariates of j observation in i study, $s\beta_{ij}$ = random term between i study and the j factors β, S_i = random term of the study, and e_{ij} = residual error ~$N(0,\sigma^2)$. The effects were deemed significant at $p < 0.05$ and tended to significant at p-value between 0.05 and 0.10 using Tukey-Kramer's test.

Table 1. Lists of studies included in the present meta-analysis study.

No	Study	Year	Exp.	Animal	Age (Months)	Status	IBW (kg)	Sources of Saponin Extract	Level (g/kg DM)	Extracted Forms
1	Valdez et al. [22]	1986	1	Dairy cows	n.d.	1st lactation (6–10 week postpartum)	n.d.	*Yucca schidigera*	0.77	Powder
2	Lu et al. [23]	1987	2–3	Sheep	n.d.	Mature wethers	49	*Medicago sativa*	20–40	Powder
3	Wu et al. [24]	1994	4–5	Dairy cows	n.d.	Lactation	650	*Yucca schidigera*	0–0.396	Powder
4	Hussain [13]	1995	6–15	Steers	n.d.	n.d.	574–658	*Yucca schidigera*	0.25	Powder
5	Wilson [25]	1998	16–17	Dairy cows		Multiparous (122 d postpartum)	640	*Yucca schidigera*	0.378	Powder
6	Sliwinski [26]	2002	18	Sheep	4.02	Castrated male lambs	35.1	*Yucca schidigera*	0.002–0.03	Powder
7	Eryavuz abd Dehoroti [8]	2004	19–22	Sheep	24–108	n.d.	186.6	*Yucca schidigera*	5–30	Liquid
8	Santoso et al. [27]	2006	23	Goat	n.d.	n.d.	20.3	*Biophytum petersianum*	0.072–0.144	Liquid
9	Wina et al. [28]	2006	24–25	Sheep	n.d.	Male sheep	16.5	*Sapindus rarak*	20.16–30.24	Powder
10	Lovett et al. [29]	2006	26–27	Dairy cows	n.d.	1st and 2nd or 3rd lactation (±39 d post calving)	585–610	*Yucca schidigera*	1.488–4.421	Powder
11	Baah et al. [3]	2007	28	Dairy cows	n.d.	Heifers	601	*Quillaja saponaria*	0–8	Powder
12	Liu et al. [30]	2007	29–34	Sheep	n.d.	Male sheep	40	*Yucca schidigera*	0.1–0.3	Powder
13	Abdelmawla [1]	2008	35	Buffalo	n.d.	4th and 5th lactation	591	*Quillaja saponaria*	0.052–0.052	Liquid
14	Benchaar et al. [6]	2008	36	Dairy cows	n.d.	Lactation (87 DIM)	730	*Yucca schidigera*	2.752	Powder
15	Singer et al. [31]	2008	37	Dairy cows	n.d.	Late lactation (298 DIM)	810	*Yucca schidigera*	2.01–6.23	Powder
16	Selcuk & Tuncer [32]	2010	38	Sheep	2–2.5	Male lamb	20.87–21.69	*Yucca schidigera*	0.2–0.4	Powder
17	Li et al. [16]	2011	39	Sheep	n.d.	Male sheep	40	*Yucca schidigera*	0.1–0.3	Powder
18	Nasri et al. [33]	2011	40–43	Sheep	5–6	Male lamb	17.8–18.8	*Quillaja saponaria*	0.1–0.09	Powder
19	Nasri et al. [34]	2012	44–55	Sheep	5–6	Female lamb	23.9–28.9	*Quillaja saponaria* *Agave americana*	0.12–0.36 0.12–0.36	Liquid Powder
20	Guyader et al. [9]	2015	56–57	Dairy cows	n.d.	Multiparous nonlactating	658	*Camellia sinensis*	5	Powder
21	Guyader et al. [10]	2017	58	Dairy cows		Primiparous & Multiparous (106 DIM)	617	*Camellia sinensis*	7.6	Powder
22	Baheg et al. [4]	2017	59–60	Sheep	43	Ewes	33.76	*Yucca schidigera*	0.2	Powder
23	Kumar et al. [14]	2017	61–63	Goat	7.03	Male kids	19.43–19.96	*Camellia sinensis*	4	Powder
25	Zhang et al. [35]	2021	65	Sheep	12	Male castrated sheep	48.37	*Camellia sinensis*	5–20	Powder
24	Yi et al. [36]	2022	64	Steers	n.d.	Steers	510.5	*Yucca schidigera*	0.198	Powder
26	Alsubait et al. [2]	2023	66	Sheep	3 and 4	Male lambs	26.26–26.97	*Yucca schidigera*	0.3–0.6	Powder

Exp. = number of experiments in the study; DIM = days in milk; IBW = initial body weight; DM = dried matter; n.d. = not determined.

3. Results

3.1. Datasets

The literature search in the present meta-analysis included 26 studies investigating the inclusion of saponin extracts supplemented in the diets of ruminants, together with information such as animal status, initial body weight during the experiment, source of saponin extract, and the range of saponin extract supplementation levels in the animal diet (Table 1). The present study observed that studies examining extracted saponins were dominated by *Yucca schidigera* and *Quillaja Saponaria* extracts. Information regarding the descriptive statistics of the datasets is presented in Table 2. The descriptive data also showed that, overall, the ranges of values of the observed parameters were within the expected values, although high variability within the studies was also identified by standard errors of the means (SEM).

Table 2. Descriptive statistics of the dataset used in the meta-analysis.

Parameters	Unit	N	Mean	SEM	Min	Max
			Feed Intake			
DMI	g/kg $BW^{0.75}$	67	113.1	4.917	50.49	187.0
OMI	g/kg $BW^{0.75}$	12	80.25	9.159	48.82	150.2
NDFI	g/kg $BW^{0.75}$	12	36.22	3.146	18.12	56.62
			Gain Performance			
ADG	g/d	18	150.8	20.20	59.6	275.0
	g/kg $BW^{0.75}$	18	13.18	2.209	5.25	27.80
			Milk production and composition			
Milk yield	kg/d	21	25.23	2.071	7.01	33.85
	g/kg $BW^{0.75}$	19	194.0	16.33	58.48	264.8
	g/kg DMI	21	1.20	0.080	0.46	1.55
Milk fat	g/kg	21	44.49	3.103	31.3	74.6
Milk protein	g/kg	17	34.22	1.358	28.0	47.9
Milk lactose	g/kg	12	47.56	0.785	43.8	52.2
			Digestibility			
DMD	g/kg	40	698.3	8.769	628	813.0
OMD	g/kg	57	683.9	11.47	405	827.0
CPD	g/kg	44	636.2	10.46	531	793.7
NDFD	g/kg	47	559.0	14.89	295	730.3
ADFD	g/kg	27	489.9	15.79	323	660.1
			Rumen fermentation parameters			
pH		148	6.30	0.030	5.51	7.1
NH_3	mg/dL	146	18.77	1.028	4.12	81.74
Total VFA	mmol/L	74	103.7	14.21	48.9	125.0
Acetate	%	71	62.23	1.305	5.73	78.4
Propionate	%	71	22.52	0.835	3.14	41.74
Butyrate	%	71	11.65	0.432	1.17	19.3
Valerate	%	29	1.27	0.141	0.16	3.68
A:P ratio		71	3.10	0.151	0.79	7.47
Protozoa	log_{10}/mL	91	5.42	0.044	4.19	6.52
			N Balance			
Urine N/MBW	g/kg $BW^{0.75}$	35	585.4	51.2	174.7	1099
Fecal N/MBW	g/kg $BW^{0.75}$	35	550.2	46.5	199.8	1494
N retention/MBW	g/kg $BW^{0.75}$	23	463.6	49.8	52.3	759.8
			Blood parameters			
Plasma NH_3	µg/dL	40	1.90	0.386	0.64	10.85

Table 2. *Cont.*

Parameters	Unit	N	Mean	SEM	Min	Max
Total protein	g/dL	20	6.97	0.255	5.32	10.31
Albumin	g/dL	13	3.01	0.105	2.55	3.84
Globulin	g/dL	10	3.72	0.158	2.97	4.66
PUN	mg/dL	72	17.22	1.20	3.59	42.0
Cholesterol	g/dL	21	75.43	8.087	28.0	153.7
Creatinine	µmol/L	28	80.0	3.814	35.4	106.7
Glucose	mg/dL	20	55.17	3.524	30.96	86.0
ALP	IU/L	10	146.6	19.837	103.7	272.9

N = sample size; SEM = standard error of the mean; DMI = Dry Matter intake; OMI = Organic matter intake; NDFI = neutral detergent fiber intake; $BW^{0.75}$ = metabolic body weight; ADG = Average Daily Gain; DMD = dry matter digestibility; OMD = Organic matter digestibility; CPD = Crude Protein digestibility; NDFD = neutral detergent fiber digestibility; ADFD = acid detergent fiber digestibility; VFA = volatile fatty acids; NH_3 = ammonia; A:P = acetate to propionate ratio; PUN = Plasma urea-N; ALP = alkaline phosphatase.

3.2. Relationship between Dietary Saponins Levels on Observed Parameters

Meta-regression indicated that the levels of saponins in the diets had no significant effect on DMI, OMI, and NDFI expressed as g/kg $BW^{0.75}$ (Table 3). However, the interaction between the level of saponin extract and the type of animal was significant for DMI and NDFI ($p < 0.05$), but tended to be significant for OMI ($p = 0.079$). In Figure 2, the pattern of DMI (g/kg $BW^{0.75}$) only showed a significant model for small ruminants in a quadratic manner ($p = 0.036$; $R^2 = 0.258$), where high accuracy on the predicted models was adjusted ($R^2 = 0.912$). Moreover, an interaction between the level of saponin extract and the type of saponin source was also observed for NDFI and OMI ($p < 0.05$).

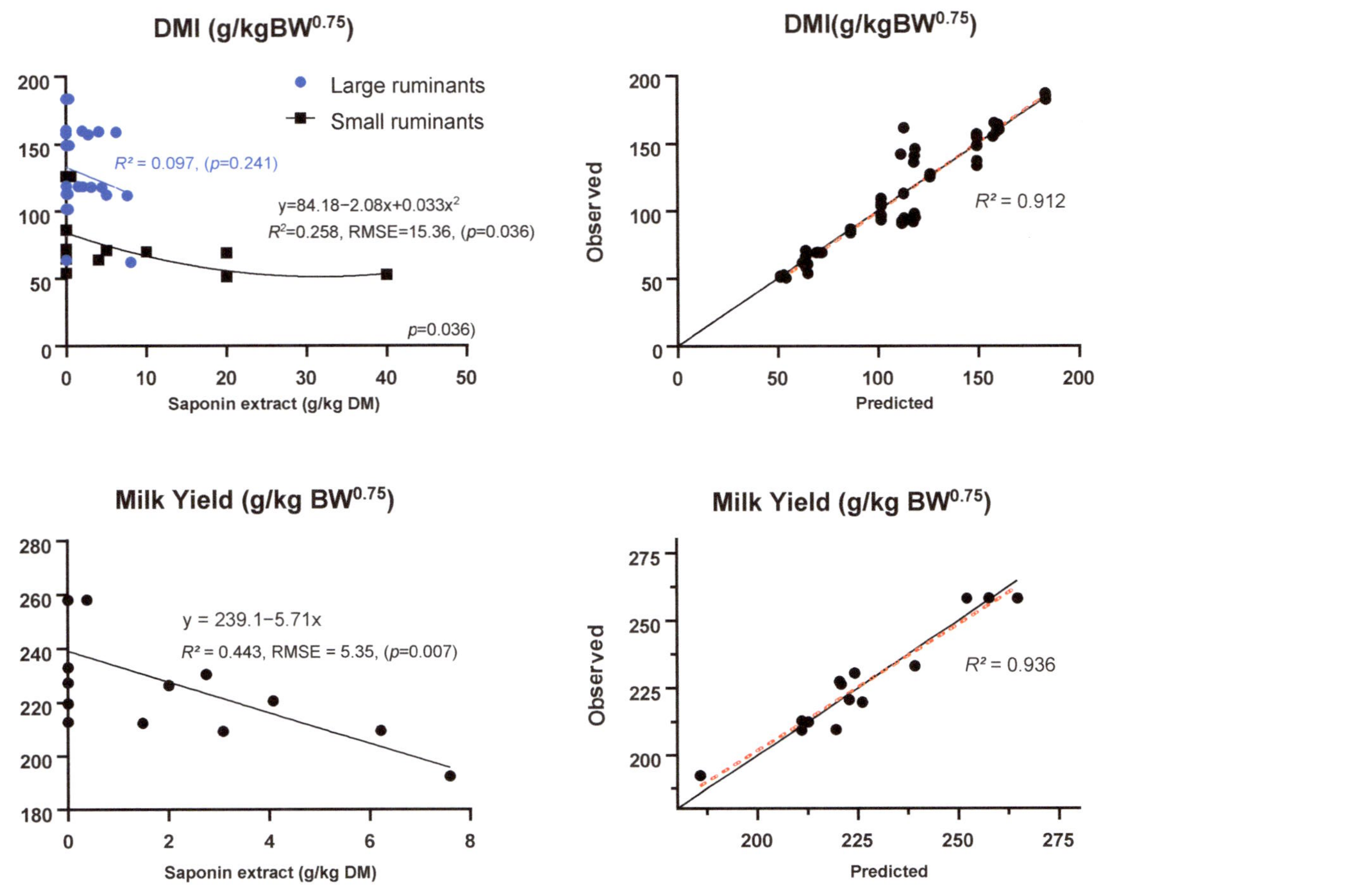

Figure 2. Meta-regression results representing the relationships between dietary saponin sources' intake (g/kg DM) on dry matter intake and milk yield per metabolic body weight ($BW^{0.75}$). The right figures represent the model performance evaluation according to the observed vs. predicted values.

Table 3. Results of meta-regression analysis of the effects of dietary saponin levels on ruminants.

Estimated Variables	Unit	N	Model	Intercept	$SE_{intercept}$	Slope	SE_{slope}	*p*-Value	RMSE	AIC	L × Animal	L × Source
						Feed Intake						
DMI	g/kg $BW^{0.75}$	67	L	112.1	7.822	−0.015	0.103	0.889	3.676	519.5	0.016	0.660
OMI	g/kg $BW^{0.75}$	12	L	80.85	13.59	−0.216	0.725	0.778	4.728	97.8	0.079	0.005
NDFI	g/kg $BW^{0.75}$	12	L	36.32	4.690	−0.036	0.288	0.905	1.878	77.98	0.032	0.021
						Gain performance						
ADG	g/d	18	L	170.8	54.36	0.183	0.538	0.739	96.11	143	0.739	0.021
	g/kg $BW^{0.75}$	18	L	12.01	4.198	0.009	0.061	0.880	0.705	83.1	0.009	0.854
						Milk production and composition						
Milk yield	kg/d	21	L	25.98	3.932	−0.364	0.109	0.005	11.31	93.9	0.016	0.001
	g/kg $BW^{0.75}$	19	L	211.6	29.18	−2.938	0.912	0.008	5.339	158.4	<0.001	0.295
	g/kg DMI	21	L	1.253	0.134	−0.002	0.005	0.698	0.030	−21.1	<0.001	0.220
Milk fat	g/kg	21	L	42.48	5.761	0.257	0.233	0.290	77.10	118	0.001	0.001
Milk protein	g/kg	17	L	33.70	2.485	−0.091	0.384	0.817	79.74	92.5	0.035	0.082
Milk lactose	g/kg	12	L	47.59	1.567	0.005	0.225	0.981	36.44	56	0.197	0.341
						Digestibility						
DMD	g/kg	40	L	702.0	17.07	0.973	0.473	0.048	31.45	382	0.015	0.425
OMD	g/kg	57	L	687.2	26.60	0.373	0.371	0.320	59.26	543	0.002	0.359
CPD	g/kg	44	L	640.2	24.20	2.017	1.546	0.201	36.40	439	0.206	0.568
NDFD	g/kg	47	L	564.8	32.90	−1.345	0.806	0.104	48.43	471	0.064	0.795
ADFD	g/kg	27	L	500.7	33.10	−3.977	1.764	0.036	39.99	275	0.005	0.235
						Rumen Fermentation Profile						
pH		148	L	6.301	0.068	−0.0004	0.005	0.945	0.176	114	0.793	0.394
NH_3	mg/dL	146	L	23.11	3.102	−0.116	0.110	0.293	10.46	982	0.361	0.653
Total VFA	mmol/L	74	Q	107.9	25.50	18.44 −0.440	6.010 0.150	0.003	78.17	897	<0.001	<0.001
Acetate	%	71	Q	62.59	3.010	−0.640 0.010	0.320 0.008	0.049	6.880	494	0.001	<0.001
Propionate	%	71	L	21.17	1.848	−0.059	0.067	0.382	4.440	416	0.001	<0.001
Butyrate	%	71	L	11.28	0.943	−0.031	0.042	0.463	2.014	348	0.038	0.001
Valerate	%	29	Q	1.370	0.260	0.190	0.080	0.001	0.355	32.5	<0.001	0.001

Table 3. *Cont.*

Estimated Variables	Unit	N	Model	Intercept	$SE_{intercept}$	Slope	SE_{slope}	*p*-Value	RMSE	AIC	L × Animal	L × Source
			L			−0.040	0.010					
A:P ratio		71	L	3.321	0.358	0.002	0.008	0.826	0.796	140	0.949	0.786
Protozoa	log_{10}/mL	91	Q	5.390	0.130	0.020	0.010	0.030	0.939	38.6	0.689	<0.001
						−0.001	0.000					
N Partitioning												
Urine N	g/kg $BW^{0.75}$	35	Q	614.1	83.99	−5.524	3.315	0.057	38.02	430.9	0.371	0.934
						0.173	0.085					
Fecal N	g/kg $BW^{0.75}$	35	L	573.96	70.65	−0.462	0.878	0.605	31.28	417.7	0.009	0.959
N retention	g/kg $BW^{0.75}$	23	L	476.43	83.87	−0.914	2.207	0.686	40.00	272	-	0.976
Blood parameters												
Total protein	g/dL	20	L	6.790	0.498	−0.027	0.255	0.915	1.261	59.2	0.562	0.993
Albumin	g/dL	13	L	2.976	0.222	0.044	0.050	0.400	3.484	6.8	0.156	0.329
Globulin	g/dL	10	L	3.756	0.358	−0.057	0.096	0.576	7.530	15.7	0.243	0.226
PUN	mg/dL	72	L	18.23	2.317	0.105	0.559	0.852	1.426	378.5	0.156	0.535
Cholesterol	mg/dL	21	L	86.18	23.49	−1.849	3.145	0.565	94.65	160	0.679	0.055
Creatinine	µmol/L	28	Q	77.54	7.510	51.85	19.62	0.014	58.79	218	0.884	<0.001
						−13.37	4.960					
Glucose	mg/dL	20	L	49.85	8.310	−1.027	3.707	0.785	115.6	153	0.785	0.083
ALP	IU/L	10	L	166.8	37.75	0.514	2.830	0.861	98.59	76.9	0.861	0.065

DMI = Dry matter intake; OMI = Organic matter intake; NDFI = neutral detergent fiber intake; $BW^{0.75}$ = metabolic body weight; ADG = Average Daily Gain; DMD = dry matter digestibility; OMD = Organic matter digestibility; CPD = Crude Protein digestibility; NDFD = neutral detergent fiber digestibility; ADFD = acid detergent fiber digestibility; NH_3 = ammonia; VFA = volatile fatty acids; A:P ratio = acetate propionate ratio; N = nitrogen; PUN = Plasma urea nitrogen; ALP = alkaline phosphatase; L = linear term; Q = quadratic term; N = sample size; $SE_{intercept}$ = standard errors of intercept; SE_{slope} = standard error of the slope; AIC = Akaike information of criterion; RMSE = root mean square error; L × Animal = interaction effects between levels of saponin extracts and type of animals; L × Source = interaction effects between levels and sources of saponin extracts.

The ADG expressed as g/d and g/kg $BW^{0.75}$, was unaffected by the increased level of saponin extract. Nonetheless, the ADG expressed as g/kg $BW^{0.75}$ showed a significant interaction between the level of saponin extract and the type of animals ($p = 0.009$), in which, results only represented by small ruminants. On the other hand, the meta-regression of milk yield expressed on kg/d and g/kg $BW^{0.75}$ was linearly decreased ($p < 0.005$) with increasing levels of saponin extract supplementation. Moreover, supplementation with saponin extract and type of animal showed an interaction effect on milk yield expressed as kg/d, g/kg $BW^{0.75}$, and g/kg DMI ($p < 0.05$), but the interaction between the level and type of saponin extracts was shown only when milk yield was expressed as kg/d ($p = 0.01$). The regression model in Figure 2 also confirmed that the milk yield, expressed as g/kg $BW^{0.75}$, linearly decreased with the increased level of saponin extracts ($p = 0.007$, $R^2 = 0.443$), with high accuracy of adjusted determination of the predicted model ($R^2 = 0.936$). Increasing levels of saponin extract in the diet did not affect milk fat, milk protein, or milk lactose proportion in dairy ruminants. However, the interaction between the level of saponin extract on the type of animals and the type of saponin source was shown in milk fat proportion ($p < 0.05$). In contrast, the milk protein proportion showed a significant interaction between the level of saponin extract and the type of animals ($p < 0.01$). However, it tended to be significant when interacting with the type of saponin source ($p = 0.082$).

Interestingly, dry matter digestibility (DMD) showed a linear increase in response to the increased levels of saponin extract (Table 3; $p = 0.048$) respected with the interaction on the type of animal ($p = 0.05$), although the crude protein digestibility (CPD) was unaffected. No effects were shown by the increased level of saponin extract on organic matter digestibility (OMD), although it showed an interaction with animal type. Nevertheless, the present results also revealed a significant decrease ($p = 0.036$) in acid detergent fiber digestibility (ADFD), followed by an interaction with animal type ($p = 0.05$). However, there was no significant effect of the increased level of saponin extract on neutral detergent fiber digestibility (NDFD), although it showed a tendency to significantly interact with animal type ($p = 0.064$).

The inclusion of saponin extract appeared to influence ruminal fermentation parameters on total volatile fatty acid (VFA) production and individual VFA composition, followed by interaction with animal type ($p < 0.05$) and saponin source type ($p < 0.001$), except for pH and ammonia concentration that showed no significant effects. Moreover, the ruminal protozoa population significantly increased saponin extract levels in a quadratic pattern, including a significant interaction effect on saponin sources ($p < 0.001$).

The absence of effects of saponin extract supplementation on fecal N and N retention was also observed, but fecal N showed an interaction with animal type ($p = 0.009$). Meanwhile, Urine N tended to decrease with increasing levels of saponin extract supplementation ($p = 0.057$), with no interaction. Blood plasma biochemical parameters, including total protein, albumin, globulin, plasma urea nitrogen, cholesterol, glucose, and alkaline phosphatase (ALP), did not affect the increased level of saponin extract supplementation. However, cholesterol, glucose, and ALP tended to interact with type saponin sources ($0.05 < p < 0.10$). Nonetheless, blood creatinine concentration was increased in a quadratic manner ($p = 0.014$), followed by an interaction with the type of saponin source ($p < 0.001$).

3.3. Comparative Analysis of Saponins' Sources

The comparative effects of different saponin sources on ruminants are presented in Table 4 and Figure 1. The results showed that the inclusion of various saponin extracts in the diets did not affect ($p > 0.05$) DMI, OMI, and NDFI but showed significant differences in the interaction between the type of saponin extract and the type of animals ($p < 0.05$). Moreover, different effects on ADG were observed; both the AA and QS extracts had relatively higher ADG, and the YS extract was relatively lower compared to the control group (CON). As displayed in Figure 3, dietary inclusion of YS extract reduced ($p < 0.05$) the ADG of ruminants. At the same time, QS and AA favorably increased ($p < 0.05$) the ADG of ruminants, where the ADG parameter in the database was only shown for the small ruminant type.

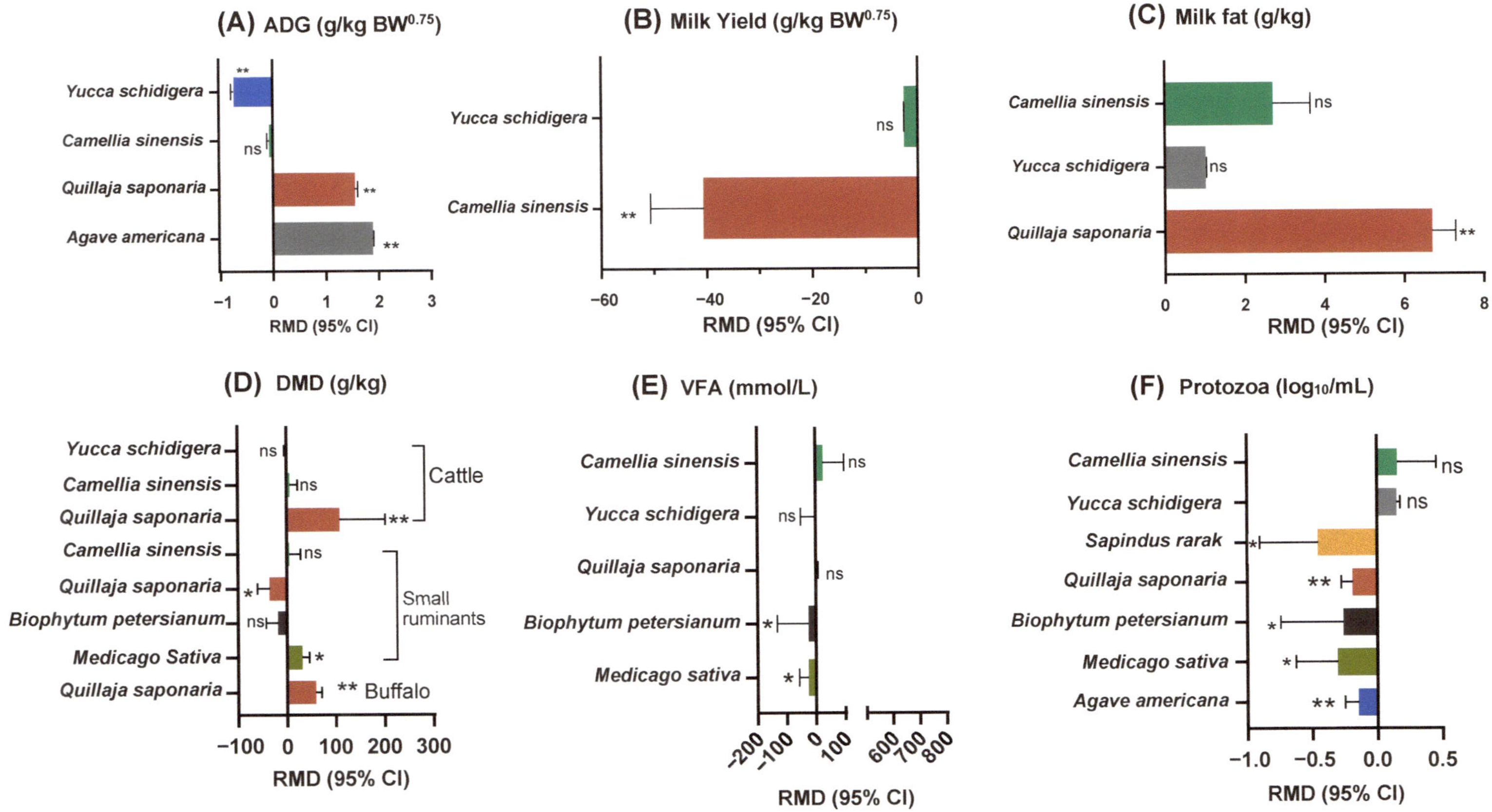

Figure 3. The effects of different sources of saponin extracts on average daily gain, dry matter intake, total volatile fatty acids production, and rumen protozoa population in ruminants. Results are presented as raw mean difference at 95% confidence intervals with control diets as a comparator. *: $p < 0.05$; **: $p < 0.01$; ns: no significance.

Table 4. Results of meta-analysis based on the sources of saponins.

Estimated Variables	Unit	*n*	Experimental Groups								SEM	*p*-Value	Source × Animal
			CON	MS	QS	CS	YS	AA	BP	SR			
Feed Intake													
DMI	g/kg $BW^{0.75}$	67	112.8	113.6	111.1	112.4	111.2	-	-	-	4.92	0.892	0.015
OMI	g/kg $BW^{0.75}$	12	84.92	-	57.26	84.77	-	-	-	-	9.16	0.776	0.043
NDFI	g/kg $BW^{0.75}$	12	39.53	-	18.54	39.99	-	-	-	-	3.15	0.142	0.017
Gain performance													
ADG	g/d	18	171.1 ^{ab}	-	188.0 ^{a}	169.9 ^{ab}	164.8 ^{b}	189.4 ^{a}	-	-	44.43	0.005	-
	g/kg $BW^{0.75}$	18	11.707 ^{b}	-	13.25 ^{a}	11.63 ^{b}	10.97 ^{c}	13.58 ^{a}	-	-	2.21	0.025	-
Milk production and composition													
Milk yield	kg/d	21	25.80 ^{a}	-	26.72 ^{a}	20.81 ^{b}	25.55 ^{a}	-	-	-	2.97	0.001	-
	g/kg $BW^{0.75}$	19	210.4 ^{a}	-	218.4 ^{a}	170.0 ^{b}	208.0 ^{a}	-	-	-	16.33	0.001	-
	g/kg DMI	21	1.247	-	1.311	1.155	1.255	-	-	-	0.08	0.146	-
Milk fat	g/kg	21	41.52 ^{b}	-	48.22 ^{a}	44.23 ^{ab}	42.54 ^{b}	-	-	-	3.48	<0.001	-
Milk protein	g/kg	17	32.97	-	39.84	31.84	32.55	-	-	-	3.08	0.084	-
Milk lactose	g/kg	12	46.92	-	50.11	48.08	46.64	-	-	-	2.19	0.339	-
Digestibility													
DMD	g/kg	40	700.4	730.5	717.02	705.7	697	-	683.4	-	29.41	0.491	0.014
OMD	g/kg	57	687.5	717	699.1	692.9	682.8	703.2	677.9	663.8	33.11	0.408	0.003
CPD	g/kg	44	644.7	-	638.1	645.5	647.3	673.8	628.2		38.59	0.658	0.321
NDFD	g/kg	47	563.2	-	568.3	550.6	560.8	569.7	551.3	547.6	47.46	0.972	0.321
ADFD	g/kg	27	494.9	-	-	484	493.8	-	496	-	48.29	0.956	0.269
Rumen fermentation parameters													
pH		148	6.28	6.09	6.26	6.11	6.31	6.14	6.8	-	0.26	0.341	0.440
NH_3	mg/dL	146	23.6	27.19	19.14	21.83	23.26	23.66	17.42	19.78	6.64	0.386	0.224
Total VFA	mmol/L	74	104.9 ^{a}	60.12 ^{b}	116.5 ^{a}	106.9 ^{a}	111.1 ^{a}	-	63.79 ^{b}	-	57.83	<0.001	0.224
Acetate	%	71	63.34	64.14	29.99	66.45	63.75	-	60.36	66.32	10.27	0.106	0.798
Propionate	%	71	20.41	-	17.52	19.57	21.47	-	26.58	21.66	-	0.665	0.156
Butyrate	%	71	11.96	13.59	7.41	10.84	11.5	-	9.01	8.87	2.95	0.295	0.541
Valerate	%	29	1.34	-	0.9	-	1.44	-	1,17	-	0.42	0.782	0.126
A:P ratio		71	3.53	3.48	1.76	3.58	3.49	-	2.67	3.22	1.47	0.521	0.240
Protozoa	log_{10}/mL	91	5.48 ^{a}	5.17 ^{b}	5.29 ^{b}	5.63 ^{a}	5.63 ^{a}	5.33 ^{b}	5.22 ^{b}	5.02 ^{b}	0.25	0.009	0.287

Table 4. *Cont.*

Estimated Variables	Unit	n	Experimental Groups								SEM	p-Value	Source × Animal
			CON	MS	QS	CS	YS	AA	BP	SR			
N Partitioning													
Urine N	g/kg $BW^{0.75}$	35	619.9	622.8	521.9	637.1	615.0	597.5	-	570.84	51.20	0.614	0.245
Fecal N	g/kg $BW^{0.75}$	35	572.6	557.2	590.1	567.9	574.8	584.8	-	554.91	46.47	0.997	0.001
N retention	g/kg $BW^{0.75}$	23	463.2	-	537.3	368.5	486.2	523.5	-	459.46	49.82	0.188	-
Blood parameters													
Total protein	g/dL	20	6.66	-	7.4	6.63	6.2	7.85	-	-	0.95	0.279	0.689
Albumin	g/dL	13	2.92	-	3.24	3.1	2.91	-	-	-	0.22	0.308	-
Globulin	g/dL	10	3.64	-	4.20	3.51	3.32	-	-	-	0.35	0.228	-
PUN	mg/dL	72	18.07 ^{b}	-	17.1335 ^{b}	19.40 ^{b}	17.90 ^{b}	23.04 ^{a}	-	-	1.20	0.001	0.345
Cholesterol	mg/dL	21	87.87	-	81.65	84.14	91.55	73.57	-	-	27.63	0.287	0.711
Creatinine	µmol/L	28	80.16 ^{b}	-	87.67 ^{b}	72.35 ^{b}	80.14 ^{b}	122.0 ^{a}	-	-	12.29	<0.001	-
Glucose	mg/dL	20	53.65 ^{a}	-	31.03 ^{b}	53.56 ^{a}	49.90 ^{a}	31.62 ^{b}	-	-	13.02	0.005	-
ALP	IU/L	10	170.8	-	-	175.9	161.9	-	-	-	32.69	0.388	-

DMI = Dry Matter intake; OMI = Organic matter intake; NDFI = neutral detergent fiber intake; $BW^{0.75}$ = metabolic body weight; ADG = Average Daily Gain; DMD = dry matter digestibility; OMD = Organic matter digestibility; CPD = Crude Protein digestibility; NDFD = neutral detergent fiber digestibility; ADFD = acid detergent fiber digestibility; NH_3 = ammonia; VFA = volatile fatty acids; A:P ratio = acetate propionate ratio; N = nitrogen; PUN = plasma urea nitrogen; ALP = alkaline phosphatase; SEM = standard error of the means; CON = Control diet, MS = *Medicago sativa*, QS = *Quilaja Saponaria*, CS = *Camelia sinensis*, YS = *Yucca schidigera*, AA = *Agave americana*, BP = *Biophytum petersianum*, SR = *Sapindus rarak*; ^{a,b} Different superscript within the row significant at $p < 0.05$.

The effects of CS extract were consistent in dairy animals, whereas CS decreased ($p < 0.05$) milk yield (expressed as kg/d and g/kg $BW^{0.75}$ units ($p < 0.05$; Figure 3B). Moreover, a higher ($p < 0.05$) milk fat proportion was observed in the QS extract group than in the CON group ($p < 0.001$; Figure 3C). In general, there were no effects of various saponin sources on digestibility parameters compared to CON, but when they showed an interaction between sources and type of animals, DMD and OMD were significant ($p < 0.05$). For example, in Figure 3D, there were showed that the digestibility among different types of ruminants showed an increased DMD on QS for large ruminants, i.e., buffalo and cattle, compared to the CON group ($p < 0.05$). However, in small ruminants, the supplementation of QS lowered DMD compared to the CON group, whereas MS had a higher DMD than the control ($p < 0.05$).

Regarding rumen fermentation parameters, only supplementation with MS and BP extracts resulted in a lower total VFA concentration than the CON group ($p < 0.01$). Nonetheless, supplementation with MS, QS, AA, BP, and SR extracts ($p < 0.05$) significantly diminished ruminal protozoa population, except for the CS and YS extracts. No effects of various types of saponin extract on N partitioning parameters were observed, except for the interaction of sources with animal types on Fecal N ($p = 0.001$). However, the AA extract had higher PUN and creatinine concentrations in the blood than the CON group ($p < 0.001$). Moreover, QS and AA extracts significantly reduced blood glucose concentration in blood than the CON ($p = 0.005$).

4. Discussion

Many studies have been conducted over the past few decades to investigate the potential benefits and drawbacks of saponins on the health and productivity of ruminants. Researchers have primarily focused on identifying natural sources of saponins and observing their positive effects on livestock health, production performance, and enteric methane emissions from ruminants [9,10,24]. Most studies have widely recognized the use of saponins in ruminants, either in their whole plant form or as an extracted defaunation agent, which can improve nutrient utilization efficiency and consequently affect production and product quality, as well as environmental impacts such as methane emissions [37–39]. However, the current study's hypothesis is focused on systematically determining the influence of saponin utilization in the extracted form, which consists of pure saponin compounds rather than the whole plant parts that considered as a source of saponins. The effects of saponin extract levels on observed parameters were analyzed in a meta-regression analysis, which was then compared to the type of saponin extracts. The results revealed the association between the levels of saponin extract and the type of animals or sources of saponin extract on the observed parameters. A comparative meta-analysis was performed to determine the association between the type of saponin extract and the influencing parameters in ruminants.

4.1. *Influence of Saponin Extract Utilization on Ruminant Performance, Digestibility, Rumen Fermentation and Health Parameters*

The current study revealed that various levels of saponin extract up to 40 g/kg DM did not negatively affect feed intake, indicating that the palatability of the diets was generally unaffected. However, it was found that the type of animal and the dietary intake of saponin extract interacted, which led to an increased interest in the differences in feed intake between large and small ruminants. Additionally, factors such as the weight and size of ruminants can affect the efficiency of nutrient uptake from diets containing saponin extracts. The quadratic regression model in Figure 2 shows that small ruminants had a reduced palatability pattern (DMI, g/kg $BW^{0.75}$) due to an increased level of dietary saponin extract (g/kg DM; $p < 0.036$), whereas there was no significant effect on larger ruminants. Some studies have suggested that saponins may decrease feed intake owing to their bitter taste [40,41]. Therefore, it is suggested that large ruminants have a more diverse palatability to tolerate the bitter taste of saponins than small ruminants [9,42]. This evidence indicates that large ruminants may naturally mitigate the adverse effects of saponin extract and have more efficient energy utilization when fed diets containing saponin extract [43].

Moreover, the interaction effects between the level of saponin extract with extract sources and animal types on nutrient digestibility were also reported in this study. DMD and OMD rates were increased ($p < 0.05$), while adverse effects were observed on NDFD and ADFD rates, where saponin extract levels interacted with ruminant type ($0.05 < p < 0.10$ and $p = 0.005$, respectively). These findings show the complexity of the noticeable influence of saponin extract mode of action that is associated with ruminant types, rumen microbiota, and digestion kinetics, where large ruminants possess a larger rumen volume that may extend their environment for microbial activity, although the bioactivity of saponin might also influence the fermentation process [38,43,44]. Furthermore, impaired digestibility might further enforce performance production in ruminants.

Despite the fact that the application of dietary saponin extract at various levels did not produce any discernible effect, the average daily weight gain (ADG) of ruminants that consumed saponin at different levels was found to be influenced by both the type of animal and the source of saponin extract. The results revealed that there was a significant interaction between these two factors, with the type of animals expressed as g/d ($p = 0.021$) and the type of saponin extract source expressed as g/kg $BW^{0.75}$ ($p = 0.009$). However, the current findings cannot be relied upon to draw definitive conclusions regarding the effect of increased saponin extract levels on daily weight gain, as the available evidence may be limited. The ADG data in the current study were collected only for small ruminants, and the inclusion of saponin extract in the meta-regression model negatively affected their natural metabolism. These findings underscore the potential differential bioactivity of saponin compounds, which may depend on the physiological characteristics of ruminants, as previously reported in the literature [9,44].

Decreased milk yield and altered composition indicate a complex interplay of factors during ruminal fermentation and nutrient absorption, which is influenced by the mode of action of the saponin extract ($p < 0.01$). The effect of various saponin extract levels on milk yield, expressed as g/kg $BW^{0.75}$, was evident, with a linear decrease observed as saponin extract supplementation increased in the diet ($p < 0.01$; Figure 2). Although the influence of supplementation levels and different types of saponin extracts on ruminants may vary, the antimicrobial activity of saponins against certain protozoa and bacteria could result in a decrease in volatile fatty acid (VFA) synthesis, which is essential for the metabolizable energy required for dairy ruminants to produce milk [33,34,45]. In the present study, the total VFA concentration, as well as the proportion of acetates and valerates, decreased due to increased saponin extract levels ($p < 0.05$). However, the interaction between levels and source saponin extract, as well as between levels and type of animals, was confirmed for milk fat ($p < 0.01$) and protein proportion ($p < 0.10$). Some studies have demonstrated the effects of saponins on rumen microbial activity by inhibiting the lipid biohydrogenation (BH) process in the rumen [10,46]. Therefore, inhibiting the ruminal BH process of long fatty acids (FA) may further enhance the increased fat proportion in milk.

The link between increased saponin extract supplementation and altered ruminal fermentation products has been well documented in previous research. In these studies, the majority of saponins have demonstrated antiprotozoal and methanogenic effects [17,28,34]. However, a decrease in VFA and acetate proportion could be related to a reduction in fiber-degrading microorganisms, such as cellulolytic bacteria and rumen protozoa, which play a critical role in fiber degradation [36]. Currently there have been noticed that *Ruminococcus* sp. and *Bacillota* bacteria genera are responsible for the degradation of cellulose and hemicellulose [36]. On the other hand, other studies also assured that rumen protozoa produce enzymes essential for the breakdown of complex carbohydrates in plant material; thus, the diversity and abundance of these enzymes contribute to the breakdown and fermentation of fiber [47]. Additionally, previous studies have found that dietary saponins improve nutrient digestibility, particularly of fiber, by selectively inhibiting protozoa, which the increased the growth of fiber-degrading bacteria [47]. As a result, the degradation rate of feed particles in the rumen is influenced by a decrease in certain protozoan families. However, the present study did not analyze rumen bacterial activity because of the limited number of studies on saponin extract supplementation. Despite this, the impact of the antimicrobial activity of saponin extract on milk production can be explained by the role of digested nutrient metabolic pathways, which influence nutrient deposition in mammary glands [48].

The population of protozoa increased quadratically with saponin extract supplementation, with some types exerting an interaction effect ($p < 0.05$ and $p < 0.01$, respectively). Previous studies have shown that saponin can increase rumen microflora, such as cellulolytic bacteria, and inhibit protozoan activity, with effects closely related to increased ammonia and VFA modulation in the rumen [17,36]. However, the results of the present study contradict these findings. Different saponin sources may have varying effects on microbial populations. For example, *Sapindus rarak* and *Quilaja saponaria* extracts effectively reduced protozoa and bacterial numbers [28,35], whereas certain levels of *Yucca schidigera* and *Camelia sinensis* increased protozoa and bacterial populations [8,9]. These findings highlight the different susceptibility rates of direct and indirect effects of various types and levels of saponin extracts on ruminal microbes [28,48,49]. These inconsistencies in microbial populations could be due to the type of saponin source, dietary ration, and duration of supplementation [36]. Additionally, reduced ruminal microbial activity, which is often associated with increased milk production and metabolism, is linked to reduced enteric methane production [50]. However, the present meta-analysis did not provide substantial evidence regarding the influence of different types and levels of saponin extract supplementation on enteric methane emissions from ruminants in vivo. Therefore, further research is needed to investigate the long-term effects of saponin extracts from different sources or at various levels, and to determine their direct impact on enteric methane mitigation in ruminants.

The results of this study suggest that the saponin extract has a modulatory effect on the metabolic health of ruminants. The levels of urinary N excretion tended to decrease according to the quadratic model ($p = 0.08$), with no significant changes in N retention or fecal N. However, the effect of saponin extract on the N cycle could not be determined. These findings are supported by Wina et al. [28], who found that ruminal protozoa defaunated by *Sapindus rarak* extract supplementation resulted in decreased protein degradation. Tea saponin extract was also confirmed to lower ammonia levels and thus reduce urinary-N excretion. Hence, the effect of saponin extract might attributed to its ability to diminished protozoa whereas consequently enhance ruminal fiber-degrading bacteria, and increased microbial protein synthesis. Available dietary nitrogen then is captured by microbial biomass and passed to the intestine for absorption, rather than being lost as ammonia and excreted as urinary-N and fecal N [51].

In contrast, a higher blood creatinine concentration was identified in a quadratic model ($p = 0.014$) followed by the interaction model between levels and source of saponin extracts ($p < 0.001$). Additionally, there was a tendency for an interaction between levels and sources of saponin extracts in cholesterol, glucose, and alkaline phosphatase in blood plasma ($0.05 < p < 0.10$). It appears that the ruminal microbial activity in ruminants fed specific types of saponin extracts affects the post-ruminant nutrient absorption process, which further absorbs metabolizable nutrients that are also modulated in the blood. Lowering cholesterol and glucose levels and increasing creatinine in blood serum in ruminants fed with saponin extract might indicate favorable health conditions. Previous studies have confirmed that changes in blood serum parameters, such as decreases in serum urea, creatinine, cholesterol, and liver enzyme activities, suggest that saponin's anti-inflammatory and antioxidant modes of action could positively affect kidney function and metabolic health [1,52]. However, increased creatinine levels in the present meta-analysis might demonstrate the inconsistencies and high variability of experimental studies in ruminants fed with saponin extract. Hence, future studies may raise concerns regarding the effects of saponins on ruminant health and organ function, such as the kidneys. Previous clinical studies have confirmed that increased creatinine in the blood serum of living organisms might indicate kidney failure [53,54].

Supplementation with saponin extract has been associated with modifications in multiple performance parameters in ruminants, including feed intake, daily weight gain, milk yield, and milk nutrient composition. It has been observed that large and small ruminants exhibit distinct effects on milk yield and nutrient composition, particularly when supplemented with saponins [42,44]. Although the digestibility rates, ruminal fermentation effectivity, N utilization, and blood parameters between large and small ruminants also displayed intricacy, the present meta-analysis provides robust evidence regarding the effects of dietary saponin extracts on ruminant production, such as ADG and Milk yield, relative to ruminant metabolic weights (Figure 2). Moreover, a comparative analysis of different saponin extracts can elucidate the varying effects of each extract on the observed parameters in ruminants.

4.2. The Relationship between Ruminant Production Health and Metabolism by the Divergence Sources of Saponin Extract

The daily weight gain (ADG) of ruminants, expressed as g/kg $BW^{0.75}$, significantly decreased when supplemented with YS extract, while positive effects were observed with QS and AA extracts (Figure 3A; $p < 0.05$) compared to the control (CON), which refers to small ruminants. Only three studies reported a positive effect of QS extract and AA on ADG parameters [3,33,34]. However, these corresponded to only one study that assessed the effect of QS extract on dairy buffaloes [1]. Similarly, only one study has reported the effect of QS extract on ADG parameters. Other studies have reported adverse effects of YS extract [2,32], whereas only one study reported that CS extract supplementation had no effect on ruminant daily weight gain [35]. Due to the small sample size, YS, QS, and AA extracts may not be statistically sufficient to provide general implications for observed ADG and thus warrant further investigation.

The results indicated that only the tea saponin extract (CS) significantly reduced milk yield, expressed as g/d and g/kg $BW^{0.75}$ (Figure 3B; $p = 0.001$), compared to the control treatment. The strong bitter taste of the CS extract might impair ruminant palatability [35,38]. Similar to QS, CS contains triterpenoid saponins, a glycosides group that are well known for their remarkable bioactive diversity with multiple therapeutic benefits as anti-inflammatory, anti-microbial, and antioxidant properties [37]. In the studies using CS as the source of saponins, the results have been consistent to reduce milk yield [9–11,39]. Hence, the potent bioactivity of the CS extract could negatively affect rumen microflora, thereby long-term exposure of CS extract reducing the beneficial metabolizable energy to transporting nutrients [33,55] that are synthesized in milk concentrated in the mammary glands of dairy ruminants.

A meta-analysis by Yanza et al. [17] also confirmed that dietary CS extract (tea saponin) negatively affects ruminal protozoa and reduces digested dried matter, but positively influences VFA concentration and effectively mitigates methane emissions. In this study, digestibility was not reduced, indicating that the various saponin extracts had no negative impact on the degradation of fibrous nutrients. However, when digestibility (DMD) results were segregated based on ruminant type, QS extract was positively influenced in cattle and buffalo compared to the control ($p < 0.001$). Moreover, the QS extract negatively influenced small ruminants, whereas the MS extract positively affected DMD ($p < 0.05$). These findings suggest that only the MS extract (known as Lucerne or alfalfa) is susceptible to small ruminants, whereas the QS extract may have a detrimental effect, and further frequent uptake may lead to metabolic disorders. This evidence shows that the QS extract has stronger triterpene glucoside compounds than other saponin extract sources, which are toxic to small ruminants, especially when supplemented in a long term and at high levels [55]. However, large ruminants such as cattle and buffalo are more susceptible to saponin extract supplementation, and sophisticated results on digestibility have been obtained with QS extract supplementation. Therefore, different saponin extract sources are suitable only for specific types of livestock ruminants.

The results of this study indicate that supplementation with MS and AA extracts led to a reduction in rumen VFA concentrations compared to the control treatment ($p < 0.001$; Figure 3E). Although CS extract may positively influence the modulation of fiber-degrading bacteria, which could increase ruminal VFA concentration and fiber digestion, it may not be beneficial for improving lactating performance in dairy ruminants, specifically milk yield and its nutrient compounds. In contrast, the population of rumen protozoa was significantly decreased in response to MS, QS, AA BP, and SR extract supplementation ($p < 0.01$; Figure 3F), while no significant influence of CS, and YS extracts on ruminal protozoa population (Table 4). Correspondingly, although some sources of saponin extracts are well documented on exhibited a defaunating effect, the maximum protozoal reduction was only approximately 8% in this meta-analysis (SR vs. CON; 5.02 vs. 5.48 log10/mL; Table 4). Typically, negative effects on ruminal fermentation and nutrient digestibility occurred when the reduction of protozoal count was higher than 50%, as observed in previous in vitro studies using whole part of saponin sources levels as high as 40 g/kg DM [12,55], which was 10-fold higher than the inclusion levels in the studies we used. According to those studies, diets containing up to 0.4% of the whole part of saponin sources do not cause unfavorable effects on ruminal fermentation and nutrient digestion in ruminants.

The antimicrobial effect of saponin to protozoa has been described previously, because saponins form sterol-membrane complex damage and disintegrate the protozoa membrane and thus suppress the ruminal protozoa population [17,37]. However, the impact of CS extract on ruminal protozoa was inconsistent. The absence effect of CS extract on protozoal count, as reported by [10,11], coupled with the other studies reporting an increase in protozoal counts [9,17,41] suggested possible adaptation of ruminal protozoa against CS extract saponin. Perhaps, long-term exposure of ruminal protozoa to various saponin sources needs to be investigated. Generally, some previous study showed, that whole part saponin sources have been observed to have differing effects depending on their sources; however, their mode of action depends on their direct or indirect effect on the microorganisms involved in rumen fermentation [49,56]. Moreover, a high-forage diet for dairy cows containing high levels of feed like lucerne silage (source of saponin) may reduce protozoa population without negatively affecting the basic fermentation parameters and keeping milk production at a certain amount as in control group [50].

Only the QS extract supplementation resulted in a higher fat proportion in milk ($p < 0.001$). Although CS and YS extract might potentially increase fat proportion in milk but not significantly higher than the CON treatment (Figure 3F). This finding is aligned with previous studies reporting no effect of CS on milk fat and fatty acids profile [5,39]. This might be associated with the associated relationship between supplementary saponin extract with the increased activity of fiber degrading bacteria in producing acetate. Because acetate is known as the main precursor of milk fat biosynthesis [10]. Moreover, saponin mode of action might inhibit lipolytic bacteria on fatty acids BH process, hence, deposited essential unsaturated fatty acids may increase in milk fat [50]. Although no significant comparison compared to CON in the present results (Figure 3C), the positive effect of YC on milk fat synthesis has also been reported [29,31]. These evidences can be attributed to the effect of saponin extracts on modulating bacteria and protozoa activity those who responsible on producing VFAs and modulating fatty acids through BH process in the rumen [34,47,50]. Because the most noticeable effects can be seen from the decreased ciliate protozoal counts [23].

The administration of AA extract was associated with elevated levels of plasma urea-N and creatinine ($p = 0.001$). However, supplementation with QS or AA extract resulted in reduced blood glucose levels ($p < 0.005$). Elevated levels of urea and creatinine can be indicative of impaired kidney function or kidney disease, as the kidneys may not effectively filter these substances out of the blood [57]. Urea is a metabolic waste product that should be excreted in the urine, but the ability of AA extract to reduce protein synthesis can affect the levels of urea and creatinine in the bloodstream. The low levels of blood creatinine observed in ruminants supplemented with AA extract (1.22 μmol/L or 1.4 mg/dL) are still within the normal range (0.8 to 2.0 mg/dL) [58,59], suggesting no significant impact on kidney function.

Incorporating triterpenoid glycosides, such as AA and QS extracts, into the diet of ruminants may reduce the population of defaunated ruminal protozoa. These extracts form insoluble complexes, with hederagenin as the aglycone, that are effective in lowering blood glucose and cholesterol levels. Glucose is directly attached to hederagenin, while rhamnose and arabinose are linked to saponins in the rumen. These chemical linkages may be harmful to ruminal microbes, hindering the degradation and absorption of glucose and leading to changes in the bloodstream. The metabolism of plasma blood suggests that the bioactivity of saponin compounds can either promote or inhibit health conditions and growth performance, depending on the physiological characteristics of the animal. The effects of saponin extracts on ruminant health are dependent on the source of the saponins, as indicated by the results of the current study on the magnitude of the effects.

5. Conclusions

This meta-analysis found that the utilization of saponin extract in ruminants up to a certain threshold of 40 g/kg DM generally did not negatively impact feed intake and, therefore, did not compromise the palatability of ruminant diets. However, the type of ruminant and the level of saponin extract interaction suggests differential effects on nutrient absorption efficiency based on size. Specifically, small ruminants showed reduced palatability at higher saponin levels, indicating a species-specific tolerance threshold for the bitter taste associated with saponin extracts. This finding suggests that larger ruminants may possess adaptive mechanisms to mitigate the adverse taste effects of saponin extracts, potentially optimizing energy utilization from saponin extracts used in ruminant diets. The relationship between saponin extracts and various health and fermentation parameters is complex and shows that saponin supplementation has been linked to notable increases in average daily weight gain and milk yield, depending on the type of saponin extract and the animal's physiological condition. However, certain saponin extracts have been shown to have adverse effects on milk yield and composition due to alterations in ruminal fermentation and nutrient absorption. The antimicrobial properties of saponins, which can reduce methane emissions and improve nutrient digestibility, can also disrupt the ruminal microbial ecosystem. This disruption can lead to a decrease in volatile fatty acid synthesis, which is essential for energy metabolism in dairy ruminants, ultimately influencing the production performance of dairy ruminants.

Some studies suggest that saponins can lower blood glucose and cholesterol levels, potentially indicating a positive shift in metabolic health, others have raised concerns about the potential impacts on kidney function. In particular, increased blood creatinine levels associated with certain saponin extracts, such as *Agave americana*, necessitate a cautious approach to dietary supplementation. The meta-analysis conducted in this study highlights the role of different saponin extracts in ruminant nutrition and environmental management, advocating for a distinct understanding of their various bioactive properties. However, among the saponin sources, saponin extracted from *Quilaja saponaria* is more likely to improve large ruminant production performance while maintaining ruminant health and metabolism, but negatively affect small ruminants. Further research is needed to unravel the intricate effects of different saponin sources on ruminant health and productivity, emphasizing the importance of tailored dietary strategies that consider the unique physiological and metabolic characteristics of the target livestock.

Author Contributions: Conceptualization, methodology, data curation, original draft preparation, data analysis, project administration, resources, review and editing, Y.R.Y.; Conceptualization, methodology, data curation, original draft preparation, data analysis, visualization, A.I.; resources, data curation, data analysis, investigation, A.N.R. and F.R.; data curation, visualization, review and editing, supervision, A.J. and V.N.; data curation, visualization, review and editing, A.C. and M.S.-S.; data curation, data analysis and figure illustration, C.H. and A.F.; visualization and data curation, R.H. and U.H.T. All authors have read and agreed to the published version of the manuscript.

Funding: This study was funded by partial funding from Padjadjaran University through internal research grant, combined with partial funding from IPB University postdoctoral grant nr. IPB/SK/Rektor/74/2024, and partial funds from SSHN IFI, Government of France Nr. 757/IFI/2023.

Institutional Review Board Statement: Not applicable.

Informed Consent Statement: Not applicable.

Data Availability Statement: The dataset underpinning the results in the present study can be obtained from the corresponding author and will be considerable to share upon reasonable request.

Acknowledgments: The first author is also thankful to UMR Herbivore, INRAE, SSHN IFI-Government of France, PHC-Nusantara 2023 research grant nr. 15819/IT3.D10/PT.01.02/P/T/2023, on facilitating the postdoctoral internship in France to YRY hence this article has been written. The first author also thankful to IPB University who provide the postdoctoral research grant scheme 2024, grant nr. IPB/SK/Rektor/74/2024 under the supervision of AJ. All authors are also grateful to the Animal Feed and Nutrition Modelling (AFENUE) Research Group, Department of Nutrition and Feed Technology, Faculty of Animal Science, IPB University, Indonesia, for providing a collaborative platform across various institutions.

Conflicts of Interest: The author declares no conflict of interest.

References

1. Abdelmawla, S. Impacts of Natural Saponin Plant Extract without or with Fresh Baker's Yeast in Buffaloe Heifers Rations on Nutrients Digestibility, Some Ruminal Fermentation and Blood Serum Parameters. *J. Anim. Poult. Prod.* **2008**, *33*, 3303–3315. [CrossRef]
2. Alsubait, I.S.; Alhidary, I.A.; Al-Haidary, A.A. Effects of Different Levels of Yucca Supplementation on Growth Rates, Metabolic Profiles, Fecal Odor Emissions, and Carcass Traits of Growing Lambs. *Animals.* **2023**, *13*, 755. [CrossRef] [PubMed]
3. Baah, J.; Ivan, M.; Hristov, A.N.; Koenig, K.M.; Rode, L.M.; McAllister, T.A. Effects of potential dietary antiprotozoal supplements on rumen fermentation and digestibility in heifers. *Anim. Feed Sci. Technol.* **2007**, *137*, 126–137. [CrossRef]
4. Baheg, R.; El-Bahrawy, K.; El-Azrak, K.; Samak, M.; Sallam, S. Effect of Condensed Tannins and Saponin Supplementation on Reproductive Performance in Barki Ewes. *Egypt J. Nutr. Feeds.* **2017**, *20*, 197–210. [CrossRef]
5. Benchaar, C.; Chouinard, P.Y. Short communication: Assessment of the potential of cinnamaldehyde, condensed tannins, and saponins to modify milk fatty acid composition of dairy cows. *J. Dairy Sci.* **2009**, *92*, 3392–3396. [CrossRef] [PubMed]
6. Benchaar, C.; McAllister, T.A.; Chouinard, P.Y. Digestion, ruminal fermentation, ciliate protozoal populations, and milk production from dairy cows fed cinnamaldehyde, quebracho condensed tannin, or Yucca schidigera saponin extracts. *J. Dairy Sci.* **2008**, *91*, 4765–4777. [CrossRef]
7. Cieślak, A.; Szumacher-Strabel, M.; Stochmal, A.; Oleszek, W. Plant components with specific activities against rumen methanogens. *Animal* **2013**, *7*, 253–265. [CrossRef]
8. Eryavuz, A.; Dehority, B.A. Effect of Yucca schidigera extract on the concentration of rumen microorganisms in sheep. *Anim. Feed Sci. Technol.* **2004**, *117*, 215–222. [CrossRef]
9. Guyader, J.; Eugène, M.; Doreau, M.; Morgavi, D.P.; Gérard, C.; Loncke, C.; Martin, C. Nitrate but not tea saponin feed additives reduce methane in dairy cows. *J. Anim. Sci.* **2015**, *93*, 5367–5377. [CrossRef]
10. Guyader, J.; Eugène, M.; Doreau, M.; Morgavi, D.P.; Gérard, C.; Martin, C. Tea saponin reduced methanogenesis in vitro but increased methane yield in lactating dairy cows. *J. Dairy Sci.* **2017**, *100*, 1845–1855. [CrossRef]
11. Gunun, P.; Cherdthong, A.; Khejornsart, P.; Wanapat, M.; Polyorach, S.; Kang, S.; Kaewwongsa, W.; Gunun, N. The Effect of Phytonutrients in Terminalia chebula Retz. on Rumen Fermentation Efficiency, Nitrogen Utilization, and Protozoal Population in Goats. *Animals* **2022**, *12*, 2022. [CrossRef] [PubMed]
12. Hu, W.L.; Liu, J.X.; Ye, J.A.; Wu, Y.M.; Guo, Y.Q. Effect of tea saponin on rumen fermentation in vitro. *Anim. Feed Sci. Technol.* **2005**, *120*, 333–339. [CrossRef]
13. Hussain, I.; Cheeke, P.R. Effect of dietary Yucca schidigera extract on rumen and blood profiles of steers fed concentrate- or roughage-based diets. *Anim. Feed Sci. Technol.* **1995**, *51*, 231–242. [CrossRef]
14. Jayanegara, A.; Sujamoko, T.U.P.; Ridla, M.; Kondo, M.; Kreuzer, M. Silage quality as influenced by concentration and type of tannins present in the material ensiled: A meta-analysis. *J. Anim. Physiol. Anim. Nutr.* **2019**, *103*, 456–465. [CrossRef]

15. Kumar, M.; Kannan, A.; Bhar, R.; Gulati, A.; Gaurav, A.; Sharma, V.K. Nutrient intake, digestibility and performance of Gaddi kids supplemented with tea seed or tea seed saponin extract. *Asian-Australas. J. Anim. Sci.* **2017**, *30*, 486–494. [CrossRef]
16. Li, Z.; Liu, D.; Ma, H.; Fu, B.; Wang, L.; Gou, Z.; Sun, J.; He, X.; Liu, C. Effects of Yucca on the sheep ruminal fermentation and some serum parameters. In Proceedings of the ITME 2011—IEEE International Symposium on IT in Medicine and Education, Cuangzhou, China, 9–11 December 2011; Volume 2, pp. 541–545. [CrossRef]
17. Yanza, Y.R.; Mahmudah, N.; Hidayat, C.; Jayanegara, A.; Fitri, A.; Ramadhani, F.; Syarifah, H.; Basri Jumin, H.; Andrian, D.; Melissa; et al. Tea Saponin Modulates In Vitro Rumen Fermentation Profile and Reduces Methane Production: A Meta-Analysis. *Biotropia* **2023**, *30*, 82–90. [CrossRef]
18. Respati, A.N.; Yanza, Y.R.; Yano, A.A.; Astuti, D.; Ningsih, N.; Triswanto; Purnamayanti, L.; Gading, B.M.; Wardani, W.W.; Jayanegara, A.; et al. Meta-analysis of the effects of dietary sources of selenium on lactational performance and oxidative status of dairy cows. *Anim. Feed Sci. Technol.* **2023**, *305*, 115782. [CrossRef]
19. Yanza, Y.R.; Fitri, A.; Suwignyo, B.; Elfahmi; Hidayatik, N.; Kumalasari, N.R.; Irawan, A.; Jayanegara, A. The Utilisation of Tannin Extract as a Dietary Additive in Ruminant Nutrition: A Meta-Analysis. *Animals* **2021**, *11*, 3317. [CrossRef] [PubMed]
20. St-Pierre, N.R. Invited review: Integrating quantitative findings from multiple studies using mixed model methodology. *J. Dairy Sci.* **2001**, *84*, 741–755. [CrossRef]
21. Yanza, Y.R.; Szumacher-Strabel, M.; Jayanegara, A.; Kasenta, A.M.; Gao, M.; Huang, H.; Patra, A.K.; Warzych, E.; Cieslak, A. The Effects of Dietary Medium-Chain Fatty Acids on Ruminal Methanogenesis and Fermentation In Vitro and In Vivo: A Meta-analysis. *J. Anim. Physiol. Anim. Nutr.* **2021**, *105*, 874–889. [CrossRef]
22. Valdez, F.R.; Bush, L.J.; Goetsch, A.L.; Owens, F.N. Effect of steroidal sapogenins on ruminal fermentation and on production of lactating dairy cows. *J. Dairy Sci.* **1986**, *69*, 1568–1575. [CrossRef] [PubMed]
23. Lu, C.D.; Jorgensen, N.A. Alfafa saponins affect site and extent of nutrient digestion in ruminants. *J. Nutr.* **1987**, *117*, 919–927. [CrossRef] [PubMed]
24. Wu, Z.; Sadik, M.; Sleiman, F.T.; Simas, J.M.; Pessarakli, M.; Hubers, J.T. Influence of yucca extract on ruminal metabolism in cows. *J. Anim. Sci.* **1994**, *72*, 1038–1042. [CrossRef] [PubMed]
25. Wilson, R.C.; Overton, T.R.; Clark, J.H. Effects of Yucca shidigera extract and soluble protein on performance of cows and concentrations of urea nitrogen in plasma and milk. *J. Dairy Sci.* **1998**, *81*, 1022–1027. [CrossRef] [PubMed]
26. Sliwiński, B.J.; Kreuzer, M.; Wettstein, H.R.; Machmuller, A. Rumen fermentation and nitrogen balance of lambs fed diets containing plant extracts rich tannins and saponins, and associated emissions of nitrogen and methane. *Anim. Nutr.* **2002**, *56*, 379–392. [CrossRef] [PubMed]
27. Santoso, B.; Kilmaskossu, A.; Sambodo, P. Effects of saponin from Biophytum petersianum Klotzsch on ruminal fermentation, microbial protein synthesis and nitrogen utilization in goats. *Anim. Feed Sci. Technol.* **2006**, *137*, 58–68. [CrossRef]
28. Wina, E.; Muetzel, S.; Becker, K. Effects of daily and interval feeding of Sapindus rarak saponins on protozoa, rumen fermentation parameters and digestibility in sheep. *Asian-Australas. J. Anim. Sci.* **2006**, *19*, 1580–1587. [CrossRef]
29. Lovett, D.K.; Stack, L.; Lovell, S.; Callan, J.; Flynn, B.; Hawkins, M.; Mara, F.P.O. Effect of feeding Yucca schidigera extract on performance of lactating dairy cows and ruminal fermentation parameters in steers. *Livest. Sci.* **2006**, *102*, 23–32. [CrossRef]
30. Liu, C.; Li, Z.; Juan, D.; An-shan, S. The Effect of Yucca schidigera extract on ruminal fermentation and parameters traits in sheep. *J. Agr. Sci.* **2007**, *6*, 121–128. [CrossRef]
31. Singer, M.D.; Robinson, P.H.; Salem, A.Z.M.; Depeters, E.J. Impacts of rumen fluid modified by feeding Yucca schidigera to lactating dairy cows on in vitro gas production of 11 common dairy feedstuffs, as well as animal performance. *Anim. Feed Sci. Technol.* **2008**, *146*, 242–258. [CrossRef]
32. Selcuk, Z.; Tuncer, S.D. The effects of different levels of Yucca schidigera added to the lamb's diets containing urea on growth performance, carcass characteristic, some rumen, and blood parameters. *Anim. Veter Adv.* **2010**, *9*, 654–660. [CrossRef]
33. Nasri, S.; Ben Salem, H.; Vasta, V.; Abidi, S.; Makkar, H.; Priolo, A. Effect of increasing levels of Quillaja saponaria on digestion, growth, meat quality of barbarine lamb. *Anim. Feed Sci. Technol.* **2011**, *164*, 71–78. [CrossRef]
34. Nasri, S.; Ben Salem, H. Effect of oral administration of Agave americana or Quillaja saponaria extracts on digestion and growth of barbarine female lamb. *Livest. Sci.* **2012**, *147*, 59–65. [CrossRef]
35. Zhang, F.; Li, B.; Ban, Z.; Liang, H.; Li, L.; Zhao, W.; Yan, X. Evaluation of origanum oil, hydrolysable tannins and tea saponin in mitigating ruminant methane: In vitro and In vivo methods. *Anim. Physiol. Anim. Nutr.* **2021**, *105*, 630–638. [CrossRef] [PubMed]
36. Yi, X.; Wu, B.; Ma, J.; Cui, X.; Deng, Z.; Hu, S.; Li, W.; Runa, A.; Li, X.; Meng, Q.; et al. Effects of dietary capsaicin and Yucca schidigera extracts as feed additives on rumen fermentation and microflora of beef cattle fed with a moderate-energy diet. *J. Ferment.* **2022**, *9*, 30. [CrossRef]
37. Wang, J.K.; Ye, J.A.; Liu, J.X. Effects of tea saponins on rumen microbiota, rumen fermentation, methane production and growth performance—A review. *Trop. Anim. Health Prod.* **2012**, *44*, 697–706. [CrossRef] [PubMed]
38. Pen, B.; Sar, C.; Mwenya, B.; Kuwaki, K.; Morikawa, R.; Takahashi, J. Effects of Yucca schidigera and Quillaja saponaria extracts on in vitro ruminal fermentation and methane emission. *Anim. Feed Sci. Technol.* **2006**, *129*, 175–186. [CrossRef]
39. Wang, B.; Tu, Y.; Zhao, S.P.; Hao, Y.H.; Liu, J.X.; Liu, F.H.; Xiong, B.H.; Jiang, L.S. Effect of tea saponins on milk performance, milk fatty acids, and immune function in dairy cow. *J. Dairy Sci.* **2017**, *100*, 8043–8052. [CrossRef]
40. Wina, E.; Muetzel, S.; Becker, K. The impact of saponins or saponin-containing plant materials on ruminant production--a review. *J. Agric. Food Chem.* **2005**, *53*, 8093–8105. [CrossRef]

41. Zhou, L.; Shi, Y.; Guo, R.; Liang, M.; Zhu, X.; Wang, C. Digital Gene-Expression Profiling Analysis of the Cholesterol-Lowering Effects of Alfalfa Saponin Extract on Laying Hens. *PLoS ONE* **2014**, *9*, e98578. [CrossRef]
42. Provenza, F.D.; Villalba, J.J.; Dziba, L.E.; Atwood, S.B.; Banner, R.E. "Linking herbivore experience, varied diets, and plant biochemical diversity. *Small Rumin. Res.* **2015**, *64*, 157–166. [CrossRef]
43. Kamra, D.N. Rumen microbial ecosystem. *Curr. Sci.* **2011**, *101*, 845–858. Available online: http://www.jstor.org/stable/24110438 (accessed on 29 March 2024).
44. Clauss, M.; Hume, I.D.; Hummel, J. Evolutionary adaptations of ruminants and their potential relevance for modern production systems. *Animal* **2010**, *8*, 979–992. [CrossRef]
45. Mao, H.L.; Wang, J.K.; Zhou, Y.Y.; Liu, J.X. Effects of addition of tea saponins and soybean oil on milk performance, milk fatty acid composition, and nutrient digestibility in dairy cows. *Livest. Sci.* **2010**, *129*, 56–62. [CrossRef]
46. Tavendale, M.H.; Meagher, L.P.; Pacheco, D.; Walker, N.; Attwood, G.T.; Sivakumaran, S. Methane production from in vitro rumen incubations with *Lotus pedunculatus* and *Medicago sativa*, and effects of extractable condensed tannin fractions on methanogenesis. *Anim. Feed Sci. Technol.* **2005**, *123–124*, 403–419. [CrossRef]
47. Sophie Comtet-Marre, N.; Parisot, P.; Lepercq, F.; Chaucheyras-Durand, P.; Mosoni, E.; Peyretaillade, A.; Bayat, K.; Shingfield, P.P.; Forano, E. Metatranscriptomics Reveals the Active Bacterial and Eukaryotic Fibrolytic Communities in the Rumen of Dairy Cow Fed a Mixed Diet. *Front. Microbiol.* **2017**, *8*, 67. [CrossRef]
48. Anantasook, N.; Wanapat, M.; Cherdthong, A.; Gunun, P. Effect of tannins and saponins in Samanea saman on rumen environment, milk yield and milk composition in lactating dairy cows. *J. Anim. Physiol. Anim. Nutr.* **2015**, *99*, 335–344. [CrossRef]
49. Kozłowska, M.; Cieślak, A.; Jóźwik, A.; El-Sherbiny, M.; Stochmal, A.; Oleszek, W.; Kowalczyk, M.; Filipiak, W.; Szumacher-Strabel, M. The effect of total and individual alfalfa saponins on rumen methane production. *J. Sci. Food Agric.* **2020**, *100*, 1922–1930. [CrossRef]
50. Kozłowska, M.; Cieślak, A.; Jóźwik, A.; El-Sherbiny, M.; Gogulski, M.; Lechniak, D.; Gao, M.; Yanza, Y.R.; Vazirigohar, M.; Szumacher-Strabel, M. Effects of partially replacing grass silage by lucerne silage cultivars in a high-forage diet on ruminal fermentation, methane production, and fatty acid composition in the rumen and milk of dairy cows. *Anim. Feed Sci. Technol.* **2021**, *277*, 114959. [CrossRef]
51. Patra, A.; Saxena, J. The effect and mode of action of saponins on the microbial populations and fermentation in the rumen and ruminant production. *Nutr. Res. Rev.* **2009**, *22*, 204–219. [CrossRef]
52. Lu, C.; Fan, G.; Wang, D. Akebia Saponin D ameliorated kidney injury and exerted anti-inflammatory and anti-apoptotic effects in diabetic nephropathy by activation of NRF2/HO-1 and inhibition of NF-KB pathway. *Int. Immunopharmacol.* **2020**, *84*, 106467. [CrossRef]
53. Du, Y.; Wang, L.; Qian, J.; Zhang, K.; Chai, K. Panax notoginseng saponins protect kidney from diabetes by up-regulating silent information regulator 1 and activating antioxidant proteins in rats. *Chin. J. Integr. Med.* **2016**, *22*, 910–917. [CrossRef]
54. Chen, G.; Wu, S.; Na, S.; Li, L. Effect of total saponin of dioscorea on uric acid excretion indicators in chronic hyperuricemia rats. *Chin. J. Integr. Tradit. West. Med.* **2014**, *341*, 75–80.
55. Francis, G.; Kerem, Z.; Makkar, H.P.; Becker, K. The biological action of saponins in animal systems: A review. *Brit. J. Nutr.* **2002**, *88*, 587–605. [CrossRef]
56. Wang, B.; Ma, M.; Diao, Q.; Tu, Y. Saponin-Induced Shifts in the Rumen Microbiome and Metabolome of Young Cattle. *Front. Microbiol.* **2019**, *10*, 356. [CrossRef]
57. Constable, P.D.; Hinchcliff, K.W.; Done, S.H.; Grünberg, W. *Veterinary Medicine: A Textbook of the Diseases of Cattle, Horses, Sheep, Pigs, and Goats*, 11th ed.; Elsevier: Amsterdam, The Netherlands, 2017.
58. Kaneko, J.J.; Harvey, J.W.; Bruss, M.L. *Clinical Biochemistry of Domestic Animals*, 6th ed.; Academic Press: Cambridge, MA, USA, 2008.
59. Thrall, M.A.; Weiser, G.; Allison, R.W.; Campbell, T.W. *Veterinary Hematology and Clinical Chemistry*; Wiley-Blackwell: Hoboken, NJ, USA, 2012.

animals

Article

Effect of Fermented Concentrate on Ruminal Fermentation, Ruminal and Fecal Microbiome, and Growth Performance of Beef Cattle

Seul Lee *,†, Chae Hwa Ryu †, Youl Chang Back, Sung Dae Lee and Hyeran Kim

Animal Nutrition and Physiology Division, National Institute of Animal Science, Rural Development Administration, Wanju 55365, Republic of Korea; chryu0629@korea.kr (C.H.R.)
* Correspondence: tabababy@korea.kr; Tel.: +82-63-238-7470
† These authors contributed equally to this work.

Simple Summary: Fermented feed is utilized to promote the digestion and absorption of nutrients in animal diets while enhancing the host's immune system and overall health. Fermented feed has been shown to increase the efficiency of nutrient digestion and absorption in livestock, thereby reducing waste production. Additionally, fermentation helps eliminate pathogenic microorganisms that may be present in the feed. The effects of fermented feed, which often contains beneficial microorganisms such as probiotics and yeast, have been extensively studied in various animal species. In ruminant animals, research has predominantly focused on the effects of fermented silage, with limited studies on the feeding effects of fermented concentrate. In this study, we show that fermenting formulated concentrate can have a positive impact on the growth and health of ruminant livestock.

Abstract: The impact of fermented concentrate on the growth and rumen health of beef cattle remains an area of emerging research. This study aimed to assess the influence of a fermented concentrate (TRT) compared to a conventional concentrate (CON) on the growth, rumen fermentation characteristics, and microbiota composition in Korean cattle. Using a crossover design, eight cattle were alternately fed TRT and CON diets, with subsequent analysis of feed components, rumen fermentation parameters, and microbial profiles. TRT and CON diets did not differ significantly in their effect on animal growth metrics. However, the TRT diet was associated with reduced digestibility of rapidly degradable carbohydrates and modified rumen fermentation patterns, as evidenced by an elevated pH and increased acetate-to-propionate ratio ($p < 0.05$). Furthermore, the TRT diet increased the abundance of lactic acid bacteria, *Bacillus*, and yeast and organic acid levels in the rumen ($p < 0.05$). Moreover, Lachnospiraceae and Bacteroidales populations in the rumen and fecal *Akkermansia* abundance increased in the TRT group compared to the CON group. These microbial changes suggest a potential enhancement of the immune system and overall health of the host. Further research on the long-term implications of incorporating fermented concentrate into cattle diets is warranted.

Keywords: fermented concentrate; beef cattle; growth performance; rumen fermentation; microbial diversity

Citation: Lee, S.; Ryu, C.H.; Back, Y.C.; Lee, S.D.; Kim, H. Effect of Fermented Concentrate on Ruminal Fermentation, Ruminal and Fecal Microbiome, and Growth Performance of Beef Cattle. *Animals* **2023**, *13*, 3622. https://doi.org/10.3390/ani13233622

Academic Editor: Lizhi Wang

Received: 10 October 2023
Revised: 17 November 2023
Accepted: 19 November 2023
Published: 23 November 2023

1. Introduction

Use of fermented feed in livestock aims to enhance digestion and absorption by processing low-quality feed and supplying it to animals, ultimately promoting the well-being of the host. When fermented feed is used, costs are significantly reduced by maximizing feed efficiency and reducing the amount of feces produced due to high nutrient digestion and absorption [1]. Fermentation is the process by which microorganisms convert starch and sugar into fermentation products, such as lactic acid, organic acids, and alcohols. At the beginning of fermentation, acidity is high because the levels of lactic acid bacteria, yeast, and lactic acid are generally low. During the stabilization period, levels of lactic acid bacteria (LAB) such as *Lactobacillus* increase and acidity decreases [2]. Feed fermentation is also an effective method used to kill residual pathogenic microorganisms that may remain in feed [3]. In addition, fermented feed can lower gastric acidity, promote pancreatic juice secretion, and improve digestion and absorption of nutrients [4]. Silage and fermented total mixed ration (TMR), which are types of fermented feed, have been used in ruminant breeding management programs [5]. When producing fermented feed, microorganisms such as LAB and yeast are used that typically improve feed quality. The addition of microorganisms to fermented feeds alters the fermentation properties of the rumen [6] and improves the crude protein (CP) and fiber degradation rate [7]; moreover, it improves digestion [8] and increases palatability and intake [5]. In particular, because fungi produce various protein- and carbohydrate-decomposing enzymes, such as amylase, xylanase, and protease, their cultures are widely used in fermented feeds [9]. Enzymes produced by fungi improve feed value by increasing substrate availability [10]. In addition, they supplement the action of endogenous enzymes in animals to promote the absorption of nutrients into the body [11]. However, because aerobic spoilage is usually initiated by yeasts, care must be taken to prevent spoilage when yeast is added [12].

Although previous studies have acknowledged the nutritional benefits of fermented feeds [1], there has been a predominant research focus on monogastric animals, including pigs and poultry, often within the realms of silage or TMR. In contrast, investigations into the effects of fermentation on feed, especially concentrate formulations for ruminant species, are comparatively underrepresented in the literature. This research aims to contribute to this underexplored area by examining the quality of fermented concentrates and assessing their influence on growth performance, ruminal fermentation patterns, and microbial diversity in beef cattle.

2. Materials and Methods

2.1. Experimental Design and Animal Management

All of the animal experimental procedures were reviewed and approved by the Institutional Animal Care and Use Committee of the National Institute of Animal Science (No. 2021-501).

Eight Hanwoo steers (25 months old, 603 ± 89.4 kg) participated in a 110-d feeding trial at a beef cattle farm of the National Institute of Animal Science, Rural Devel-opment Administration. Two cattle with similar body weights (BWs) were grouped and housed in a pen. Each pen (5 m × 5 m) was equipped with a feed bin. The animals were fed twice daily at 09:00 and 16:00 with each experimental diet, the amount of which was adjusted to achieve 10% refusal based on previous intake. The forage utilized a blend of hay produced by mixing Tall Fescue and Kentucky Bluegrass in a 7:3 ratio. Forage contained 90.1% dry matter (DM), 8.5% CP, 6.0% crude fat, 69.1% neutral detergent fiber analyzed using heat-stable amylase and expressed inclusive of residual ash (aNDF), 45.0% acid detergent fiber (ADF), 15.7% non-fiber carbohydrates (NFCs), and 8.1% ash. Drinking water was freely accessible.

The cattle were randomly assigned to one of the two treatment sequences in a switchback design with three 30-d periods and a 10-d washout period. Each period consisted of 28 d for diet adaptation and data collection (feed intake) and 2 d for sample collection (rumen, feces) and data collection (BW). The experimental treatments comprised conventional concentrate (CON) and fermented concentrate (TRT). Both the CON and TRT groups were given the same forage, with the difference lying in the concentrate feed. The CON group was administered concentrate in its powdered form, whereas the TRT group was given the same concentrate that had been fermented for three days in a fermentation concentrate manufacturing machine made of stainless steel (EO-1500, EO tech, Gwangju, Republic of Korea) set at 35 °C, with water added to achieve a moisture content of 40%. In our research, we used the EO-1500 fermentation machine, which has a 1500-L capacity and is constructed from durable stainless steel (SUS304). Its design includes a triple-layer structure for efficient heat retention and a double ribbon mixer for effective content circulation. The machine also features a dual temperature control system and a unique convection-based drying method, ensuring optimal fermentation conditions. During production of the fermented concentrate, we utilized a microbial complex product (*Bacillus subtilis* 1.0×10^6 cfu/g, *Enterococcus faecium* 1.0×10^6 cfu/g, *Saccharomyces cerevisiae* 1.0×10^6 cfu/g; Bio 5050, Nonghyup Feed Gunsan Bio, Gunsan, Republic of Korea). Ultimately, a probiotic was added at a level of 0.25% of the final fermented concentrate weight. All cattle were placed completely randomly. Experimental animals received both diets throughout the three experimental periods in the order TRT-CON-TRT or CON-TRT-CON (switchback design) [13,14]. Daily DM intake was measured throughout the feeding trial. Individual daily feed intake was recorded by measuring the feed offered and refusals. Before providing the diet to the animals, we measured the moisture content and feed weight of the raw feed. Each morning, we collected the uneaten refusal diet and measured the moisture content and weight of the raw refusal diet. Moisture measurements of both the diet given to the animals and the remaining diet were taken after sub-sampling following thorough mixing, with three replicates. BW was measured at the start and end of the period before morning feeding.

2.2. Formula and Chemical Composition of Experimental Diets

The formulas and chemical compositions of the experimental diets are summarized in Tables 1 and 2. The vitamin complex used in this study consisted of 2,650,000 IU vitamin A, 530,000 IU vitamin D3, 1050 IU vitamin E, 10 g nicotinic acid, 4.4 g manganese sulfate, 4.4 g zinc sulfate, 13.2 g ferrous sulfate and ferric oxide, 2.2 g cupric sulfate, 0.44 g calcium iodate, and 0.44 g cobaltous carbonate per kg. The fermented concentrate was prepared by adding probiotics and moisture to initiate fermentation.

Table 1. Concentrate feed ingredients.

Items	Ingredients (% as-Fed Basis)
Corn flakes	35.7
Wheat bran	19.6
Corn gluten feed	12.5
Wheat	8.8
Soybean meal	7.8
Palm meal	6.3
Soybean hull	3.7
Lupin flake	3.7
Limestone	0.9
Salt	0.4
Sodium bicarbonate	0.4
Vitamin complex	0.1
Total	100.0

Table 2. Chemical composition of experimental diets.

Items	CON Diet	TRT Diet	SEM	*p*-Value
Dry matter, %	90.39	61.59	3.018	<0.05
% dry matter				
CP	18.33	18.93	0.301	0.375
Soluble CP	7.60	8.40	0.342	0.286
Neutral detergent-insoluble CP	1.52	1.52	0.088	1.000
Acid detergent-insoluble CP	0.83	0.92	0.033	0.185
Ether extract	2.94	2.74	0.139	0.520
Neutral detergent fiber (1)	28.13	26.50	1.316	0.595
Acid detergent fiber	12.50	10.80	0.753	0.307

Table 2. *Cont.*

Items	CON Diet	TRT Diet	SEM	p-Value
Non-fiber carbohydrate	46.90	47.50	1.416	0.858
Ash	5.24	5.84	0.220	0.203

CON diet, conventional concentrate without microbial additive; CP, crude protein; TRT diet, fermented concentrate with microbial additive; SEM, standard error of the mean. (1) NDF, neutral detergent fiber analyzed using heat-stable amylase and expressed inclusive of residual ash.

All feed samples used in the experiment were dried at 60 °C for 48 h and ground in a cyclone mill (Foss, Hillerød, Denmark) fitted with a 1-mm screen. The DM, ADF, ash, and ether extract (EE) were analyzed using the procedure described by Horwitz [15]. NDF was analyzed using heat-stable amylase and was expressed inclusive of residual ash [16]. Total nitrogen content was measured via the Dumas combustion method [17] using an elemental combustor (Vario Max Cube, Elementar Gmbh, Frankfurt, Germany), and CP content was calculated as 6.25 times the nitrogen content. The acid detergent-insoluble CP (ADICP) and neutral detergent-insoluble CP (NDICP) levels in each sample were determined as described by Licitra et al. [18]. NFC level was calculated based on the guidelines provided by the National Research Council [19].

$$\text{NFC (\%DM)} = 100 - \text{ash} - \text{EE} - \text{CP} - (\text{aNDF} - \text{NDICP}) \quad (1)$$

2.3. *Calculation of Degradable Carbohydrate and Protein Fractions*

The formulae for calculating carbohydrate (CHO) and protein digestibility in the rumen are as follows [20]:

$$\text{CA (\%CHO)} = \text{CHO} - \text{CB1} - \text{CB3} - \text{CC} \quad (2)$$

$$\text{CB1 (\%CHO)} = \text{starch} \quad (3)$$

$$\text{CB2 (\%CHO)} = \text{NFC} - \text{CA} - \text{CB1} \quad (4)$$

$$\text{CB3 (\%CHO)} = \text{aNDF} - \text{NDICP} \times \text{CP}/1000 - \text{CC} \quad (5)$$

$$\text{CC (\%CHO)} = \text{aNDF} \times \text{Lignin} \times 2.4/1000 \quad (6)$$

CA: carbohydrate A fraction, instantaneously degradable carbohydrates.
CB1: carbohydrate B1 fraction, starch.
CB2: carbohydrate B2 fraction, intermediately degradable carbohydrates.
CB3: carbohydrate B3 fraction, slowly degradable carbohydrates.
CC: carbohydrate C fraction, unavailable cell wall.

$$\text{PA (\%CP)} = \text{non-protein nitrogen (NPN) (\%SOLP)} \times 0.01 \times \text{SOLP (\%CP)} \quad (7)$$

$$\text{PB1 (\%CP)} = \text{SOLP (\%CP)} - \text{PA} \quad (8)$$

$$\text{PB2 (\%CP)} = 100 - \text{PA} - \text{PB1} - \text{PB3} - \text{PC} \quad (9)$$

$$\text{PB3 (\%CP)} = \text{NDICP (\%CP)} - \text{ADICP (\%CP)} \quad (10)$$

$$\text{PC (\%CP)} = \text{ADICP (\%CP)} \quad (11)$$

PA: protein A fraction, instantaneously solubilized protein.
SOLP: soluble CP.
PB1: protein B1 fraction, rapidly degradable protein.
PB2: protein B2 fraction, intermediately degradable protein.

PB3: protein B3 fraction, slowly degradable protein.
PC: protein C fraction, completely undegradable protein.

2.4. Fermentation Feed Quality Analysis

The feed samples (10 g) and distilled water (90 mL) were mixed and homogenized for 2 min in a stomacher (Wisemix®, Daihan, Republic of Korea). After centrifugation (2300× *g*; 4 °C; 20 min), the supernatant was used for pH and ammoniacal nitrogen content analyses. The pH of the samples was analyzed using a pH meter (Seven Easy; Mettler Toledo®, Columbus, OH, USA). Ammoniacal nitrogen content was determined according to the method described by Chaney and Marbach [21]. The phenol color reagent (1 mL) and alkali hypochlorite reagent (1 mL) were mixed with 20 μL of the supernatant. After reacting at 37 °C for 15 min, the absorbance was measured at 630 nm using a spectrophotometer (Optizen UV2120, Mecasis, Republic of Korea). Each fermentation sample (10 g) was mixed with 90 mL peptone water (Difco, Detroit, MI, USA). After homogenizing for 2 min using a stomacher (Wisemix®, Daihan, Republic of Korea), the homogenate was subjected to organic acid analysis via HPLC (Varian, Palo Alto, CA, USA) utilizing a C18 column. Sample preparation was performed by dissolving 0.1 g of the sample in 20 mL of 0.4% hydrochloric acid, followed by ultrasonication. HPLC conditions included isocratic pumping, a mobile phase of 520 mM H_2SO_4, UV detection at 210 nm, a flow rate of 1.0 mL/min, and a 20 μL injection volume. Quantification was based on the formula: Organic acid (%) = (sample peak area/standard peak area) × (concentration of standard solution (g/50 mL)/sample weight (g)) × 100. Viable cell counts were determined as described by Miller and Wolin [22]. LAB and *Bacillus* were cultured for 48 h in an incubator at 37 ± 1 °C using De Man, Rogosa, and Sharpe (Difco, Detroit, MI, USA) and Luria–Bertani (Difco, Detroit, MI, USA) media, respectively. The yeast cells were cultured for 48 h in malt chloramphenicol. Thereafter, the number of viable cells was determined by counting the number of colonies formed on each plate.

2.5. Sample Collection and Analysis

Fresh fecal samples were obtained from the cattle on day 29 of each experimental period. The samples were immediately stored at −80 °C until metagenomic DNA extraction.

On day 30 of each experimental period, representative rumen fluid samples were collected via a stomach tube approximately 3 h after feeding [23,24]. Between samples, the stomach tube was thoroughly washed with warm water to prevent cross-contamination from the previous animal [25,26]. The first 200 mL of ruminal fluid was discarded to reduce contamination by saliva. Immediately after collection, the pH of the sampled inoculum was measured using a pH meter (Pinnacle pH meter M540; Corning, NY, USA). The ruminal fluid was then sealed in a tube and frozen in liquid nitrogen. The samples were stored at −80 °C until the analysis of volatile fatty acid (VFA) and ammonia nitrogen (NH_3-N) levels and metagenomic DNA extraction. The VFA and NH_3-N concentrations were determined as described by Erwin et al. [27] and Chaney and Marbach [21] with minor modifications.

The ruminal fluid was centrifuged at 6000× *g* for 15 min at 20 °C. The supernatant was used for VFA analysis. For VFA analysis, a 25% metaphosphoric acid solution was added to the ruminal fluid at 10% of the total volume. The supernatant was injected into a gas chromatograph (TRACE 1610, Thermo Fisher Scientific, Waltham, MA, USA) equipped with a flame ionization detector and capillary column (Nukol™, fused silica capillary column 15 m × 0.53 mm × 0.5 μm; Supelco Inc., Bellefonte, PA, USA). The oven, injector, and detector temperatures were 110 °C, 250 °C, and 250 °C, respectively. A standard curve was generated using a VFA standard solution (catalog number. 46975-U; Sigma-Aldrich, St. Louis, MO, USA). The mixtures of inoculum and 25% metaphosphoric acid were centrifuged at 14,000× *g* for 5 min at 4 °C for NH_3-N analysis. After centrifugation, 20 μL of each supernatant was mixed with 1 mL of phenol color reagent (50 g/L of phenol plus 0.25 g/L of nitroferricyanide) and 1 mL of alkali hypochlorite reagent (25 g/L of sodium hydroxide and 16.8 mL/L of 4–6% sodium hypochlorite). The mixture was incubated in a water bath for color development at 37 °C for 15 min; thereafter, 8 mL of distilled water was added, and the NH_3-N concentration was determined by measuring the absorbance at 630 nm using a UV spectrophotometer (Bio-Rad, US/benchmark plus, Tokyo, Japan). All analyses were repeated three times, and the mean values are presented.

2.6. Microbial Diversity Analysis

Metagenomic DNA was extracted from the collected fecal and ruminal fluid samples. For metagenome community analysis, the V3–V4 region of the 16S rRNA was used as a phylogenetic marker. The amplicons were generated using the 337 F and 805R 16S V3–V4 rRNA universal primers (GACTCCTACGGGAGGCWGCAG and GACTACCAGGGTATCTAATC). Sequencing libraries were constructed using Herculase II Fusion DNA Polymerase Nextera XT Index Kits by Agilent Technologies (Santa Clara, CA, USA). All library construction processes were conducted following the 16S Metagenomic Sequencing Library Preparation guidelines by Illumina. Before conducting analysis, initial quality control was performed using Trimmomatic [28] to remove low-quality base calls and sequencing artifacts. The parameters used were ILLUMINACLIP:TruSeq3-PE.fa:2:30:10:2:True LEADING:5 TRAILING:20 MINLEN:250. Microbial community analysis was performed using the QIIME2 [29] pipeline with a DADA2 [30] and SILVA [31] pre-trained Naive Bayes classifier based on the SILVA full-length 16S rRNA database. A Kruskal–Wallis test and linear discriminant analysis effect size [32] were used to compare the effect of the fermented concentrate by treatment on the rumen and intestine.

2.7. Statistical Analysis

Data for the experimental diets, growth performance, and rumen fermentation parameters satisfied the conditions of normality and homoscedasticity for each group and were analyzed using descriptive statistics and the *t*-test in SPSS (Version 26, IBM, Armonk, NY, USA). Statistical significance was set at $p < 0.05$.

3. Results and Discussion

TRT diet with water added for fermentation had a significantly lower DM content than that in the CON diet ($p < 0.05$; Table 1). However, there were no significant differences between the treatments in terms of the composition of CP, crude fat, aNDF, and others on a DM basis. The digestibility of carbohydrates and proteins was calculated based on the chemical composition of the experimental diets (Table 3). The proportion of instantaneously degradable carbohydrates (CA), which are immediately digested in the rumen, was significantly lower in the TRT diet than that in the CON diet ($p < 0.05$). This difference may be attributed to the fact that microorganisms had already decomposed some carbohydrates during the concentration process in TRT. There was no significant difference between the treatment groups in terms of the degradable carbohydrate fraction, which is classified according to the carbohydrate decomposition rate, and the unavailable cell wall fraction in the rumen. Furthermore, there was no significant difference in protein digestibility among the instantaneously solubilized protein, degradable protein, and completely undegradable protein fractions, regardless of the rumen digestion rate.

In general, feed fermentation can be determined by microbial population and organic acids [33]. *Enterococcus faecium*, a homofermentative lactic acid bacterium, is used as an additive that can increase lactic acid production in silage [34]. Owing to the presence of microbial additives, the TRT diet had significantly higher abundances of LAB, *Bacillus*, and yeast than the CON diet ($p < 0.05$, Table 4). The pH was significantly lower in the TRT group than that in the CON diet ($p < 0.05$). This finding is consistent with results of previous studies reporting that the addition of the homofermentative LAB lowers the pH during silage fermentation [35,36]. pH and acetic acid levels are used as indicators to confirm fermentation stability in fermented feeds [37]. The acetic acid level was significantly higher in the TRT diet than that in the CON diet ($p < 0.05$). In addition, the TRT diet exhibited significantly higher levels of organic acids, such as lactic acid, propionic acid, and butyric acid, than the CON diet did, indicating a stable fermentation in the former. Ammonia nitrogen production was also significantly higher in the TRT diet than that in the CON diet ($p < 0.05$). During the fermentation of the concentrate, proteins and other nitrogen-containing compounds are broken down by microbes, resulting in the conversion to NH_3-N. The concentration of NH_3-N during this process can serve as a critical indicator of the fermentation quality. Maintaining appropriate levels of ammonia nitrogen in fermented feeds is important because it affects protein metabolism and energy balance within the rumen. By ensuring that these levels are kept within optimal ranges, the productivity and health of ruminants can be optimized [38].

Table 3. Degradable carbohydrate and protein fractions of experimental diets (dry matter basis).

Items	CON Diet	TRT Diet	SEM	*p*-Value
		Carbohydrate fractions [(1)], %CHO		
CA	4.96	1.24	0.932	<0.05
CB1	43.64	46.26	2.733	0.683
CB2	15.17	18.03	1.002	0.173
CB3	28.59	27.65	1.443	0.783
CC	7.68	6.79	0.534	0.468
		Protein fractions [(2)], %CP		
PA + PB1	41.46	44.34	1.466	0.383
PB2	50.28	47.60	1.336	0.374
PB3	3.73	3.18	0.530	0.662
PC	4.54	4.88	0.191	0.443

CON diet, conventional concentrate without microbial additive; TRT diet, fermented concentrate with microbial additive; SEM, standard error of mean. [(1)] CHO, carbohydrate; CA, instantaneously degradable carbohydrates; CB, degradable carbohydrates; CB1, starch; CB2, intermediately degradable carbohydrates; CB3, slowly degradable carbohydrates; CC, unavailable cell wall. [(2)] CP, crude protein; PA, instantaneously solubilized protein; PB, degradable protein; PB1, rapidly degradable protein; PB2, intermediately degradable protein; PB3, slowly degradable protein; PC, completely undegradable protein.

Table 4. Fermentation and microbial profile of the experimental diets (as-fed basis).

Items	CON Diet	TRT Diet	SEM	*p*-Value
	Microbial profile, log10 cfu/g			
Lactic acid bacteria	4.26	8.46	0.579	<0.05
Bacillus	4.37	5.14	0.110	<0.05
Yeast	4.00	6.41	0.439	<0.05
	Fermentation profile			
pH	5.69	4.40	0.150	<0.05
Lactic acid, %	0.12	0.24	0.028	<0.05
Acetic acid, mM	0.34	18.31	2.897	<0.05
Propionic acid, mM	0.00	0.66	0.109	<0.05
Butyric acid, mM	0.00	0.26	0.067	<0.05
Ammonia nitrogen, mg/dL	0.63	2.61	0.259	<0.05

CON diet, conventional concentrate without microbial additive; TRT diet, fermented concentrate with microbial additive; Forage, mixed hay with Tall Fescue and Kentucky Bluegrass in a 7:3 ratio; SEM, standard error of mean.

After the experimental trial period, there was no significant difference in the final BW, daily weight gain, feed intake, or feed conversion ratio between the treatments (Table 5). Although the mean daily weight gain was higher in the TRT group (0.96) than in the CON group (0.67), the difference was not statistically significant. The large standard error associated with the weight gain data (0.105) indicates a considerable variability within the groups, thereby precluding a definitive conclusion regarding the effect of the treatment on weight gain. Fermentation did not affect the chemical composition of the feed or the amount of feed consumed; thus, it had no effect on nutrient intake.

Table 5. Effects of fermented concentrate on growth performance of Hanwoo steers (dry matter basis).

Items	CON	TRT	SEM	*p*-Value
	Growth performance			
Initial body weight, kg	655.66	666.09	33.506	0.888
Final body weight, kg	687.66	690.41	34.343	0.971
Average daily gain, kg/d	0.96	0.67	0.105	0.169
Feed intake, kg/d	11.61	12.18	0.479	0.568
Concentrate intake, kg/d	9.98	10.53	0.425	0.558
Forage intake, kg/d	1.63	1.65	0.124	0.929
Feed conversion ratio	16.13	22.99	2.311	0.175
	Nutrient intake, kg/d			
Crude protein	1.97	2.21	0.076	0.114
Ether extract	0.31	0.31	0.010	0.785
Neutral detergent fiber	3.93	4.02	0.118	0.719
Non-fiber carbohydrate	4.94	5.46	0.186	0.173
Gross energy	51.39	55.82	1.784	0.233

CON, conventional concentrate without microbial additive; TRT, fermented concentrate with microbial additive; SEM, standard error of mean.

Ruminal pH was significantly higher in the TRT group than in the CON group ($p < 0.05$; Table 6). There were no significant differences in the NH_3-N and total VFA levels between treatments; however, acetic acid, propionic acid, and butyric acid production significantly differed between the groups ($p < 0.05$). The acetic acid and lactic acid levels were higher and propionic acid levels were lower in the TRT group than those in the CON group ($p < 0.05$). The acetate to propionate ratio (AP) ratio was significantly higher in the TRT group than that in the CON group ($p < 0.05$). In accordance with our research findings, a study that fed fermented soybean meal to Holstein cows also reported an increase in the acetate percentage, pH, and AP ratio in the rumen fluid [39]. Conversely, in another study, Holsteins fed fermented soybean meal showed no changes in the ruminal pH and an increase in the AP ratio [40]. In light of our research results, the higher rumen pH and elevated AP ratio observed in the TRT group may be correlated, aligning with the findings of Amin [39] and Russell [41]. In the context of beef cattle experiencing a rapid increase in BW and marbling score just before slaughter due to excessive concentrate feeding, a sharp decline in rumen pH can lead to an increased risk of acidosis. In our study, despite a relatively high proportion of concentrate intake within the total diet, accounting for 90.6% in the CON group and 86.5% in the TRT group, the fermented concentrate resulted in an elevation of the rumen pH. Therefore, using fermented concentrate during periods of excessive concentrate feeding may raise the rumen pH, potentially preventing acidosis.

Table 6. Effects of fermented concentrate on rumen fermentation characteristics of Hanwoo steers.

Items	CON	TRT	SEM	*p*-Value
pH	6.27	6.60	0.056	<0.05
Ammonia nitrogen, mg/dL	11.13	16.33	3.019	0.348
Total volatile fatty acid, mM	113.82	103.83	6.347	0.566
Acetate, %	60.67	64.00	1.074	<0.05
Propionate, %	25.56	19.18	1.677	<0.05
Butyrate, %	10.35	13.03	0.533	<0.05
Valerate, %	3.42	3.79	0.173	0.227
Acetate-to-propionate ratio	2.44	3.39	0.242	<0.05

CON, conventional concentrate without microbial additive; TRT, fermented concentrate with microbial additive; SEM, standard error of the mean.

In response to the administration of fermented concentrate, there was a statistically significant increase in the alpha diversity of rumen microbiota, as indicated by the higher Shannon entropy values, in the TRT group compared to the CON group ($p < 0.05$, Figure 1). This finding suggests that the intake of fermented concentrate results in increased microbial diversity within the rumen. Conversely, when comparing the alpha diversity of the fecal microbiome across different treatments, no statistically significant differences were observed. In terms of beta diversity analysis, rumen microbiota displayed relatively distinct clustering patterns among the groups, highlighting clear separations. In contrast, fecal microbiota did not exhibit discernible clustering patterns among the groups. In summary, the use of fermented concentrate exerts a more pronounced impact on rumen microbiota than that on the post-rumen digestive intestinal. Fermented feed primarily affects the rumen microbiome in ruminants, enhancing the microbial activity, which is crucial for digestion [42,43]. This modulation can improve feed efficiency and nutrient uptake. The fecal microbiome is less directly impacted by fermented feed, reflecting post-digestive processes. However, diet can influence both the rumen and fecal microbiome, with more pronounced changes in the former.

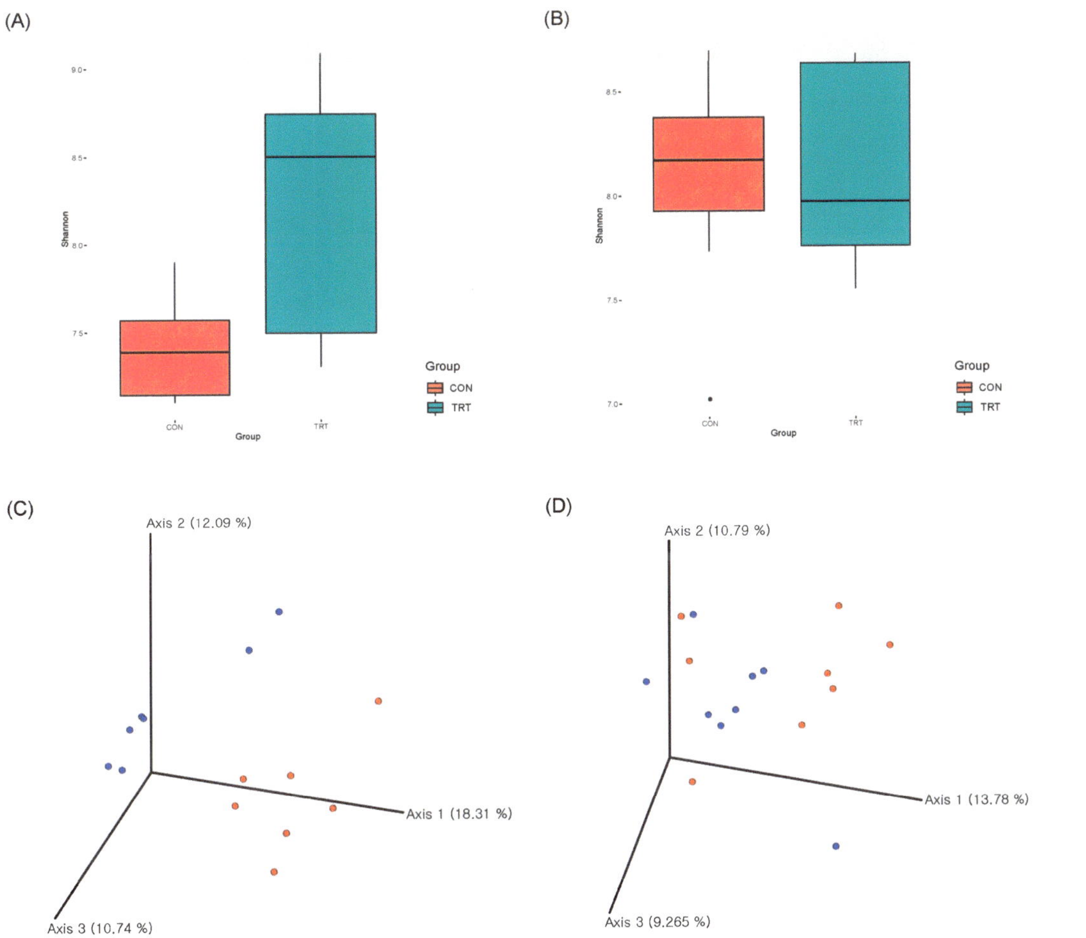

Figure 1. Ruminal and fecal bacterial community diversities. Alpha diversity in the ruminal bacterial community (**A**) and fecal bacterial community (**B**). Principal coordinate analysis plots for the ruminal bacterial community (**C**) and fecal bacterial community (**D**); each dot in the plots represents a cluster. Samples were collected from Hanwoo steers fed conventional concentrate (CON; $n = 8$, represented by red dots) and fermented concentrate (TRT; $n = 8$, represented by blue dots).

The abundance of rumen microbial clusters belonging to the family Lachnospiraceae increased in cattle administered the fermented concentrate (Figure 2). Lachnospiraceae is commonly found in cattle rumen and plays a role in fermenting plant polysaccharides to produce short-chain fatty acids and alcohol. It is also associated with disease resistance because it is involved in the production of butyric acid, a critical substance for microbe–host epithelial cell growth, which can mitigate intestinal inflammation and maintain the intestinal barrier [44]. However, there are conflicting findings regarding the impact of Lachnospiraceae on livestock productivity. Previous analyses of rumen microbes in cattle with low nitrogen efficiency showed high levels of Lachnospiraceae [44,45]. In contrast, the abundance of the genus *Moryella* belonging to Lachnospiraceae, which has been linked to improved feed efficiency in livestock, was found to be increased in the TRT group. *Moryella* was the dominant genus in the rumen of calves with low residual feed intake. Additionally, inoculation of the rumen with *Moryella* led to a significant increase in its abundance, potentially enhancing propionate production capacity and improving feed efficiency [46,47]. In this study, the altered abundance of previously reported taxa did not directly impact weight gain or feed efficiency.

In the TRT group, an increase in the abundance of ruminal microorganisms related to the breakdown of feed nutritional components was observed. Genera belonging to the order Bacteroidales showed a significant increase in the TRT group compared to the CON group. Bacteroidales are involved in the decomposition and fermentation of plant fibers, as well as the breakdown of complex polysaccharides and other complex fibers into simple organic compounds [48]. Furthermore, the abundance of Bradymonadales and *Desulfovibrio*, belonging to the phylum Termodesulfobacteriota, which is known for its involvement in sulfate reduction and digestion of organic matter, increased in the TRT group compared to the CON group [49]. Bradymonadales, although relatively less known, preferentially prey on Bacteroidetes and Proteobacteria [50].

Analysis of fecal microbiota revealed an increase in the abundance of *Akkermansia* in the TRT group compared to the CON group (Figure 2). *Akkermansia*, a slender-shaped bacterium, is commonly found in bovine feces and is particularly abundant in individuals consuming a high forage diet [51–53]. Furthermore, *Akkermansia* maintains a healthy gut microbiota and enhances immune function, potentially aiding in disease prevention. Moreover, it plays a positive role in preventing metabolic disorders as a member of the gut microbiota and is often highly abundant in healthy humans [54,55]. However, additional research is needed to understand the potential immunomodulatory and disease-preventing effects of fecal *Akkermansia* in cattle, particularly regarding its impact on host health.

Our study acknowledges the limitation of single time point collection in reflecting the rumen's dynamic nature, and suggests the potential for more comprehensive insights through multi-time point sampling in future research. We recognize the possible influence of using a stomach tube on rumen fluid characteristics, and have detailed our mitigating measures in the Methods section, acknowledging the need for careful consideration of these factors in interpreting our results.

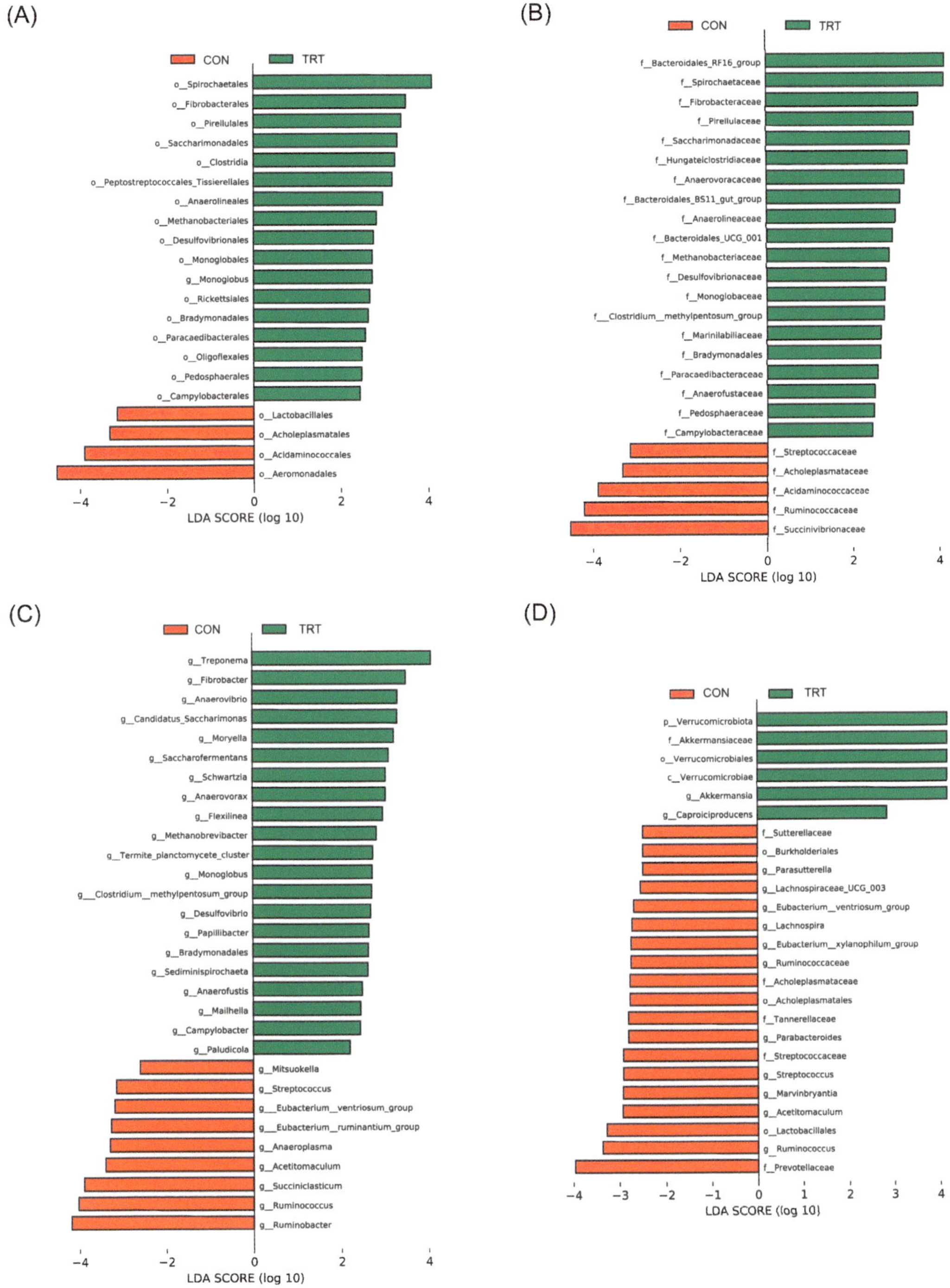

Figure 2. Linear discriminant analysis of the (**A–C**) rumen and the (**D**) feces. Samples were collected from Hanwoo steers fed conventional concentrate (CON; n = 8) and fermented concentrate (TRT; n = 8).

4. Conclusions

This study investigated the effect of fermented concentrate feed on its nutritional composition and on beef cattle. Fermentation led to a decrease in the CA fraction of the concentrate but no significant changes in the levels of other components. Microbial abundance increased during fermentation. Stable fermentation resulted in higher organic acid levels and a lower pH, potentially improving rumen digestibility. Fermented concentrate treatment did not significantly affect weight gain, feed conversion ratio, or final BW, but did improve the rumen fermentation characteristics, including pH and acetic acid levels, indicating stable rumen fermentation. Our findings suggest that fermented concentrate primarily affects the rumen microbial community rather than the gastrointestinal tract. Fermented concentrate treatment increases the abundance of microorganisms enhancing the feed digestibility, immunity, and health of the host. While fermented concentrate does not impact cattle growth and digestibility in the short term, it may have long-term effects on cattle growth and farm economics. Future research with a greater number of sampling days is needed to explore these potential long-term benefits.

Author Contributions: Conceptualization, methodology, validation, S.L.; formal analysis, S.L. and C.H.R.; investigation, S.L. and H.K.; resources, S.D.L. and Y.C.B.; data curation, C.H.R. and Y.C.B.; writing—original draft preparation, S.L. and C.H.R.; writing—review and editing, S.L.; supervision, S.L. All authors have read and agreed to the published version of the manuscript.

Funding: This study was supported by the Cooperative Research Program for Agriculture Science and Technology Development (Project No. PJ015647012023) of the Rural Development Administration of Korea and by the 2023 RDA Fellowship Program of the National Institute of Animal Science, Rural Development Administration, Republic of Korea.

Institutional Review Board Statement: The experiments conformed to the guidelines of the Animal Ethics Committee of the National Institute of Animal Science (NIAS. 2021-501, approved on 30 March 2021).

Informed Consent Statement: Animals were sourced from the National Institute of Animal Science, and PIC approval was not required.

Data Availability Statement: The data presented in this study are available on request from the first author. The data are not publicly available due to restrictions by the research group.

Conflicts of Interest: The authors declare no conflict of interest.

References

1. Park, B.K.; Gil, J.M.; Kim, J.B.; Hong, B.J.; Ra, C.S.; Shin, J.S. Effects of fermented feedstuff with wet brewer's grain and soybean on fattening performance and carcass grade in Hanwoo steers. *J. Anim. Sci. Technol.* **2003**, *45*, 397–408. [CrossRef]
2. Eun, J.B.; Jin, T.Y.; Wang, M.H. The effect of waxy glutinous rice degree of milling on the quality of Jinyangju, a Korean traditional rice wine. *Korean J. Food Sci. Technol.* **2007**, *39*, 546–551.
3. Niba, A.T.; Beal, J.D.; Kudi, A.C.; Brooks, P.H. Bacterial fermentation in the gastrointestinal tract of non-ruminants: Influence of fermented feeds and fermentable carbohydrates. *Trop. Anim. Health Prod.* **2009**, *41*, 1393–1407. [CrossRef]
4. Scholten, R.H.; van der Peet-Schwering, C.M.; Verstegen, M.W.; Den Hartog, L.A.; Schrama, J.W.; Vesseur, P.C. Fermented co-products and fermented compound diets for pigs: A review. *Anim. Feed Sci. Technol.* **1999**, *82*, 1–19. [CrossRef]
5. Lee, S.J.; Shin, N.H.; Jung, H.S.; Moon, Y.H.; Lee, S.S.; Lee, S.S. Effects of supplemental Synbiotics composed of anaerobic bacteria, yeast and mold on the change of chemical composition and fermentation characteristics of total mixed ration for cattle. *J. Life Sci.* **2009**, *19*, 241–248. [CrossRef]
6. Martin, S.A.; Nisbet, D.J. Effects of Aspergillus oryzae fermentation extract on fermentation of amino acids, Bermudagrass and starch by mixed ruminal microorganisms in vitro. *J. Anim. Sci.* **1990**, *68*, 2142–2149. [CrossRef]
7. Williams, P.E.; Tait, C.A.; Innes, G.M.; Newbold, C.J. Effects of the inclusion of yeast culture (Saccharomyces cerevisiae plus growth medium) in the diet of dairy cows on milk yield and forage degradation and fermentation patterns in the rumen of steers. *J. Anim. Sci.* **1991**, *69*, 3016–3026. [CrossRef]
8. Harrison, G.A.; Hemken, R.W.; Dawson, K.A.; Harmon, R.J.; Barker, K.B. Influence of addition of yeast culture supplement to diets of lactating cows on ruminal fermentation and microbial populations. *J. Dairy Sci.* **1988**, *71*, 2967–2975. [CrossRef]
9. Gomez-Alarcon, R.A.; Dudas, C.; Huber, J.T. Influence of cultures of Aspergillus oryzae on rumen and total tract digestibility of dietary components. *J. Dairy Sci.* **1990**, *73*, 703–710. [CrossRef]

10. Pettersson, D.; Graham, H.; Åman, P. Enzyme supplementation of low or high crude protein concentration diets for broiler chickens. *Anim. Sci.* **1990**, *51*, 399–404. [CrossRef]
11. Graham, H.; Lowgren, W.; Pettersson, D.; Aman, P. Effect of enzyme supplementation on digestion of a barley/pollard-based pig diet. *Nutr. Rep. Int.* **1988**, *38*, 1073–1079.
12. Woolford, M.K.; Cook, J.E. A note on the effect on the aerobic deterioration of maize silage of the manipulation of the microflora by means of antibiotics. *Anim. Feed Sci. Technol.* **1978**, *3*, 89–94. [CrossRef]
13. Donkin, S.S.; Velez, J.C.; Totten, A.K.; Stanisiewski, E.P.; Hartnell, G.F. Effects of feeding silage and grain from glyphosate-tolerant or insect-protected corn hybrids on feed intake, ruminal digestion, and milk production in dairy cattle. *J. Dairy Sci.* **2003**, *86*, 1780–1788. [CrossRef]
14. Fessenden, S.W.; Hackmann, T.J.; Ross, D.A.; Block, E.; Foskolos, A.; Van Amburgh, M.E. Rumen digestion kinetics, microbial yield, and omasal flows of nonmicrobial, bacterial, and protozoal amino acids in lactating dairy cattle fed fermentation by-products or urea as a soluble nitrogen source. *J. Dairy Sci.* **2019**, *102*, 3036–3052. [CrossRef]
15. Horwitz, W. *Official Methods of Analysis of AOAC (Association of Official Analytica Chemists) International, United States of America*, 17th ed.; Methods 2020; AOAC International: Gaithersburg, MD, USA, 2020; Volume 942, p. 930.915.
16. Van Soest, P.J.; Robertson, J.B.; Lewis, B.A. Methods for dietary fiber, neutral detergent fiber, and nonstarch polysaccharides in relation to animal nutrition. *J. Dairy Sci.* **1991**, *74*, 3583–3597. [CrossRef]
17. Simonne, A.H.; Simonne, E.H.; Eitenmiller, R.R.; Mills, H.A.; Cresman, C.P. Could the Dumas method replace the Kjeldahl digestion for nitrogen and crude protein determinations in foods? *J. Sci. Food Agric.* **1997**, *73*, 39–45. [CrossRef]
18. Licitra, G.; Hernandez, T.M.; Van Soest, P.J. Standardization of procedures for nitrogen fractionation of ruminant feeds. *Anim. Feed Sci. Technol.* **1996**, *57*, 347–358. [CrossRef]
19. NRC (National Research Council). *Nutrient Requirements of Dairy Cattle*; National Academies Press: Washington, DC, USA, 2001.
20. Fox, D.G.; Tedeschi, L.; Tylutki, T. The Cornell Net Carbohydrate and Protein System model for evaluating herd nutrition and nutrient excretion. *Anim. Feed Sci. Technol.* **2004**, *112*, 29–78. [CrossRef]
21. Chaney, A.L.; Marbach, E.P. Modified reagents for determination of urea and ammonia. *Clin. Chem.* **1962**, *8*, 130–132. [CrossRef]
22. Miller, T.L.; Wolin, M.J. A serum bottle modification of the Hungate technique for cultivating obligate anaerobes. *Appl. Microbiol.* **1974**, *27*, 985–987. [CrossRef]
23. De Evan, T.; Carro, M.D.; Fernández Yepes, J.E.; Haro, A.; Arbesú, L.; Romero-Huelva, M.; Molina-Alcaide, E. Feeding mango wastes to dairy goats: Effects on diet digestibility, ruminal fermentation, and milk yield and composition. *Anim. Feed Sci. Technol.* **2022**, *286*, 115252. [CrossRef]
24. Kidane, A.; Gregersen Vhile, S.; Ferneborg, S.; Skeie, S.; Olsen, M.A.; Torunn Mydland, L.; Øverland, M.; Prestl Kken, E. Cyberlindnera jadinii yeast as a protein source in early-to mid-lactation dairy cow diets: Effects on feed intake, ruminal fermentation, and milk production. *J. Dairy Sci.* **2022**, *105*, 2343–2353. [CrossRef]
25. Steiner, S.; Neidl, A.; Linhart, N.; Tichy, A.; Gasteiner, J.; Gallob, K.; Baumgartner, W.; Wittek, T. Randomised prospective study compares efficacy of five different stomach tubes for rumen fluid sampling in dairy cows. *Vet. Rec.* **2015**, *176*, 50. [CrossRef]
26. Paz, H.A.; Anderson, C.L.; Muller, M.J.; Kononoff, P.J.; Fernando, S.C. Rumen bacterial community composition in Holstein and Jersey cows is different under same dietary condition and is not affected by sampling method. *Front. Microbiol.* **2016**, *7*, 1206. [CrossRef]
27. Erwin, E.S.; Marco, G.J.; Emery, E.M. Volatile fatty acid analyses of blood and rumen fluid by gas chromatography. *J. Dairy Sci.* **1961**, *44*, 1768–1771. [CrossRef]
28. Bolger, A.M.; Lohse, M.; Usadel, B. Trimmomatic: A flexible trimmer for Illumina sequence data. *Bioinformatics* **2014**, *30*, 2114–2120. [CrossRef]
29. Bolyen, E.; Rideout, J.R.; Dillon, M.R.; Bokulich, N.A.; Abnet, C.C.; Al-Ghalith, G.A.; Caporaso, J.G. Reproducible, interactive, scalable and extensible microbiome data science using QIIME 2. *Nat. Biotechnol.* **2019**, *37*, 852–857. [CrossRef]
30. Callahan, B.J.; McMurdie, P.J.; Rosen, M.J.; Han, A.W.; Johnson, A.J.A.; Holmes, S.P. DADA2: High-resolution sample inference from Illumina amplicon data. *Nat. Methods* **2016**, *13*, 581–583. [CrossRef]
31. Quast, C.; Pruesses, E.; Yilmaz, P.; Gerken, J.; Schweer, T.; Yarza, P.; Glöckner, F.O. The SILVA ribosomal RNA gene database project: Improved data processing and web-based tools. *Nucleic Acids Res.* **2012**, *41*, 590–596. [CrossRef]
32. Segata, N.; Izard, J.; Waldron, L.; Gevers, D.; Miropolsky, L.; Garrett, W.S.; Huttenhower, C. Metagenomic biomarker discovery and explanation. *Genome Biol.* **2011**, *12*, R60. [CrossRef]
33. Lee, A.; Shin, S.J.; Yang, J.; Cho, S.; Choi, N.J. Effect of lactic acid bacteria and enzyme supplementation on fermentative patterns of ensiling silages, their in vitro ruminal fermentation, and digestibility. *J. Korean Soc. Grassl. Forage Sci.* **2016**, *36*, 7–14. [CrossRef]
34. Heinl, S.; Wibberg, D.; Eikmeyer, F.; Szczepanowski, R.; Blom, J.; Linke, B.; Goesmann, A.; Grabherr, R.; Schwab, H.; Pühler, A.; et al. Insights into the completely annotated genome of Lactobacillus buchneri CD034, a strain isolated from stable grass silage. *J. Biotechnol.* **2012**, *161*, 153–166. [CrossRef]
35. Filya, I. The effect of Lactobacillus buchneri and Lactobacillus plantarum on the fermentation, aerobic stability, and ruminal degradability of low dry matter corn and sorghum silages. *J. Dairy Sci.* **2003**, *86*, 3575–3581. [CrossRef]
36. Lee, H.J.; Kim, D.H.; Amanullah, S.M.; Kim, S.C.; Song, Y.M.; Kim, H.Y. Effects of bacterial inoculants and cutting height on fermentation quality of barley silage. *J. Korean Soc. Grassl. Forage Sci.* **2014**, *34*, 163–168. [CrossRef]

37. Song, T.H.; Han, O.K.; Park, T.I.; Kim, K.H.; Kim, K.J.; Yoon, S.K. Effect of pre-wilting time on change of the moisture content and its silage quality at different harvest stages of whole crop barley. *KSIA* **2009**, *1*, 316–321.
38. Huhtanen, P.; Nousiainen, J.I.; Khalili, H.; Jaakkola, S.; Heikkilä, T. Relationships between silage fermentation characteristics and milk production parameters: Analyses of literature data. *Livest. Prod. Sci.* **2003**, *81*, 57–73. [CrossRef]
39. Amin, A.B.; Zhang, L.; Zhang, J.; Mao, S. Fermented soybean meal modified the rumen microbiome to enhance the yield of milk components in Holstein cows. *Appl. Microbiol. Biotechnol.* **2022**, *106*, 7627–7642. [CrossRef]
40. Wang, Z.; Yu, Y.; Li, X.; Xiao, H.; Zhang, P.; Shen, W.; Wan, F.; He, J.; Tang, S.; Tan, Z.; et al. Fermented soybean meal replacement in the diet of lactating Holstein dairy cows: Modulated rumen fermentation and ruminal microflora. *Front. Microbiol.* **2021**, *12*, 625857. [CrossRef]
41. Russell, J.B. The importance of pH in the regulation of ruminal acetate to propionate ratio and methane production in vitro. *J. Dairy Sci.* **1998**, *81*, 3222–3230. [CrossRef]
42. Uyeno, Y.; Shigemori, S.; Shimosato, T. Effect of probiotics/prebiotics on cattle health and productivity. *Microbes Environ.* **2015**, *30*, 126–132. [CrossRef]
43. Kibegwa, F.M.; Bett, R.C.; Gachuiri, C.K.; Machuka, E.; Stomeo, F.; Mujibi, F.D. Diversity and functional analysis of rumen and fecal microbial communities associated with dietary changes in crossbreed dairy cattle. *PLoS ONE* **2023**, *18*, e0274371. [CrossRef]
44. Meehan, C.J.; Beiko, R.G. A phylogenomic view of ecological specialization in the Lachnospiraceae, a family of digestive tract-associated bacteria. *Genome Biol. Evol.* **2014**, *6*, 703–713. [CrossRef]
45. Gomes Carvalho Alves, K.L.; Granja-Salcedo, Y.T.; Messana, J.D.; Carneiro de Souza, V.; Generoso Ganga, M.J.; Detogni Colovate, P.H.; Kishi, L.T.; Berchielli, T.T. Rumen bacterial diversity in relation to nitrogen retention in beef cattle. *Anaerobe* **2021**, *67*, 102316. [CrossRef]
46. Hernandez-Sanabria, E.; Goonewardene, L.A.; Wang, Z.; Durunna, O.N.; Moore, S.S.; Guan, L.L. Impact of feed efficiency and diet on adaptive variations in the bacterial community in the rumen fluid of cattle. *Appl. Environ. Microbiol.* **2012**, *78*, 1203–1214. [CrossRef]
47. Yu, S.; Zhang, G.; Liu, Z.; Wu, P.; Yu, Z.; Wang, J. Repeated inoculation with fresh rumen fluid before or during weaning modulates the microbiota composition and co-occurrence of the rumen and colon of lambs. *BMC Microbiol.* **2020**, *20*, 29. [CrossRef]
48. Huws, S.A.; Edwards, J.E.; Creevey, C.J.; Rees Stevens, P.; Lin, W.; Girdwood, S.E.; Pachebat, J.A.; Kingston-Smith, A.H.; Kingston-Smith, A.H. Temporal dynamics of the metabolically active rumen bacteria colonizing fresh perennial ryegrass. *FEMS Microbiol. Ecol.* **2016**, *92*, fiv137. [CrossRef]
49. Leng, H.; Wang, Y.; Zhao, W.; Sievert, S.M.; Xiao, X. Identification of a deep-branching thermophilic clade sheds light on early bacterial evolution. *Nat. Commun.* **2023**, *14*, 4354. [CrossRef]
50. Mu, D.S.; Wang, S.; Liang, Q.Y.; Du, Z.Z.; Tian, R.; Ouyang, Y.; Wang, X.P.; Zhou, A.; Gong, Y.; Chen, G.J.; et al. Bradymonabacteria, a novel bacterial predator group with versatile survival strategies in saline environments. *Microbiome* **2020**, *8*, 126. [CrossRef]
51. Kim, M.; Kim, J.; Kuehn, L.A.; Bono, J.L.; Berry, E.D.; Kalchayanand, N.; Freetly, H.C.; Benson, A.K.; Wells, J.E. Investigation of bacterial diversity in the feces of cattle fed different diets. *J. Anim. Sci.* **2014**, *92*, 683–694. [CrossRef]
52. Zhang, Z.; Yang, L.; He, Y.; Luo, X.; Zhao, S.; Jia, X. Composition of fecal microbiota in grazing and feedlot Angus beef cattle. *Animals* **2021**, *11*, 3167. [CrossRef]
53. Sim, S.; Lee, H.; Yoon, S.; Seon, H.; Park, C.; Kim, M. The impact of different diets and genders on fecal microbiota in Hanwoo cattle. *J. Anim. Sci. Technol.* **2022**, *64*, 897–910. [CrossRef]
54. Keshavarz Azizi Raftar, S.; Hoseini Tavassol, Z.; Amiri, M.; Ejtahed, H.S.; Zangeneh, M.; Sadeghi, S.; Ashrafian, F.; Kariman, A.; Khatami, S.; Siadat, S.D.; et al. Assessment of fecal Akkermansia muciniphila in patients with osteoporosis and osteopenia: A pilot study. *J. Diabetes Metab. Disord.* **2021**, *20*, 279–284. [CrossRef]
55. Cruz-Aguliar, R.M.; Wantia, N.; Clavel, T.; Vehreschild, M.J.G.T.; Buch, T.; Bajbouj, M.; Haller, D.; Busch, D.; Schmid, R.M.; Stein-Thoeringer, C.K.; et al. An open-labeled study on fecal microbiota transfer in irritable bowel syndrome patients reveals improvement in abdominal pain associated with the relative abundance of Akkermansia muciniphila. *Digestion* **2019**, *100*, 127–138. [CrossRef]

animals

Article

Effects of Isochlorogenic Acid on Ewes Rumen Fermentation, Microbial Diversity and Ewes Immunity of Different Physiological Stages

Shuyan Li [1,2], Xiongxiong Li [3], Yuzhu Sha [3], Shuai Qi [1,2], Xia Zhang [1,2], Huning Wang [1,2], Zhengwen Wang [1,2], Shengguo Zhao [3] and Ting Jiao [1,2,*]

1 College of Prataculural Science, Gansu Agricultural University, Lanzhou 730070, China; lisy@st.gsau.edu.cn (S.L.); idreamworker@sina.com (S.Q.); 18894012309@139.com (X.Z.); 18419165872@163.com (H.W.); gsauwangzhengwen@163.com (Z.W.)
2 Key Laboratory for Grassland Ecosystem of Ministry of Education, Gansu Agricultural University, Lanzhou 730070, China
3 College of Animal Science and Technology, Gansu Agricultural University, Lanzhou 730070, China; lxx_gsau@163.com (X.L.); 18894314093@163.com (Y.S.); zhaosg@gsau.edu.cn (S.Z.)
* Correspondence: jiaoting207@126.com; Tel.: +86-136-0930-4144; Fax: +86-0931-642-468

Simple Summary: In the context of large-scale production, it is necessary to improve ewes health status and reproductive performance to promote newborns survival rate and later fattening performance. Previous studies have indicated that phenolic acids could change the host rumen microflora and inhibit pathogenic bacteria to ensure rumen health and host body health. However, there are few studies on isochlorogenic acid (ICGA) in ruminants. Therefore, based on regulating the rumen environment of breeding ewes during the whole physiological period, the experiment was conducted to study the effects of ICGA on rumen fermentation, microbial diversity and immunity of ewes at estrus, pregnancy and lactation stages. The experimental data obtained showed that adding ICGA could regulate ewes rumen fermentation mode, optimize microbial flora of different physiological stages by increasing *Bacteroidota* relative abundance while reducing *Firmicutes* relative abundance, maintain rumen microbial homeostasis at the pregnancy stage, and increase ewes blood immuneglobulin content, thereby improving ewes health.

Citation: Li, S.; Li, X.; Sha, Y.; Qi, S.; Zhang, X.; Wang, H.; Wang, Z.; Zhao, S.; Jiao, T. Effects of Isochlorogenic Acid on Ewes Rumen Fermentation, Microbial Diversity and Ewes Immunity of Different Physiological Stages. *Animals* **2024**, *14*, 715. https://doi.org/10.3390/ani14050715

Academic Editor: Lizhi Wang

Received: 11 December 2023
Revised: 12 February 2024
Accepted: 19 February 2024
Published: 24 February 2024

Abstract: The effects of isochlorogenic acid (ICGA) on ewes rumen environment, microbial diversity, and immunity at different physiological stages (estrus, pregnancy and lactation) were studied in this experiment. Twenty healthy female Hu lambs of 1.5 months with similar body weight (17.82 ± 0.98 kg) and body condition were selected and randomly divided into two groups: the control group (CON) and the ICGA group (ICGA). The lambs of CON were fed a basal diet, while the lambs of ICGA were supplemented with 0.1% ICGA based on the basal diet. Lambs rumen fermentation characteristics, microbial diversity and immunity at estrus, pregnancy, and lactation stages were determined and analyzed, respectively. The results showed that the rumen pH in CON increased first and then decreased as lambs grew ($p < 0.05$). However, it showed the opposite change in ICGA. The content of ammonia nitrogen (NH_3-N) showed the highest at estrus stage in both groups, but it was significantly higher in ICGA than that in CON ($p < 0.05$). The Acetic acid/propionic acid (A/P) ratio at estrus stage and the volatile fatty acids (VFAs) at pregnancy stage in ICGA were significantly higher than those of the CON ($p < 0.05$). The 16S rDNA sequencing analysis showed that the Shannon, Chao 1 and ACE indexes of the ICGA were significantly higher than those of the CON both at estrus and lactation stages ($p < 0.05$), while they showed higher at the pregnancy stage in CON ($p > 0.05$). Principal component analysis (PCA) showed that there were significant differences in rumen microorganism structure between CON and ICGA at all physiological stages ($p < 0.01$). At the phylum level, compared with the CON, *Firmicutes* relative abundance of three physiological stages decreased ($p > 0.05$) while *Bacteroidota* increased ($p > 0.05$). The relative abundance of *Synergistota* at estrus stage and *Patescibacteria* at the lactation stage increased significantly too ($p < 0.05$). At the genus level, compared with the CON, the relative abundance of *Prevotella* at three stages showed the highest ($p > 0.05$), while the relative abundance of *Succiniclasticum*, *unclassified_Selenomonadaceae* and *Rikenellaceae_RC9_gut_group* showed different abundances at different physiological stages in ICGA. Compared with the CON, the lambs of the ICGA showed higher blood IgG, IgM, and TNF- α contents at three physiological stages and higher IL-6 contents at pregnancy stage ($p < 0.05$). Conclusion: Adding ICGA could regulate ewes rumen fermentation mode at different physiological stages by increasing rumen NH_3-N at estrus, VFAs at pregnancy, and the ratio of A/P at lactation. It optimizes rumen microbial flora of different physiological stages by increasing *Bacteroidota* relative abundance while reducing *Firmicutes* relative abundance, maintaining rumen microbial homeostasis at pregnant stage, increasing the number of beneficial bacteria in later lactating and ewes blood immunoglobulins content at three physiological stages.

Keywords: isochlorogenic acid; ewes; Hu lambs; physiological stages; rumen fermentation; rumen microbial diversity

1. Introduction

With the development of urbanization, animal product consumption has undergone a historic turning point in China [1]. People currently prefer mutton to meet their high-quality protein needs [2]. To meet mutton supply, sheep breeding and feeding are gradually transitioning from grazing to large-scale and standardized production [3]. In this context, sheep health status has become one of the focuses, especially ewes, which has directly impacted ewes reproductive performance, growth rate, and lambs' production performance [4]. Meanwhile, the problems of ewes low immunity, low fecundity, and the survival rate of newborns faced by the animal industry has affected the development of sheep industry seriously [5]. Therefore, it is necessary to improve ewes' health condition and reproductive performance to promote newborns survival rate and later fattening performance.

Stevia rebaudiana was native to the subtropical region of South America [6]. It has been planted widely in both southern and northern regions since it was introduced and cultivated successfully in China in 1977 [7]. In recent years, there have been reports that the stevia extract contained a variety of phenolic acids, such as chlorogenic acid (CGA) and isochlorogenic acid (ICGA), which have biological activities, such as scavenging free radicals, antibacterial and anti-inflammatory, inhibiting tumors, protecting liver and gallbladder, promoting blood circulation and reducing blood pressure, etc. [8,9]. Previous studies indicated that CGA had multiple benefits for animals gastrointestinal health [10]. It can optimize the gastrointestinal flora composition to improve animal intestinal health by increasing short-chain fatty acids (SFAs), promoting intestinal probiotics growth and inhibiting pathogenic bacteria [11–13]. ICGA has a similar structure as CGA [14], it has one more caffeoyl group than CGA in structure, and so, has stronger biological activity [15]. In our previous study, we found that adding ICGA to the sheep diet could increase rumen acetic acid (AA), butyric acid (BA), and total volatile fatty acid (TVFA) concentration, so as to regulate the rumen fermentation mode [16].

Rumen is a special digestive and absorptive organ of ruminants [17]. Microorganisms in the rumen not only play a key role in animal digestion and absorption but also play an important role in host immune response and body health [18–20]. A large number of studies on cows have shown that ruminal dysbiosis could lead to metabolic changes in rumen and other parts of the digestive tract, causing a large number of microorganisms to release toxins into gastrointestinal lumen, triggering various systemic inflammations and directly affecting the health of next generation [21]. Therefore, from the perspective of ruminant reproductive health, strengthening rumen health is particularly important [22].

Previous studies have shown that ICGA could regulate the rumen fermentation mode. So, we speculate that ICGA may change the host rumen microflora and inhibit pathogenic bacteria to ensure rumen health and host body health. Therefore, the primary aim of this study is to investigate the effects of isochlorogenic acid (ICGA) on rumen fermentation and microbial diversity in ewes across different physiological stages. Our hypothesis is that ICGA can regulate the mode of rumen fermentation and contribute to the optimization of microbial flora, thereby improving the health of Hu sheep.

2. Materials and Methods

2.1. Experimental Material

The ICGA derived from stevia is purchased from Chenguang Biotechnology Co., Ltd. (Handan, China), with a total acidity greater than 55%.

The 1.5-month-old weaned female lambs of Hu sheep came from Guanghe county, Gansu province, with similar body conditions selected as experimental animals. All of the lambs are in a state of health.

2.2. Experimental Design

Twenty 1.5-month-old healthy and homogeneous weaned female lambs with similar body weights (17.82 ± 0.98 kg) were randomly divided into two groups: the control group (CON) and ICGA group (ICGA), 10 lambs in each and housed in one pen (one pen per group) throughout the research. The lambs in CON were fed a basal diet, while the lambs in ICGA were supplemented with 0.1% ICGA based on the basal diet (air-dried basis). The amount of ICGA addition was referred to the results of our previous research of in vitro rumen fermentation [16]. It was weighed and mixed into their diet, stirred thoroughly and evenly before feeding. Lambs in each group were raised in one pen, free feeding and drinking water. After all lambs were fed to sexual maturity (7 months old), embolectomy and estrus synchronization operations were started. Pregnant mare's serum hormones were injected 10 days later. The embolus was withdrawn after 11 days, then insemination was performed 2 days later, and again 24 h later. Artificial insemination status assessment was conducted 10 days later after the insemination. Ewe pregnancy was checked using B-ultrasound 45 days later, then the successfully pregnant ewe was fed until they gave birth (lactation). Every period the ewes were fed the according diets shown in Table 1. The basal diets of two groups at different physiological stages were formulated according to Agricultural Industry Standard Mutton Sheep Feeding Standard of People's Republic of China. "Estrus ration" was used during estrus, "Pregnancy ration" during pregnancy, and "Lactation ration" during lactation. The composition and nutritional components of raw materials are shown in Table 1.

2.3. Sample Collection and Processing

The ewes rumen fluid and blood samples in each group were collected at estrus, pregnancy (120 days after pregnancy) and lactation (30 days after delivery) stages, respectively. Rumen fluid samples were collected through oral cavity using a rumen tube. Briefly, a flexible PVC tube (2 mm of wall thickness × 6 mm of internal diameter) with holes of 2.5 mm diameter in the 15 cm-probe head (Anscitech Co., Ltd., Wuhan, China) was connected to an electric vacuum pump (7 mbar) and inserted into the ewes rumen via the esophagus to collect the rumen sample. About 75 mL of rumen fluid from each ewe was collected 3 h after morning feeding. The first 5 mL of rumen fluid in each sampling was discarded to remove the potential saliva contamination and the remaining contents were collected. One part was divided into a 50 mL centrifuge tube and stored at −20 °C for fermentation parameters determination; and one part was placed in 5 mL cryotube and stored at −80 °C for rumen microbiota determination. Approximately 10 mL of fasting jugular vein blood in every period were also collected and placed in a common blood collection vessel, centrifuged at 3500 r·min^{-1} for 10 min to obtain the separated serum, then stored at −20 °C for the determination of blood immune indexes.

Table 1. Diet composition and nutritional levels of ewes at different physiological stages (air-dried basis).

Physiological Period		Estrus (7.5 Months Old)	Pregnancy (7.5–12.5 Months Old)	Lactation (12.5–14.5 Months Old)
Formula composition/%	Whole plant silage corn	20	10	22
	Wheat straw	25	25	10
	Alfalfa	25	30	33
	Corn grain	13	15	15
	Bran	2	5	5
	Sesame cake	8	7	7
	Cottonseed meal	3	4	4
	Baking soda	0.5	0.5	0.5
	Calcium hydrogen phosphate	1	1	1
	Salt	0.5	0.5	0.5
	Premix [1]	2	2	2
	Total	100	100	100
Nutritional level	Dry matter, DM (%)	80.19	85.91	82.85
	Digestible energy, DE/(MJ kg $^{-1}$)	11.01	11.53	13.03
	Metabolizable energy, ME/(MJ kg $^{-1}$)	9.07	8.85	9.12
	Calcium, Ca (%)	1.94	1.07	1.08
	Phosphorus, P (%)	0.71	0.54	0.58
	Crude protein, CP (%)	12.89	13.76	13.07

[1] The premix provided the following per kg of the diet: Cu 5.40 mg, Fe 19.80 mg, Zn 9.00 mg, Mn 14.40 mg, Se 0.11 mg, I 0.18 mg, Co 0.07 mg, VA 7200 IU, VD 1440 IU, VE 500 IU; Metabolic energy and digestible energy were calculated, and other values are measured.

2.4. Rumen Fermentation Index Determination

The pH value was measured by Sartorius acidometer (PB-10). Volatile fatty acids content was determined by the Agilent 7890B meteorological chromatograph. After the rumen fluid sample was pretreated, it was injected into a gas chromatograph to obtain the chromatogram of VFAs sample, then we calculated the content of VFAs using the peak area external standard method [23]. Ammonia nitrogen content was determined by the phenol-sodium hypochlorite colorimetric method. The rumen fluid was centrifuged at 4000 r/min for 10 min, and 2 mL of supernatant was taken into a 15 mL centrifuge tube, 8 mL of 0.2 mol/L hydrochloric acid was added, then shaken well. Then, 2 mL of solution A (0.08 g of sodium nitroso ferricyanide dissolved in 100 mL of 14% sodium salicylate solution) and 2 mL of solution B (2 mL of sodium hypochlorite solution mixed with 100 mL of 0.3 mol/L sodium hydroxide solution) were added in turn, respectively. After shaking and standing for 10 min, the absorbance value was recorded at 700 nm to calculate the ammonia nitrogen content [24].

2.5. Analysis of Rumen Microbial Diversity

From DNA extraction to sequencing steps: (1) Microbial total DNA extraction; (2) Target fragment PCR amplification; (3) Amplification products purification and recovery by magnetic beads; (4) Amplification products fluorescence quantification; (5) Sequencing library preparation; (6) High-throughput sequencing; (7) Results analysis, including preliminary screening of the original offline data of high-throughput sequencing according to the sequence quality, the problem samples retesting, then removing the primer fragments of the sequence, discarding the unmatched primers sequence, and performing quality control, denoising, splicing, chimera removal and other steps. Based on valid data, the sequences were clustered into operational taxonomic units (OTUs) with 97% homology, and species annotation analysis was performed using OTUs sequences and the RDP Classifier database. According to the results of species annotation, alpha diversity and beta diversity were calculated, and the differences between groups were compared to reveal the differences of the same flora under each treatment. The diversity index was calculated based on the OTU level. Beta diversity analysis (PCA) was used to analyze the differences in community structure among different populations.

2.6. Blood Immune Index Determination

Serum immunoglobulin A (IgA), immunoglobulin G (IgG), immunoglobulin M (IgM), interleukin-1β (IL-1β), interleukin-6 (IL-6) and Tumor necrosis factor α (TNF-α) were measured using commercial ELISA kits (Nanjing Jiancheng Biotech, Nanjing, Jiangsu, China) [25].

2.7. Statistical Analysis

Excel 2010 was used to sort out the experimental data and the independent sample *t*-test in SPSS 23.0 software was used to analyze the significance between different treatments at the same physiological stage. $p < 0.05$ indicated the difference was significant, while $p < 0.01$ showed extremely significant.

3. Results

3.1. Effects of ICGA on Ewes Rumen Fermentation at Different Physiological Stages

With the advancement of ewes physiological periods, the ewes rumen pH increased at pregnancy and then decreased at lactation in CON, while the content of NH_3-N and VFAs decreased at pregnancy and then increased at lactation, while in ICGA, the ewes rumen NH_3-N content decreased first and then increased, but the pH and VFAs content showed the opposite change (Table 2). Compared with the CON, the rumen propionic acid (PA), butyric acid (BA) and TVFA content decreased significantly, and the A/P ratio and NH_3-N content increased significantly ($p < 0.05$) at estrus stage, while acetic acid (AA), PA, isobutyric acid (IBA), BA, isovaleric acid (IVA), valeric acid (VA) and TVFA increased ($p < 0.05$) at pregnancy stage in ICGA. There were no significant differences in rumen fermentation parameters between CON and ICGA at lactation stage ($p > 0.05$) (Table 2).

Table 2. Effects of ICGA on ewes rumen fermentation at different physiological stages.

Index	Estrus			Pregnancy			Lactation		
	CON	ICGA	*p* Value	CON	ICGA	*p* Value	CON	ICGA	*p* Value
pH	6.55 ± 0.05^{b}	6.91 ± 0.02^{a}	0.002	6.73 ± 0.13^{a}	6.28 ± 0.02^{b}	0.025	6.52 ± 0.15^{a}	6.93 ± 0.04^{a}	0.053
Ammonia nitrogen (mg dL^{-1})	8.62 ± 0.19^{b}	10.50 ± 0.60^{a}	0.040	4.79 ± 0.78^{a}	4.16 ± 0.20^{a}	0.481	8.10 ± 0.94^{a}	4.73 ± 0.25^{a}	0.061
Acetic acid (mmol L^{-1})	49.39 ± 3.49^{a}	44.18 ± 1.08^{a}	0.227	26.14 ± 0.46^{b}	45.63 ± 2.14^{a}	0.009	45.30 ± 1.91^{a}	38.33 ± 3.72^{a}	0.172
Propionic acid (mmol L^{-1})	12.91 ± 0.85^{a}	9.36 ± 0.46^{b}	0.021	6.64 ± 0.67^{b}	18.99 ± 1.37^{a}	0.001	15.00 ± 1.15^{a}	10.84 ± 1.07^{a}	0.057
Isobutyric acid (mmol L^{-1})	0.59 ± 0.04^{a}	0.66 ± 0.02^{a}	0.214	0.26 ± 0.02^{b}	0.41 ± 0.03^{a}	0.011	0.55 ± 0.11^{a}	0.43 ± 0.01^{a}	0.405
Butyric acid (mmol L^{-1})	7.99 ± 0.46^{a}	4.58 ± 0.20^{b}	0.002	3.31 ± 0.11^{b}	7.26 ± 0.39^{a}	0.001	6.58 ± 0.79^{a}	5.49 ± 0.40^{a}	0.284
Isovaleric acid (mmol L^{-1})	0.93 ± 0.06^{a}	0.95 ± 0.10^{a}	0.872	0.32 ± 0.00^{b}	0.64 ± 0.02^{a}	0.003	0.88 ± 0.18^{a}	0.78 ± 0.04^{a}	0.638
Valeric acid (mmol L^{-1})	0.52 ± 0.01^{a}	0.48 ± 0.03^{a}	0.365	0.27 ± 0.02^{b}	0.57 ± 0.05^{a}	0.006	0.64 ± 0.06^{a}	0.61 ± 0.02^{a}	0.715
Acetic acid/Propionic acid	3.82 ± 0.02^{b}	4.74 ± 0.16^{a}	0.029	4.00 ± 0.34^{a}	2.42 ± 0.12^{b}	0.012	3.04 ± 0.11^{a}	3.55 ± 0.21^{a}	0.100
Total volatile fatty acids (mmol L^{-1}))	71.45 ± 3.69^{a}	59.1 ± 2.21^{b}	0.045	36.88 ± 1.07^{b}	78.31 ± 4.78^{a}	0.001	67.17 ± 3.69^{a}	56.39 ± 5.21^{a}	0.167

^{a,b} Values with different treatments in the same physiological stage are significantly different ($p < 0.05$).

3.2. Effects of ICGA on Ewes Rumen Microbial Diversity at Different Physiological Stages

3.2.1. Effect on the Number of Operational Taxonomic Units (OUT)

A total of 18,302 OTUs were obtained in this sequencing. The number of OTUs were higher at estrus and lactation stage while they were lower at pregnancy stage in ICGA than those in CON (Figure 1). There were 2171 unique OTUs in CON and 2873 unique OTUs in ICGA, with a total of 785 OTUs at estrus stage (Figure 2A), 4371 unique OTUs in CON and 3303 unique OTUs in ICGA, with a total of 955 OTUs at pregnancy stage (Figure 2B) and

3206 unique OTUs in CON and 3884 unique OTUs in ICGA, with a total of 1077 OTUs at lactation stage (Figure 2C).

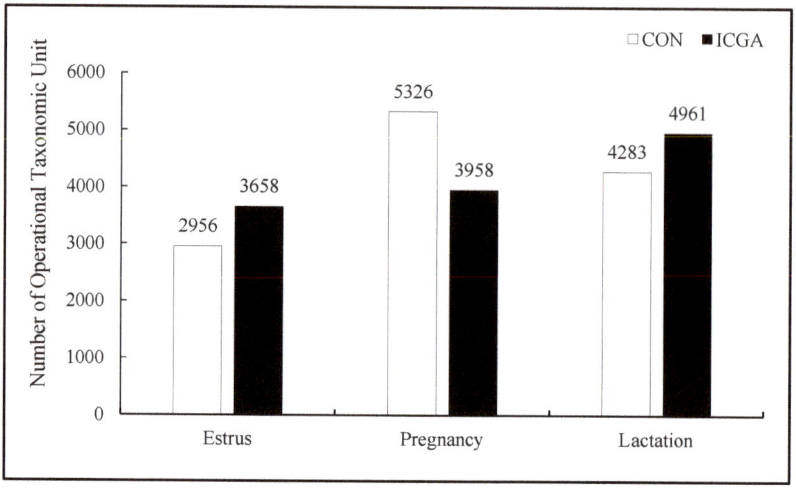

Figure 1. Operational Taxonomic Unit number distribution map.

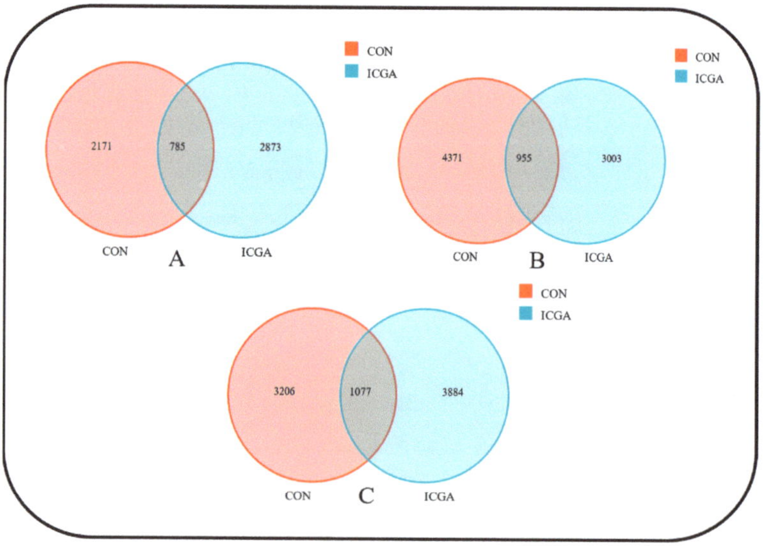

Figure 2. Venn diagram for OTU distribution of rumen microorganisms. (**A**): Estrus; (**B**): Pregnancy; (**C**): Lactation.

3.2.2. Effect on Dilution Curve

The sample dilution curves of ewes three physiological stages in both groups showed a trend of rising first and then leveling off (Figure 3), indicating that the sequencing depth was sufficient to cover all species in the sample, and the sequencing results were reliable.

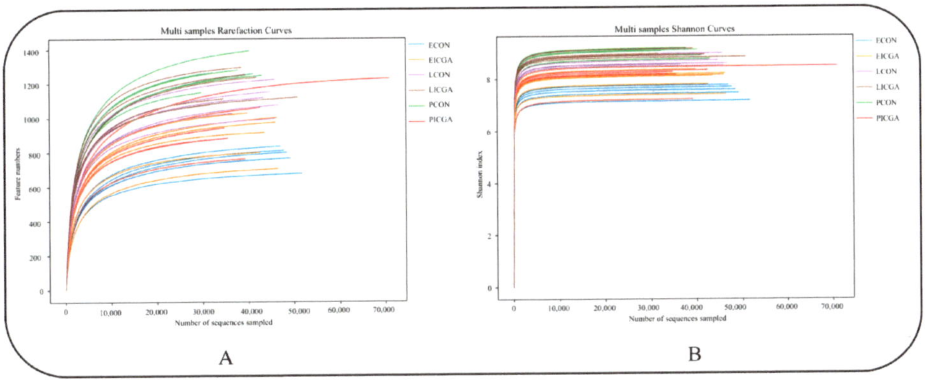

Figure 3. Sample dilution curve. (**A**): OTU number dilution curve; (**B**): Shannon index dilution curve.

3.2.3. Effect on Alpha Diversity

The Shannon, Chao 1, and ACE indexes of the ICGA at estrus stage were significantly higher and those at the pregnancy stage were significantly lower than those of the CON group ($p < 0.05$). The Shannon and Simpson indexes of the ICGA were significantly higher than those of the CON at lactation stage ($p < 0.05$) (Figure 4).

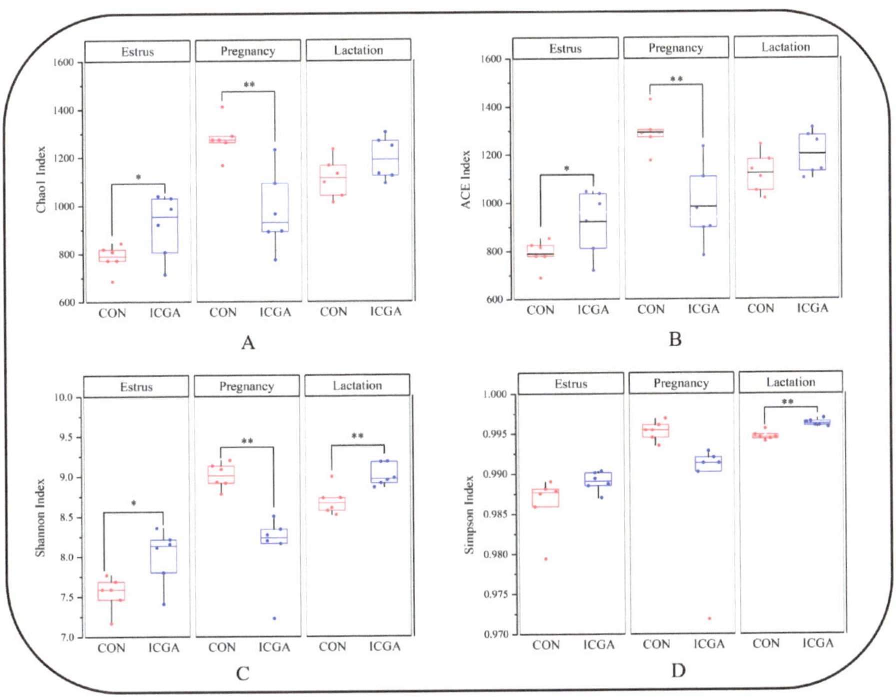

Figure 4. Alpha diversity index of rumen microbes. (**A**): Chao 1 Index; (**B**): ACE Index; (**C**): Shannon Index; (**D**): Simpson Index. * $p < 0.05$, ** $p < 0.01$.

3.2.4. Effect on Beta Diversity

There were significant differences in rumen microflora between CON and ICGA at estrus (Figure 5A), pregnancy (Figure 5B) and lactation (Figure 5C) stage ($p < 0.01$).

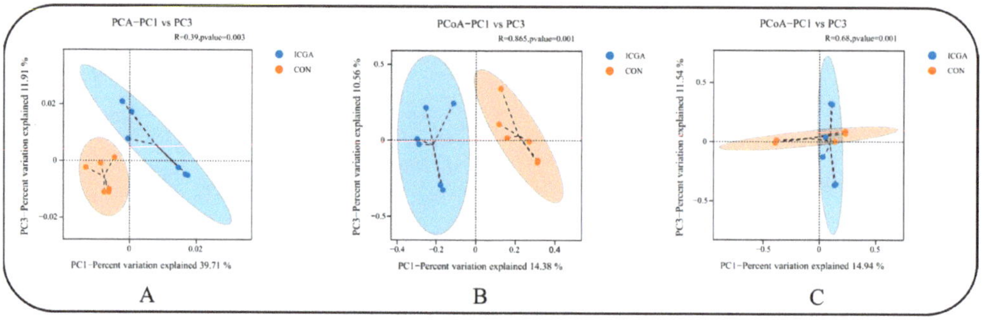

Figure 5. PCA of rumen microflora. (**A**): Estrus; (**B**): Pregnancy; (**C**): Lactation.

3.2.5. Effect on Bacterial Composition and Structure at the Phylum Level

At the phylum classification level, the top 10 species in the relative abundance of rumen microbial communities at three physiological stages of two groups were counted (Figure 6A) (Table 3). With the advancement of the physiological periods, the relative abundance of *Firmicutes* in CON decreased first and then increased, while that in ICGA decreased. The relative abundance of *Bacteroidota* in both groups increased first and then decreased. The relative abundance of *Firmicutes* in both groups were highest at three physiological stages, followed by *Bacteroidota*. Compared with the CON, the relative abundance of *Firmicutes* in ICGA was lower ($p > 0.05$), and the relative abundance of *Bacteroidota* and *Proteobacteria* were higher than those in CON ($p > 0.05$). The relative abundance of *Synergistota* in ICGA was significantly higher than that in CON at the estrus stage ($p < 0.05$), but the relative abundance of *Patescibacteria* in ICGA was significantly higher than that in CON at lactation stage ($p < 0.05$).

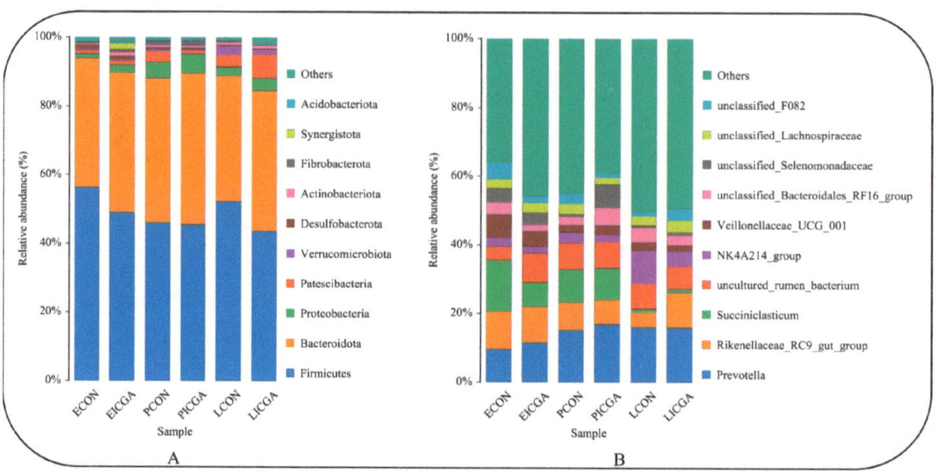

Figure 6. The relative abundance of phylum (**A**) and genus (**B**) of rumen microbiota Top 10.

Table 3. Rumen microbial relative abundance at phylum level Top 10 (%).

Phylum	Estrus			Pregnancy			Lactation		
	CON	ICGA	*p* Value	CON	ICGA	*p* Value	CON	ICGA	*p* Value
Firmicutes	56.05 ± 0.04	48.86 ± 0.02	0.99	46.42 ± 0.04	45.78 ± 0.02	1.00	51.80 ± 0.03	42.78 ± 0.04	0.93
Bacteroidota	37.79 ± 0.04	40.78 ± 0.02	1.00	42.13 ± 0.02	42.86 ± 0.02	1.00	36.92 ± 0.02	41.78 ± 0.05	0.99
Proteobacteria	1.26 ± 0.00	2.25 ± 0.01	1.00	4.78 ± 0.01	5.68 ± 0.02	1.00	2.60 ± 0.00	3.70 ± 0.00	1.00
Patescibacteria	1.09 ± 0.00	1.22 ± 0.00	1.00	3.11 ± 0.00	0.92 ± 0.00	0.06	3.51 ± 0.00	6.92 ± 0.01	0.00
Verrucomicrobiota	0.41 ± 0.00	0.28 ± 0.00	1.00	0.44 ± 0.00	0.27 ± 0.00	1.00	2.63 ± 0.01	1.26 ± 0.00	0.48
Desulfobacterota	1.28 ± 0.00	1.13 ± 0.00	1.00	0.80 ± 0.00	0.48 ± 0.00	0.99	0.38 ± 0.00	0.33 ± 0.00	1.00
Actinobacteriota	0.27 ± 0.00	1.13 ± 0.01	0.55	0.63 ± 0.00	0.66 ± 0.00	1.00	0.67 ± 0.00	0.87 ± 0.00	1.00
Fibrobacterota	0.35 ± 0.00	0.72 ± 0.00	1.00	1.16 ± 0.01	1.64 ± 0.01	1.00	0.38 ± 0.00	0.11 ± 0.00	1.00
Synergistota	0.06 ± 0.00	1.88 ± 0.01	0.00	0.04 ± 0.00	0.03 ± 0.00	1.00	0.03 ± 0.00	0.05 ± 0.00	1.00
Acidobacteriota	0.01 ± 0.00	0.35 ± 0.00	0.41	0.34 ± 0.00	0.31 ± 0.00	1.00	0.26 ± 0.00	0.40 ± 0.00	1.00

3.2.6. Effect on Bacterial Composition and Structure at the Genus Level

At the genus classification level, the top 10 species in the relative abundance of rumen microbial communities in CON and ICGA at three physiological stages were counted (Figure 6B) (Table 4). With the advancement of the physiological periods, the relative abundance of *Prevotella* in CON increased gradually, while the relative abundance of *Rikenellaceae_RC9_gut_group* and *Succiniclasticum* decreased. The relative abundance of *Prevotella* and *Succiniclasticum* in ICGA increased first and then decreased, and the relative abundance of *Rikenellaceae_RC9_gut_group* decreased first and then increased. Compared with the CON, the relative abundance of *Prevotella* in ICGA was higher at three physiological stages ($p > 0.05$), the relative abundance of *Rikenellaceae_RC9_gut_group* in ICGA at lactation stage was significantly higher ($p < 0.05$), and the relative abundance of *NK4A214_group* was significantly lower ($p < 0.05$). The relative abundance of *unclassified_Selenomonadaceae* was significantly higher at the pregnancy stage ($p < 0.05$) while the relative abundance of *Succiniclasticum* and *unclassified_F082* was significantly lower at the estrus stage in ICGA than those in CON ($p < 0.05$).

Table 4. Rumen microbial relative abundance at genus levels Top10 (%).

Genus	Estrus			Pregnancy			Lactation		
	CON	ICGA	*p* Value	CON	ICGA	*p* Value	CON	ICGA	*p* Value
Prevotella	9.67 ± 0.01	11.49 ± 0.01	1.00	14.97 ± 0.01	16.78 ± 0.01	1.00	16.09 ± 0.01	16.44 ± 0.03	1.00
Rikenel-laceae_RC9_gu-t_group	11.16 ± 0.01	10.44 ± 0.01	1.00	8.13 ± 0.01	6.89 ± 0.00	1.00	4.54 ± 0.01	10.36 ± 0.01	0.00
Succiniclasticum	14.80 ± 0.01	6.98 ± 0.01	0.00	9.45 ± 0.01	9.78 ± 0.02	1.00	0.88 ± 0.00	1.11 ± 0.00	1.00
uncultured_ru-men_bac-terium	4.00 ± 0.01	8.70 ± 0.02	0.23	7.90 ± 0.01	7.19 ± 0.02	1.00	7.53 ± 0.01	6.76 ± 0.01	1.00
NK4A214_group	2.46 ± 0.00	1.89 ± 0.00	1.00	3.08 ± 0.00	2.05 ± 0.00	1.00	9.54 ± 0.02	4.32 ± 0.00	0.00
Veillonel-laceae_UCG_001	6.76 ± 0.01	4.45 ± 0.01	0.27	2.08 ± 0.00	2.95 ± 0.01	1.00	2.4 ± 0.01	1.81 ± 0.00	1.00
unclassified_Ba c-teroid-ales_RF16_group	3.56 ± 0.01	1.81 ± 0.01	0.91	2.40 ± 0.01	5.59 ± 0.02	0.09	4.45 ± 0.01	2.79 ± 0.00	0.94
unclassified_Se-lenom-onadaceae	4.11 ± 0.01	3.57 ± 0.01	1.00	0.72 ± 0.00	6.87 ± 0.04	0.02	0.52 ± 0.00	0.94 ± 0.00	1.00
unclassified_La-chnospi-raceae	2.75 ± 0.00	2.77 ± 0.01	1.00	2.87 ± 0.00	2.00 ± 0.00	1.00	2.77 ± 0.01	3.2 ± 0.00	1.00
unclassified_F082	5.09 ± 0.01	1.93 ± 0.01	0.02	3.05 ± 0.00	1.14 ± 0.00	0.56	1.15 ± 0.00	3.56 ± 0.00	0.19

3.2.7. Effect on LEfSe Difference

LEfSe difference analysis was performed on rumen microorganisms of two groups at three physiological stages at different classification levels (Figure 7). At the estrus stage (Figure 7A), the ewes in CON had eight species groups with significant differences. They were mainly *Negativicutes, uncultured_rumen_bacterium, Acidaminococcales, Acidaminococcaeae, Succiniclasticum*, and *Christensenellaceae_R7_group*. There were four significantly different microbial groups in ICGA, mainly *Bacteroidales_BS11_gut_group, uncultured_rumen_bacterium* and *Quinella*. At the pregnancy stage (Figure 7B), there were ten significantly different species groups in CON, mainly including *Clostridia, Oscillospirales, Lachnospiraceae* and *unclassified_Rikenellaceae_RC9_gut_group*. There were seven significantly different microbial groups in ICGA, mainly for *Negativicutes, Veillonellales-selenomonadales, unclassified_Selenomonadaceae*, and *uncultured_rumen_bacterium*. At the lactation stage (Figure 7C), there were two significantly different species groups in CON, which were *unclassified_NK4A214-group* and *Prevotellaceae_UCG_001*, respectively. There were twelve significantly different microbial groups in ICGA, mainly *Rikenellaceae, Rikenellaceae_RC9_gut_group* and *uncultured_rumen_bacterium*.

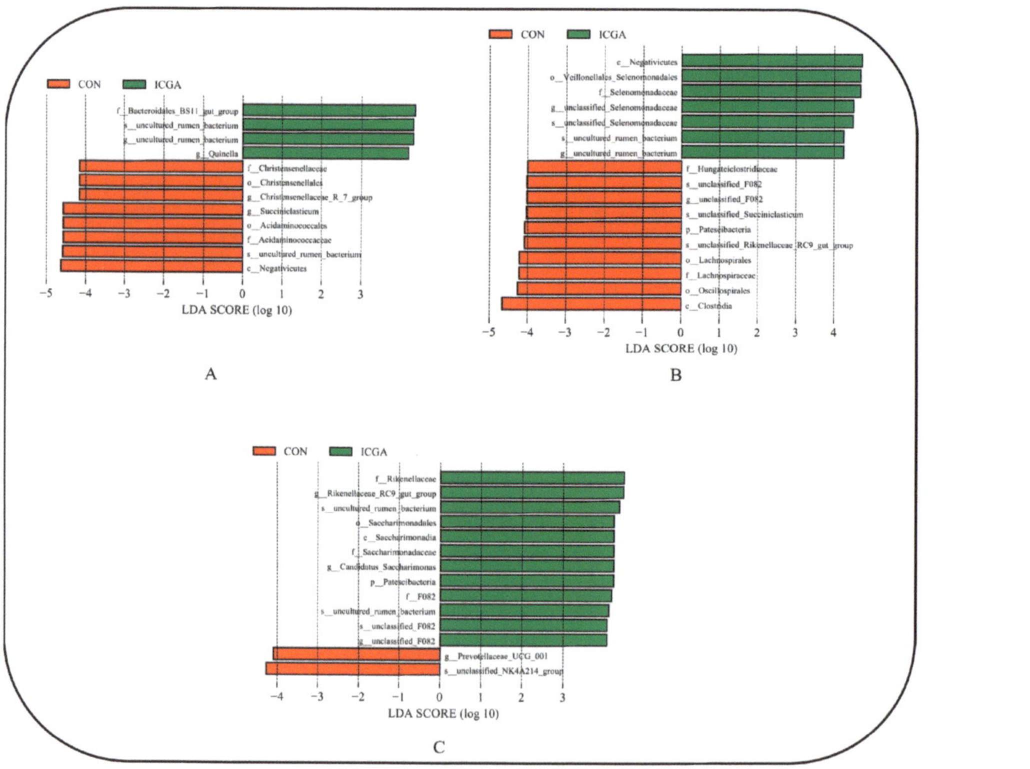

Figure 7. LDA value distribution histogram. (**A**): Estrus; (**B**): Pregnancy; (**C**): Lactation.

3.3. *Effects of ICGA on Ewes Blood Immune Index at Different Physiological Stages*

Compared with the CON, the ewes of ICGA showed higher serum IgG, IgM, TNF-α contents at three physiological stages ($p < 0.05$), while lower serum IL-6 contents at the estrus and lactation stages, but higher IL-6 at the pregnancy stage ($p < 0.05$) (Figure 8).

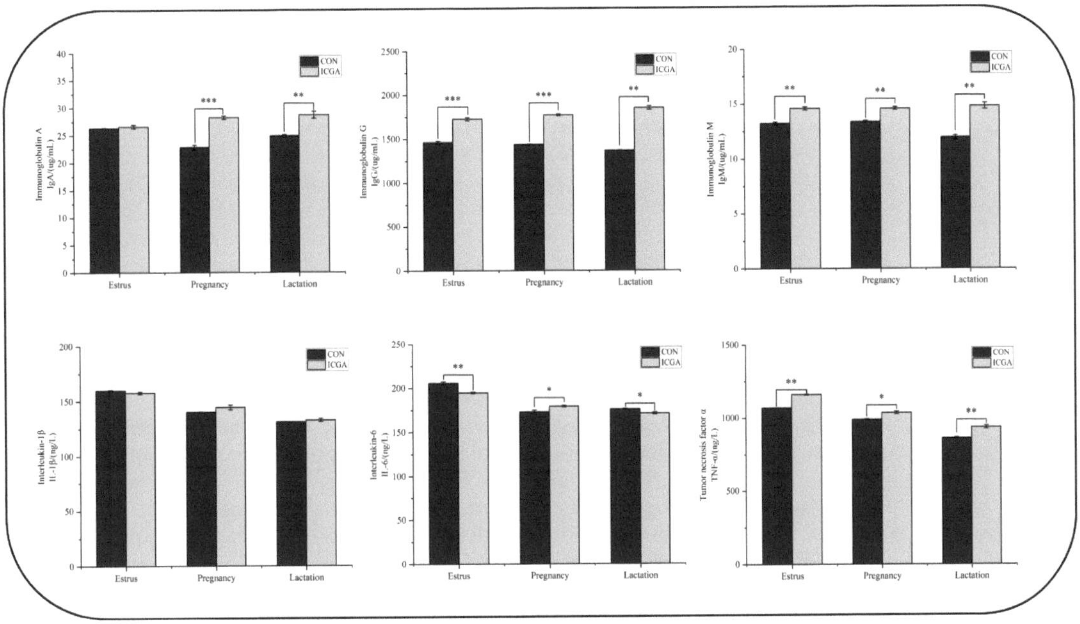

Figure 8. Effects of ICGA on ewes blood immune index at different physiological stages. * $p < 0.05$, ** $p < 0.01$, *** $p < 0.001$.

4. Discussion

4.1. Effects of ICGA on Ewes Rumen Fermentation at Different Physiological Stages

Ruminant rumen is a relatively stable anaerobic fermentation tank. NH_3-N, pH, and VFAs are the main internal environmental indicators for rumen fermentation [26]. In this experiment, the rumen pH is within the normal range of 6.3 and 7.0 at three physiological stages in both groups, meaning that there was no adverse effect on ewes rumen fermentation by ICGA. The rumen NH_3-N content had a significant increase at estrus in ICGA, which was beneficial to synthesize more microbial protein (MCP) to provide large amounts of protein for body, so improve ovulation rate and conception rate of ewes at estrus stage [27]. Weston pointed out the TVFA content decreased with the prolongation of pregnancy time [28], which may be due to ewes absorbing VFAs rapidly by rumen wall to need higher energy demand because of the rapid development of pregnant embryos, thus reduce the rumen VFAs concentration of pregnant ewes. Adding ICGA may stimulate the rumen microorganisms (*unclassified_Selenomonadaceae*) of pregnant ewes to produce VFAs, providing sufficient energy for pregnant ewes, which is beneficial to the health of ewes and embryos [29]. Due to the heavy burden and nutrient consumption at the pregnancy stage, the feed intake of lactating ewes increased, and the rumen NH_3-N and VFAs content also increased [30,31] However, the NH_3-N and VFAs contents showed the opposite in ICGA. Adding ICGA may accelerate the absorption of NH_3-N and VFAs in the rumen wall so that the ewes rumen NH_3-N and VFAs contents decrease at the lactation stage in ICGA. Liu Yongjun pointed out the high-yield group of dairy cows had a stronger absorption capacity for VFAs produced by rumen fermentation [32]. In this experiment, the A/P of the ICGA was higher than that of the CON at the estrus and lactation stages and was significantly lower than that of the CON at the pregnancy stage. AA is an important precursor for synthesis of milk fat, PA is the precursor of body fat synthesis, and the A/P ratio is significantly positively correlated with milk fat rate [33,34]. Adding ICGA could promote the synthesis of body fat in estrus ewes and accumulate energy for pregnancy, promote synthesis of glucose from propionic acid to provide more energy for pregnant ewes and synthesis of milk fat in lactating ewes to increase the milk fat rate of milk.

4.2. Effects of ICGA on Ewes Rumen Microbial Diversity at Different Physiological Stages

Through sequencing data analysis, it was found that the diversity dilution curve of rumen flora increased first and then tended to be gentle, indicating that the depth of this sequencing was sufficient and the sequencing results were reliable. The rumen microbial community is dynamic and often changes with diet, host, physiological state and environment [35]. In ICGA, the rumen microorganisms richness and diversity at estrus and lactation stages were higher than those of the CON while lower than those of the CON at pregnancy stage. It may be due to the stress response during pregnancy that the rumen fluid flora in CON was disordered, resulting in an increase of rumen species richness in CON during pregnancy. Guo pointed out the rumen flora diversity of ewes increased at 120 days of pregnancy [36]. It is speculated that adding ICGA to the diet may maintain the rumen microbial homeostasis of pregnant ewes and alleviate pregnant ewes stress response. Through Beta diversity analysis of samples, we found that the sample composition of CON and ICGA at different physiological stages was significantly separated, and there were significant differences in rumen microflora. So, adding ICGA could change the ewes rumen microbial flora structure at estrus, pregnancy and lactation stages.

The analysis of rumen microflora composition and structure at the phylum level showed that the dominant phyla in CON and ICGA were *Firmicutes* and *Bacteroidota* at three physiological stages, and the dominant phyla distribution was similar to results of other studies [37]. In this experiment, the relative abundance of *Firmicutes* in ICGA was lower than that in CON, while the relative abundance of *Bacteroidota* was higher than that in CON. High abundance of *Bacteroidota* usually has anti-inflammatory effects [38], showing ICGA may exert anti-inflammatory effects by reducing *Firmicutes* abundance and increasing *Bacteroidota* abundance, so increase the ratio of *Bacteroidota* to *Firmicutes*, the same as other results [39], which can reduce fat ewes deposition, thereby accelerating material circulation and energy flow to be benefit to maintain ewes health [40]. It is also indicated that ICGA can promote the protein and non-structural carbohydrates degradation and feed absorption and transformation in rumen by reducing *Firmicutes* abundance and increasing *Bacteroidota* abundance [41]. The relative abundance of *Synergistota* at estrus stage and the relative abundance of *Patescibacteria* at lactation stage in ICGA both were significantly increased. Studies showed that *Synergistota* isolated from ruminant foods could use arginine and histidine as substrates to degrade toxins in diet, while *Patescibacteria* has a strong ability to adapt to adverse environments [42,43]. It can be suggested that ICGA may improve ewes stress resistance on adverse environments to help host animals survive well.

The composition and structure analysis of rumen microbial flora at the genus level showed that the relative abundance of *Prevotella* in ICGA was higher than that in CON at three physiological stages. *Prevotella* is the dominant genus in mature rumen and the most abundant and common group of rumen microorganisms [44]. ICGA may promote rumen microflora maturation [45]. In this experiment, the relative abundance of *unclassified_Selenomonadaceae* in ICGA at pregnancy stage was significantly higher than that in CON, which could promote VFAs production in pregnant ewes rumen [46]. It was consistent with the status of pregnant ewes rumen fermentation. Studies showed that the *Rikenellaceae_RC9_gut_group* could promote the digestion and absorption of carbohydrates in intestine and protect mucosal barrier function by increasing butyrate levels [47,48]. Just in time, the relative abundance of *Rikenellaceae_RC9_gut_group* in ICGA was significantly higher than that in CON in the experiment. ICGA could improve lactating ewes rumen health by increasing rumen beneficial bacteria amounts and promoting rumen VFAs absorption and utilization of lactating ewes., The relative abundance of *Succiniclasticum* at the estrus stage in ICGA was significantly lower than that in CON in the experiment. *Succiniclasticum* played an important role in PA production, which may be the reason for the decrease of rumen PA concentration at estrus stage in ICGA [49].

LEfSe difference analysis showed that *Negativicutes* species were significantly enriched in CON at estrus, and *Bacteroidales_BS11_gut_group* and *Quinella* were significantly enriched in ICGA. The studies showed that the increase in *Quinella* abundance was related to a decrease in methane production [50]. It was speculated that ICGA not only promoted the ewes protein and soluble carbohydrates degradation at estrus but also inhibited methane production. At pregnancy, the species significantly enriched in CON were mostly *Clostridia*; *Negativicutes* species in ICGA were significantly enriched. Both *Clostridia* and *Negativicutes* belong to *Firmicutes*, which could promote fiber degradation to provide energy for pregnant ewes. At lactation, *Prevotellaceae_UCG_001* was significantly enriched in CON, and *Rikenellaceae* and *Rikenellaceae_RC9_gut_group* were significantly enriched in ICGA. ICGA could increase rumen beneficial bacteria of postpartum lactating ewes, promote ewes carbohydrates digestion and absorption at lactating, so maintain lactating ewes rumen health.

4.3. Effects of ICGA on Ewes Blood Immune Index at Different Physiological Stages

Immunoglobulin is a type of large molecular protein that is rapidly produced after being stimulated by foreign pathogens and can combine with antigens to enhance the body's resistance to stress, thereby promoting animal overall health [51]. In this experiment, the ewes serum IgG, IgM, and TNF-α contents increased significantly at three physiological stages in ICGA, showing that ICGA can increase ewes serum immunoglobulin contents. Once external pathogens invade the body, they quickly bind with complement, dissolve bacteria, and prevent or reduce the occurrence of stillbirth, miscarriage, and other phenomena [52]. It can also promote the proliferation of immune cells, improve immune ability, and ewes health status [53]. In this experiment, the blood IL-6 contents in ICGA ewes was significantly higher than that of the CON at pregnancy stage. IL-6 belongs to the inflammatory cytokine family and is one of the essential components in the complex cytokine network in animal. Low IL-6 value may lead to trophoblast invasion into the surface and hinder placental vascular formation, resulting in placental trophoblast ischemia [54]. Researchers found that IL-6 mRNA levels were significantly decreased in preeclamptic placentas before the development of gestational hypertension syndrome [55]. So, adding ICGA can prevent gestational hypertension syndrome in pregnant ewes and maintain maternal pregnancy health.

5. Conclusions

ICGA could regulate ewes rumen fermentation mode at different physiological stages by increasing rumen NH_3-N content at the estrus stage, the VFAs content at the pregnancy stage, and the ratio of A/P at the lactation stage, optimizing rumen microbial flora of different physiological stages by increasing *Bacteroidota* relative abundance while reducing *Firmicutes* relative abundance. This maintains the rumen microbial homeostasis at the pregnancy stage, and increases the number of beneficial bacteria in later lactating and ewes blood immunoglobulins content at three physiological stages, therefore reducing the use of antibiotics and achieving healthy breeding purposes.

Author Contributions: Conceptualization, S.Z. and T.J.; methodology, T.J.; software, X.L. and Z.W.; validation, X.L., Y.S. and S.Q.; formal analysis, X.Z. and H.W.; investigation, S.L.; resources, T.J.; data curation, T.J.; writing—original draft preparation, S.L.; writing—review and editing, T.J.; visualization, T.J.; supervision, T.J.; project administration, T.J.; funding acquisition, S.Z. and T.J. All authors have read and agreed to the published version of the manuscript.

Funding: This research was funded by Gansu Provincial Science and Technology Plan Project Technology Innovation Guidance Plan-Lugan Science and Technology Cooperation Special Topic (22CX2NA005); South Xinjiang Key Industry Innovation Development Support Program Project (2022DB017); Major Special Project of Gansu Provincial Department of Science and Technology (21ZD11NM001) and Guanghe County Cow and Sheep Industry Quality Improvement and Efficiency Enhancement Technology Demonstration Project (LXZC21520230202).

Institutional Review Board Statement: The study was conducted according to the guidelines of the Declaration of Helsinki, and approved by the Animal Committee of Gansu Agricultural University (GSAU-Eth-AST-2023-035).

Informed Consent Statement: Not applicable.

Data Availability Statement: The data presented in this study are available on request from the corresponding author.

Acknowledgments: The author would like to thank Chen Jiangang for guiding the use of experimental instruments and Jinheyuan Family Ecological Farm for providing breeding base.

Conflicts of Interest: The authors declare that the research was conducted in the absence of any commercial or financial relationships that could be construed as a potential conflict of interest.

References

1. Li, Y.C.; Zhao, Q.R.; Zhai, T.C.; Si, W. Structural transition of protein intake in urban China: Stage characteristics and driving forces. *Agribusiness* **2023**, *39*, 1559–1577. [CrossRef]
2. Ma, Z.F. *Research on the Construction and Application of Dairy Big Data Platform from the Perspective of Grassland Agriculture*; Lanzhou University: Lanzhou, China, 2020.
3. Li, X.L.; Yuan, Q.H.; Wan, L.Q.; He, F. Perspectives on livestock production systems in China. *Rangel. J.* **2008**, *30*, 211–220. [CrossRef]
4. Khanal, P.; Nielsen, M.O. Impacts of prenatal nutrition on animal production and performance: A focus on growth and metabolic and endocrine function in sheep. *J. Anim. Sci. Biotechnol.* **2017**, *8*, 75. [CrossRef]
5. Zhang, S.Z.; Zhang, S.B.; Han, J.Y.; Liu, H.B.; Chen, Q.L.; Yang, J.F.; Shi, Z.H.; Wang, Y.Z.; Li, X.L.; Song, L.G. Effects of compound Chinese herbal medicine additive on reproductive performance, some blood biochemical indexes and reproductive hormones of Hu sheep ewes. *Heilongjiang J. Anim. Sci. Vet. Med. Sci.* **2020**, *16*, 118–120.
6. Hossain, M.F.; Islam, M.T.; Islam, M.A.; Akhtar, S. Cultivation and uses of stevia (*Stevia rebaudiana* Bertoni): A review. *Afr. J. Food Agric. Nutr. Dev.* **2017**, *17*, 12745–12757. [CrossRef]
7. Angelini, L.G.; Martini, A.; Passera, B.; Tavarini, S. Cultivation of Stevia rebaudiana Bertoni and associated challenges. In *Sweeteners*; Springer International Publishing AG: Cham, Switzerland, 2018; pp. 35–85.
8. Chen, F.; Du, E.C.; Fan, Q.W.; Wei, J.T. The application and research progress of chlorogenic acids in livestock and poultry production. *Hubei Agric. Sci.* **2020**, *59*, 10–13+16.
9. Xi, L.S.; Mu, T.H.; Sun, H.N. Research progress of chlorogenic acids at home and abroad. *Acta Agric. Nucl. Sin.* **2014**, *28*, 292–301.
10. Chen, J.L.; Chen, D.W.; Yu, B.; Luo, Y.H.; Zheng, P.; Mao, X.B.; Yu, J.; Luo, J.Q.; Huang, Z.Q.; Yan, H.; et al. Chlorogenic acid attenuates oxidative stress-Induced intestinal mucosa disruption in weaned pigs. *Front. Vet. Sci.* **2022**, *9*, 806253. [CrossRef] [PubMed]
11. Ye, X.L.; Liu, Y.; Hu, J.J.; Gao, Y.Y.; Ma, Y.N.; Wen, D.L. Chlorogenic acid-induced gut microbiota improves metabolic endotoxemia. *Front. Endocrinol.* **2021**, *12*, 762691. [CrossRef] [PubMed]
12. Cao, Q. Effects of Chlorogenic Acid on Growth Performance, Gastrointestinal Tissue Morphology and Microflora of Hu Sheep. Master's Thesis, Henan Agricultural University, Zhengzhou, China, 2022.
13. Liu, Y.; Zhang, Y.; Bai, D.; Li, Y.Q.; He, X.L.; Ito, K.; Liu, K.X.; Tan, H.Q.; Zhen, B.K.; Ma, Y.B. Dietary supplementation with chlorogenic acid enhances antioxidant capacity, which promotes growth, jejunum barrier function, and cecum microbiota in broilers under high stocking density stress. *Animals* **2023**, *13*, 303. [CrossRef] [PubMed]
14. Wang, Z.Y.; Zhen, J.D.; Liu, W. Research progress on biological activity and extraction process of isochlorogenic acid. *Feed Ind.* **2021**, *42*, 54–59.
15. Li, Y.C.; Niu, Z.P.; Xu, M.L.; Tian, H.; Gao, W. Comparative analysis of chlorogenic acids in stevia and other plants. *China Food Addit.* **2021**, *32*, 1–6.
16. Li, S.Y.; Jiao, T.; Qi, S.; Zhang, X.; Wang, H.N.; Zhao, S.G. In vitro study of the effects of isochlorogenic acid supplementation on nutrient digestion and rumen fermentation in sheep diets. *J. Zhejiang Agric.* **2024**, accepted.
17. Ren, Q.M. Study on the Structure and Function of Microbial Community in Different Niches of Yak Rumen. Ph.D. Thesis, Lanzhou University, Lanzhou, China, 2022.
18. Wang, J.; Zhang, L.Q.; Wang, W.L.; Zhao, Z.W.; Ma, Q.; Wen, X.F. Effects of caragana korshinskii fermented feed on growth performance and rumen microflora of Ningxia Tan sheep. *J. Anim. Nutr.* **2023**, *35*, 1035–1045.
19. Guo, H.; Yang, H.C.; Gao, Q.X.; Xin, G.X. Rumen microbiota and its regulation. *Pratacultural Sci.* **2022**, *39*, 371–380.
20. Shen, Y.Y.; Tong, J.J.; Xiong, B.H.; Jiang, L.S. Research progress of multi-omics technology in the interaction mechanism between rumen microorganisms and host in dairy cows. *Chin. Milk Ind.* **2021**, *8*, 68–75.
21. Zhao, C.X.; Liu, G.W.; Li, X.B.; Guan, Y.; Wang, Y.Z.; Yuan, X.; Sun, G.Q.; Wang, Z.; Li, X.W. Inflammatory mechanism of rumenitis in dairy cows with subacute ruminal acidosis. *BMC Vet. Res.* **2018**, *14*, 135. [CrossRef] [PubMed]
22. Hu, X.X.; Bian, Q.; Zhang, S.P.; Wang, F.; Liu, H.L.; Zhang, J.B. The relationship between rumen health, body health and reproductive performance in dairy cows. *Hunan Feed* **2021**, *3*, 31–34.

23. Rey, J.; Otálora DD, X.; Atxaerandio, R.; Mandaluniz, N.; García-Rodríguez, A.; González-Recio, O.; Lopez-García, A.; Ruiz, R.; Goiri, A. Effect of chitosan on ruminal fermentation and microbial communities, methane emissions, and productive performance of dairy cattle. *Animals* **2023**, *13*, 2861. [CrossRef]
24. Feng, Z.C.; Gao, M. Improvement of colorimetric method for determination of ammonia nitrogen in rumen fluid. *Anim. Husb. Feed Sci.* **2010**, *31*, 37.
25. Luo, T.; Li, Y.T.; Zhang, W.Y.; Liu, J.X.; Shi, H.B. Rumen and fecal microbiota profiles associated with immunity of young and adult goats. *Front. Immunol.* **2022**, *13*, 978402. [CrossRef]
26. Li, H.T.; Yang, Z.B.; Yang, W.R.; Jiang, S.Z.; Zhang, G.G.; Zhang, Q.; Zhang, C.Y. Effects of different dietary types on rumen pH, ammonia nitrogen and volatile fatty acid concentration in Dorper sheep. *Chin. J. Anim. Husb.* **2017**, *53*, 67–72.
27. Zhang, H.T. Effects of Different Nutritional Levels on Reproductive Performance (Estrus Time) of Ultra-Early-Weaned Small-Tailed Han Ewes. Master's Thesis, Gansu Agricultural University, Lanzhou, China, 2004.
28. Weston, R.H. The effect of pregnancy and early lactation on the digestion of a medium-quality roughage. *Aust. J. Agric. Res.* **1988**, *39*, 659–669. [CrossRef]
29. Zhang, F.; Diao, Q.Y. Effects of energy on the health of ewes and their lambs during late pregnancy. *Chin. Anim. Vet.* **2017**, *44*, 1369–1374.
30. Du, H.D.; Na, R.H. Research progress on physiological metabolism and microbial changes of ruminants during gestation and lactation and their effects on offspring development. *J. Anim. Husb. Vet. Med.* **2023**, *54*, 4458–4467.
31. Kaske, M.; Groth, A. Changes in factors affecting the rate of digesta passage during pregnancy and lactation in sheep fed on hay. *Reprod. Nutr. Dev.* **1997**, *37*, 573–588. [CrossRef]
32. Liu, Y.J. Study on the Differences in Physiological Metabolism and Rumen Internal Environment of Dairy Cows with Different Milk Yields. Master's Thesis, Jilin Agricultural University, Changchun, China, 2021.
33. Cui, C.Y.; Li, W.J.; Zhang, F.; Wang, Y.L.; Wu, Q.C.; Wang, W.K.; Jiang, Y.W.; Chen, H.W.; Yang, H.J. Effects of Bacillus subtilis on rumen microbial fermentation and degradation characteristics of peanut vine, oat and alfalfa hay. *Chin. J. Anim. Husb.* **2023**, *59*, 266–271.
34. Luo, Z.J.; Ma, L.; Pu, D.P.; Qin, X. Meta-analysis of the effects of rumen fermentation product acetic acid and propionic acid ratio on production performance and milk composition of dairy cows. *Chin. Anim. Husb. Vet. Med.* **2021**, *48*, 1613–1624.
35. Xiang, K.H.; Hu, X.Y.; Li, S.; Wang, Y.; Zhang, Y.X.; Zhang, N.S. Research progress on the changes and influencing factors of rumen flora in dairy cows during perinatal period. *Vet. Med. Prog.* **2022**, *43*, 93–97.
36. Guo, Y.J. *Comparison of Rumen Outflow Rate and Microflora of Ewes during Non-Pregnancy, Pregnancy and Lactation*; Shanxi Agricultural University: Jinzhong, China, 2021.
37. Yang, D.; Yin, F.Y.; He, R.H.; Tian, R.G.; Wu, J.H. Research progress on rumen and intestinal microorganisms in ruminants. *Anim. Husb. Feed Sci.* **2021**, *42*, 28–38.
38. Chaput, N.; Lepage, P.; Coutzac, C.; Soularue, E.; Roux, K.L.; Monot, C.; Boselli, L.; Routier, E.; Cassard, L.; Collins, M.; et al. Baseline gut microbiota predicts clinical response and colitis in metastatic melanoma patients treated with ipilimumab. *Ann. Oncol.* **2017**, *28*, 1368–1379. [CrossRef]
39. Shi, A.M.; Li, T.; Zheng, Y.; Song, Y.H.; Wang, H.T.; Wang, N.; Dong, L.; Shi, H.T. Chlorogenic acid improves NAFLD by regulating gut microbiota and GLP-1. *Front. Pharmacol.* **2021**, *12*, 693048. [CrossRef]
40. Jami, E.; Israel, A.; Kotser, A.; Mizrahi, I. Exploring the bovine rumen bacterial community from birth to adulthood. *Isme J.* **2013**, *7*, 1069–1079. [CrossRef]
41. Dodd, D.; Moon, Y.H.; Swaminathan, K.; Mackie, R.; Cann, I. Transcriptomic analyses of xylan degradation by prevotella bryantii and insights into energy acquisition by xylanolytic Bacteroidetes. *Biol. Chem.* **2010**, *285*, 30261–30273. [CrossRef] [PubMed]
42. Cao, Q.N.; Zhang, Y.M.; Ao, C.J.; Zhang, T.L.; Zhang, X.F.; Bai, C.; Zhao, Y.B.; Qi, J.W. Effects of γ-polyglutamic acid on rumen fermentation and microflora structure of Small Tail Han sheep. *Chin. J. Anim. Husb.* **2021**, *57*, 136–142.
43. Long, T.H.; Ou Yang, K.H.; Fang, Y.T.; Chen, X.F.; Wang, C.B.; Li, Y.J.; Ou Yang, K.H.; Qiu, Q.H. Effects of dietary lysine supplementation on rumen fermentation characteristics and microbial community structure. *Feed Ind.* **2023**, *44*, 1–9.
44. Zhang, K. *Study on the Vertical Effect of Maternal Microorganisms on Gastrointestinal Flora and Intestinal Function of Offspring during Pregnancy-Lactation Period*; Northwest University of Agriculture and Forestry: Xianyang, China, 2022.
45. Stevenson, D.M.; Weimer, P.J. Dominance of prevotella and low abundance of classical ruminal bacterial species in the bovine rumen revealed by relative quantification real-time PCR. *Appl. Microbiol. Biotechnol.* **2007**, *75*, 165–174. [CrossRef]
46. Scheifinger, C.C.; Linehan, B.; Wolin, M.J. H2 production by selenomonas ruminantium in the absence and presence of methanogenic bacteria. *Appl. Microbiol.* **1975**, *29*, 480–483. [CrossRef] [PubMed]
47. Wei, Z.W.; Chi, Z.; Deng, M.; Tian, H.C.; Zhong, C.M.; Sun, B.L.; Guo, Y.Q. Effects of dietary concentrate and chromium propionate levels on growth performance, fecal fermentation parameters and microflora of Leizhou goats. *Ann. Anim. Nutr.* **2021**, *33*, 5781–5793.
48. Zhao, G.L.; Niu, Y.; Zhang, R.Q.; Dai, B.; Yang, C.M. Effects of yeast culture on growth performance, immune function and caecal microflora of yellow-feathered broilers. *Ann. Anim. Nutr.* **2022**, *34*, 7701–7710.
49. Van, G.N.O. Succiniclasticum ruminis gen. nov., sp. nov., a ruminal bacterium converting succinate to propionate as the sole energy-yielding mechanism. *Int. J. Syst. Bacteriol.* **1995**, *45*, 297–300.

50. Wuriliga Aladar Cui, S.; Xiao, M.; Gong, J. Effects of selenium on rumen fermentation and rumen microflora of grazing sheep under heat stress. *Chin. Feed* **2022**, *3*, 8–16.
51. Yang, L. Effects of Leonurine on Immune and Antioxidation in Broilers and the Related Mechanisms. Ph.D. Thesis, Shihezi University, Shihezi, China, 2019.
52. Gao, Y. *Effects of Feeding Acidified Milk on Growth Performance, Blood Immunologic Indicator and Fecal Microorganism Diversity in Holstein Calves*; Heilongjiang Bayi Agricultural University: Daqing, China, 2018.
53. Zhang, Y.Q.; Liu, Y.Z.; Yang, Y.F.; Li, J.G. Effects of feeding ginseng powder to sows during the perinatal period on their reproductive performance and blood immunoglobulins. *Pig Breed.* **2019**, *2*, 41–44.
54. Zhang, L.; Bai, H.; Wang, Z.T. Changes of hCG and IL-1β, IL-6, TNF-α serum level in different phase of pregnancy. *J. Armed Police Med. Coll.* **2008**, *10*, 864–868.
55. Feng, D.L.; Yin, C.X.; You, H. A study on the correlation between placental IL-6 mRNA and amniotic fluid IL-6 level with preeclampsia. *Chin. J. Matern. Child Health* **2001**, *16*, 765–767.

animals

Article

Crude Protein Degradation Kinetics of Selected Tropical Forages in Buffalo Using NorFor In Situ Standards

Roshan Riaz [1,2], Rana Muhammad Bilal [3], Mahmood Ul Hassan [4], Massimo Todaro [4], Riccardo Gannuscio [4], Fatma Inal [5], Muhammad Naveed Ul Haque [6] and Muhammad Naeem Tahir [2,*]

1 Department of Animal Nutrition and Nutritional Diseases, Faculty of Veterinary Medicine, Kafkas University, Kars 36100, Türkiye; roshan.riaz@kafkas.edu.tr
2 Department of Livestock Management, Faculty of Veterinary and Animal Sciences, The Islamia University of Bahawalpur, Bahawalpur 63100, Pakistan
3 Department of Animal Nutrition, Cholistan University of Veterinary and Animal Sciences, Bahawalpur 63100, Pakistan; muhammad.bilal@cuvas.edu.pk
4 Department of Agricultural, Food and Forestry Science (SAAF), University of Palermo, Viale delle Scienze 13, 90128 Palermo, Italy; mahmoodul.hassan@unipa.it (M.U.H.); massimo.todaro@unipa.it (M.T.); riccardo.gannuscio@unipa.it (R.G.)
5 Department of Animal Nutrition and Nutritional Diseases, Faculty of Veterinary Medicine, Selçuk University, Konya 42130, Türkiye; fatmaaksakalinal@gmail.com
6 Department of Animal Nutrition, Ravi Campus, University of Veterinary and Animal Sciences, Lahore 54000, Pakistan; muhammad.naveed@uvas.edu.pk
* Correspondence: naeem.tahir@iub.edu.pk

Simple Summary: The nutritional needs of livestock are important to the sustainability of food production; however, variations in the quality of forage and environmental conditions can affect the nutritional efficiency of animal feed. This study investigated the variations in crude protein content and its degradability among ten species of cereal and legume fodder grown at two different locations. The feedstuffs comprised six types of cereals and four leguminous crops, all cultivated under similar agronomic practices at each location, and harvested at optimum growth stages. The results of the ruminal protein degradation showed that forage type, species, and geographical location significantly affected protein availability for buffalo. In general, legumes had greater protein availability than cereals and the location of growth affected the overall degradability. Some species were quite less affected by the location of production, while others were more so. The study results will enlighten both farmers and scientists in selecting the appropriate fodder with respect to the climatic conditions to help livestock nutrition and further support sustainable means of farming, which is significant for global food security.

check for updates

Academic Editor: Donald C Beitz

Received: 7 January 2025
Revised: 3 February 2025
Accepted: 14 February 2025
Published: 18 February 2025

Citation: Riaz, R.; Bilal, R.M.; Hassan, M.U.; Todaro, M.; Gannuscio, R.; Inal, F.; Haque, M.N.U.; Tahir, M.N. Crude Protein Degradation Kinetics of Selected Tropical Forages in Buffalo Using NorFor In Situ Standards. *Animals* **2025**, *15*, 585. https://doi.org/10.3390/ani15040585

Abstract: This study aimed to evaluate the crude protein (CP) degradation kinetics and degradability (CPD) of cereal and legume fodder species grown at two geographically distant locations. Ten forage species, comprising six cereals (barley, maize, millet, oats, sorghum, and wheat) and four legumes (berseem, jantar, lucerne, and mustard), were evaluated to determine the effects of forage family, species, and location of growth on CP degradation fractions and effective CPD. The forage crops were cultivated under uniform agronomic practices at two distinct agro-ecological locations and were harvested at the booting stage (cereals) and 50% flowering stage (legumes). Dried and ground samples were incubated in the rumen of four Nili-Ravi buffalo fitted with rumen cannula. The incubation periods utilized in the experiment were 0, 4, 8, 16, 24, and 48 h, and a $4 \times 2 \times 2$ split-plot design was employed. The results showed that the CP degradation fractions and CPD were significantly affected by forage family, species, and location of growth. Wide variations in degradation kinetics and degradability existed among and within the cereal and legume fodders, with wheat and jantar ranked at the top. Legume forages had larger soluble fractions, smaller potentially degradable fractions, and rapid rates and extent of degradation of dietary proteins than cereal forages. The cooler climatic conditions at location 2 increased the rapidly degradable protein fraction and overall CPD, whereas the warmer climatic conditions enhanced the slowly degradable protein fractions, thereby reducing the overall protein degradability in tropical forages. It was quite evident that some fodder species, such as maize among the cereal fodders and mustard among the legume fodders, remained quite non-responsive to the effects of the climatic conditions. A moderately positive and linear relationship between the rate of degradation and CPD was established for cereals, whereas a strongly positive and quadratic relationship was established for legume fodders. In conclusion, forage species, family, and location of growth significantly affected the degradation fractions and degradability of tropical cereal and legume fodders.

Keywords: cattle feed; crude protein degradability; tropical forage

1. Introduction

The global livestock industry is facing challenges in meeting the increasing demand for sustainable and high-quality feed resources. This industry has further attracted criticism because of competition among humans and ruminants for resources utilized in their production or for being supplied as food to humans [1]. Ruminants have the ability to convert non-human-utilized agriculture products into high-quality animal products (milk and meat), and forages (cereal and legume) are and should be the main feed resources for ruminants [2]. Although forages are an economical feed source in tropical and subtropical regions, significant fluctuations occur in the quality and availability of these feed resources, ultimately affecting animal performance [3]. Therefore, there is an ultimate need to understand the nutritional dynamics of cereals and legume fodders to optimize their use in meeting the dietary needs of animals.

Among the key nutritional parameters in ruminant nutrition, crude protein (CP) and its degradability are critically important because they govern the nitrogen supply to ruminant microbes, modulating microbial protein synthesis, digestion efficiency, and animal performance [4,5]. The rate and extent of CP degradation in the rumen is critical for formulating the ration to provide the nutritional demands of high-performing animals [6]. Several factors affect the CP degradation kinetics of tropical forages, including agronomic practices, environmental conditions, and forage family and species [7]. Cereal forages show higher degradation due to their lower lignin content, and legume forages possess higher CP levels and rate of degradation but show resistance to fast microbial breakdown due to their high lignin content in the cell wall [5,8]. The regional climate impacts forage growth, its nutritional composition, and the nutrients available for digestion. In addition, variations in the location of the growth environment impact forage CP content and degradation, and protein availability is often reduced from regions with high temperatures and low rainfall [9].

Despite the established utilization of these forages in ruminants, particularly for buffalo, there is a scarcity of data on the CP degradation kinetics of different tropical forage families, species, and geographical locations. In addition, the current ration formulation is also based on the feed evaluation information of either temperate regions or other ruminants from that climate [7,10]. Pakistan, with its diverse agro-ecological regions and varied climatic conditions, has a scarcity of information on fodder nutritional profiles and impacts of the geographic region. There are physiological and metabolic differences between buffalo and other ruminants [11]. These factors ultimately lead to under or over-feeding, resulting in influences on animal performance, economic losses, and the environmental footprint [10]. Therefore, there is a necessity for research to fill this gap to improve feeding strategies for animal production performance and optimize resource utilization efficiency.

This study aimed to determine: (1) the CP degradation parameters of selected tropical forages in Nilli-Ravi buffalo; (2) the influence of forage family, species, and location of growth and their interactions on degradation parameters; and (3) the relationship between the in situ CP degradation rate and crude protein degradability (CPD) for cereal and legume fodders. The findings of this study could contribute to the effective utilization of high-quality feed resources and provide region-specific feeding strategies to increase livestock productive performance and support sustainable agriculture systems in subtropical regions.

2. Materials and Methods

2.1. Forage Sampling and Processing

This study evaluated ten forage species, comprising six cereals and four legumes, across two locations. Summer forages included *Pennisetum glaucum* (millet), *Sorghum bicolor* (Sorghum), *Sesbania bispinosa* (Jantar), and *Zea mays* (maize). Winter forage species consisted of *Avena sativa* (oats), *Brassica napus* (mustard), *Hordeum vulgare* (barley), *Medicago sativa* (lucerne), *Triticum aestivum* (wheat), and *Trifolium alexandrinum* (berseem). Summer crops were planted in mid-March, while winter crops were sown in late November using consistent, locally recommended agronomic practices and geographical production conditions. Each species was cultivated at two different locations: Lahore (31.55° N, 74.35° E; Location 1) for the central region and Rawalpindi (33.598° N, 73.04° E; Location 2) for the northern region of Punjab Province, Pakistan (Figure 1). Within each location, three different plots were cultivated ~100 m apart for statistical sample replication. The geographic climatic conditions and soil profiles of the production sites are presented in Table 1 and Figure 2.

Figure 1. The geographical regions of fodder cultivation.

Table 1. Growing location and soil profile characteristics [2].

Parameter	OM (%)	Soil pH	Available K (mg/kg)	Available N (mg/kg)	Available P (mg/kg)	EC (dS/m)	Soil Texture	Ht (m)	PP (mm)	Topography
Lahore	0.87	7.8	158	0.054	14.3	0.67	Clay loam	210	1000	Flat and slope toward south
Rawalpindi	1.42	7.9	112	0.062	11.12	0.36	Clay loam	530	1300	Hilly sub-terrain

OM = organic matter, EC = electrical conductivity, Ht = height from sea level, PP = precipitation (annual).

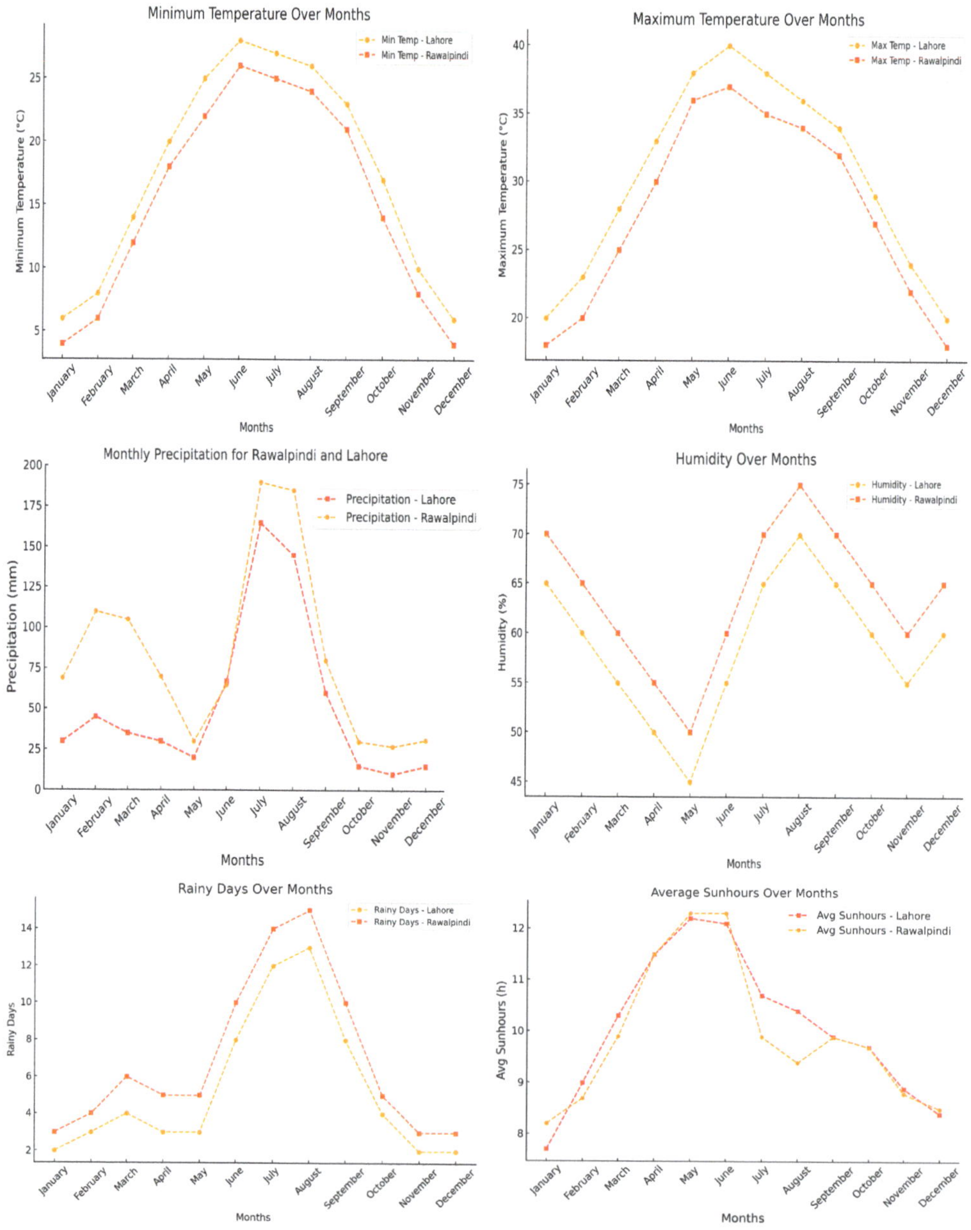

Figure 2. Mean values of climate parameters over months at the growing location.

From each site, ~10 kg of herbage was collected for each forage type from three replicate plots. The samples of each forage type collected from three replicated plots at a single location were pooled in order to obtain a more representative sample with as little variation in chemical composition as possible. This pooled sample from each location was further utilized for chemical (two statistical replicates) and in situ analysis (two experimental replicates). To ensure the optimal growth stage for analysis, the cereal and legume forages were harvested at the booting and 50% flowering stages, respectively. The freshly collected herbage was chopped to roughly 20 mm lengths using a chaff cutter (Toka 510, Patiala Agri-Industries, Faisalabad, Pakistan). Following chopping, the samples were air-dried under shade for three to seven days to achieve the recommended moisture content. After drying, the samples were ground using a hammer mill (POLYMIX PX-MFC, Kinematica GmbH, Eschbach, Germany) to two specific particle sizes: 1 mm for chemical analysis and 2 mm for in situ digestion kinetics experiment. The processed samples were stored at room temperature in small plastic containers until further analysis.

2.2. Management of Experimental Animals

A total of four rumen-cannulated (Bar Diamond, Parma, ID, USA), Nili-Ravi buffalo (average live weight [LW] = 529 ± 33.4 kg, age = 2635 ± 49.5 days, and parity 3) were used in this study. All animals received a standardized diet formulated to meet maintenance requirements according to the NorFor standards for rumen-cannulated animals [11]. The diet consisted of freshly chopped forage, commercial concentrate, lucerne hay, and cotton seed cake with a forage to concentrate ratio of 85:15. The composition of the diet is presented in Table 2. The animals were individually housed in 2 × 2.5 m stalls, fed separately, and had access to fresh water at all times.

Table 2. Ingredients and chemical composition of the diet offered to rumen-cannulated animals.

No. of Samples = 5, No. of Statistical Replicates = 2, Total No. of Observations per Feed = 10	
	Mean ± SD
Ingredients (g/kg on fresh matter)	
Sorghum forage	807.5 ± 10
Lucerne hay	141.5 ± 3
Cotton seed cake	15.0 ± 1
Concentrate mixture	35.5
Forage to concentrate ratio (on DM basis)	85:15
Chemical composition (g/kg DM)	
DM (as fed)	331 ± 10
aNDF	575 ± 12
CP	59.0 ± 3
EE	17.5 ± 2
NFC	234 ± 7
Ash	113 ± 6

DM = dry matter; aNDF = amylase-treated neutral detergent fiber; CP = crude protein; EE = ether extract; NFC = non-fiber carbohydrates. Adapted from Tahir et al. [7].

2.3. In Situ Incubation and Degradation Profiles

The CP degradation characteristics of the forage samples were studied according to NorFor guidelines using an in situ incubation technique [12]. Approximately 1 g of each sample was incubated under standard conditions of 15 mg per cm^2 density relative to the bag surface area. The samples were then enclosed in bags and adhesive-sealed in Dacron-type polyester pouches (Sefar AG, Heiden, Switzerland), measuring 11 × 8.5 cm (effective dimensions of 10 × 7.5 cm). The incubation bags had 33 μm pores and were made of PES material (140/37), with 25% free space for optimum entry of rumen fluid. Using the all-in system, the bags with the samples were simultaneously placed into the rumen of each rumen-cannulated animal and removed after 0, 4, 8, 16, 24, and 48 h. The bags, once removed, were washed with tap water and stored at −18 °C until the final incubation time was completed. After the entire incubation batch, the bags were thawed and washed twice with tap water at 25 °C. At the completion of each batch, the intact samples and residues were analyzed for CP content using the Dumas method [13].

2.4. Chemical Analyses

Samples of the diet and individual feeds were collected after a fortnight and pooled against each feed. For freshly chopped forage, the dry matter (DM) and ether extract (EE) contents were determined in accordance with AOAC guidelines [13]. The DM content of the dry feed was determined by drying the feed at 105 °C for 16 h in order to remove almost all moisture and to establish a better comparison of the chemical analyses among various forage species. The ash content was measured by the incineration of samples at 525 °C for 6 h (method 923.03). The ether extract (EE) content was evaluated through a 6 h extraction utilizing petroleum ether (method 7.062). The amylase-treated neutral detergent fiber (aNDF) content was determined utilizing an amylase-treated procedure based on the method of Van Soest et al. [14], as modified by Mertens et al. [15]. The non-fiber carbohydrate content was calculated as follows:

$$\left[NFC\left(\frac{g}{kg}DM\right) = 1000 - (CP + EE + aNDF + ash)\right]$$

2.5. Data Analysis and Curve Fitting

The in situ degradation data were categorized into two main components: the washable fraction (a), which comprised the initial loss after washing at 0 h, and the non-washable fraction. The latter was further subdivided into a potentially degradable fraction (b) and an indigestible fraction, representative of the degradation and residues obtained at the final incubation time point, respectively, using the procedures of Ørskov and McDonald [16]. A first-order kinetic model using Table Curve 2D software (version 5.0, Systat Software Inc., San Jose, CA, USA) was fitted to the degradation data. This model assumes that the conditions of degradation and passage are in a steady state and is expressed as follows:

$$Yt = a + b(1 - \exp(-Kdt))$$

In this equation, Y_t represents the gradient fraction at time t, K_d is the fractional degradation rate of fraction b, and t denotes the incubation time in hours. The effective ruminal CPD was determined using Ørskov and McDonald's model, which assumes a fractional passage rate (K_p) of 0.05/h for forages, equivalent to a 20 h rumen retention time [16]. This method facilitated the estimation of effective degradability under assumed rumen kinetics for the forage samples:

$$CPD = \alpha + \left[b \times \frac{Kd}{Kd + Kp}\right]$$

2.6. Statistical Analysis

Statistical analyses were performed using the GLM procedure in SAS® software (Statistical Analyses Software, version 9.2). The data on the in situ parameters were analyzed considering each buffalo as an experimental replicate using the following ANOVA model:

$$Y_{ijk} = \mu + S(F)_{ij} + F_j + L_k + (F \times L)_{jk} + \varepsilon_{ijk}$$

in which Y_{ijk} is the dependent variable, μ is the overall mean, S_i shows the effect of the ith forage species and F_j shows the effect of the jth forage family (cereal vs. legume), L_k shows the effect of the kth location (Lahore vs. Rawalpindi), $(F \times L)_{jk}$ is the interaction between the ith forage species and kth location, and ε_{ijk} is the residual error. The least-squares means were compared using p-values adjusted according to the Tukey–Kramer multiple comparison test. Results were considered significant when $p \leq 0.05$ and are presented as the least square mean with the standard error of the mean.

3. Results

3.1. Chemical Composition of Forage Family

The chemical composition of the forage species is presented in Table 3. The CP values were consistently lower in the cereals compared to the legumes, ranging from 50.6 g/kg DM (sorghum) to 74.5 g/kg DM (wheat) for cereal forages and from 110.3 g/kg DM (mustard) to 135.5 g/kg DM (lucerne) for legume forages. The CP values did not differ across Location 1 and Location 2, as shown by Tahir et al. [7].

Table 3. Effect of forage family, forage species, and location of growth on the chemical composition. The p-values are presented in this table, where $p \leq 0.05$ shows a significant effect. The values are presented as the least square mean (forage species) with the standard error of the mean unless otherwise stated.

Factor	DMfresh	DMdry	aNDF	CP	EE	Ash	NFC
	No. of Samples = 2, No. of Statistical Replicates = 2, Total No. of Observations/Feed = 4						
Species (Family)							
Cereal fodders							
Barley	15.4 ^{d}	95.3 ^{bc}	639.0 ^{a}	46.8 ^{e}	14.3 ^{bcd}	119.6 ^{bcd}	180.4 ^{b}
Maize	19.6 ^{e}	95.2 ^{bc}	621.1 ^{a}	55.8 ^{e}	9.0 ^{d}	94.9 ^{de}	219.2 ^{b}
Millet	19.8 ^{e}	94.4 ^{ab}	611.3 ^{a}	71.2 ^{d}	15.3 ^{bc}	104.5 ^{cde}	197.8 ^{b}
Oat	16.5 ^{d}	95.4 ^{bc}	612.8 ^{a}	52.8 ^{e}	17.6 ^{bc}	91.8 ^{de}	225.0 ^{b}
Sorghum	21.9 ^{f}	94.0 ^{a}	607.9 ^{a}	53.3 ^{e}	17.3 ^{bc}	86.7 ^{e}	234.8 ^{b}
Wheat	15.4 ^{b}	95.9 ^{cd}	527.7 ^{b}	76.8 ^{d}	15.5 ^{bc}	146.3 ^{b}	233.7 ^{b}
Legume fodders							
Berseem	8.3 ^{a}	96.8 ^{d}	429.7 ^{c}	110.3 ^{b}	13.5 ^{cd}	204.6 ^{a}	241.9 ^{b}
Jantar	16.8 ^{d}	94.4 ^{ab}	480.9 ^{bc}	93.7 ^{c}	20.2 ^{b}	79.0 ^{e}	326.3 ^{a}
Lucerne	13.2 ^{bc}	95.2 ^{bc}	183.6 ^{bc}	131.5 ^{a}	14.7 ^{bcd}	148.0 ^{b}	222.1 ^{b}
Mustard	14.9 ^{cd}	95.5 ^{bc}	529.5 ^{b}	80.4 ^{cd}	28.1 ^{a}	127.3 ^{bc}	234.7 ^{b}
SEM	0.25	0.44	16.28	4.26	1.73	8.24	17.70
Family							
Cereals	17.5	95.0	603.3 ^{a}	59.5 ^{b}	14.8 ^{b}	102.3 ^{b}	215.1 ^{b}
Legumes	13.3	95.5	480.9 ^{b}	104.0 ^{a}	19.1 ^{a}	139.7 ^{a}	256.2 ^{a}
SEM	0.22	0.12	7.39	1.94	0.77	3.74	8.03
Location							
Lahore	17.0	95.1	563.8 ^{a}	80.8	17.1	124.3	213.9 ^{b}
Rawalpindi	14.6	95.3	520.4 ^{b}	82.6	16.8	122.7	247.5 ^{a}
SEM	0.19	0.11	7.43	1.94	0.79	3.76	8.08
Family*Location							
Cereals Lahore	19.2	95.2	644.4 ^{b}	54.7 ^{c}	14.7	105.3	180.9
Cereals Rawalpindi	15.9	94.9	562.2 ^{c}	64.2 ^{b}	14.9	109.3	249.3
Legumes Lahore	13.9	95.4	483.3 ^{a}	106.9 ^{a}	19.6	143.4	246.8
Legumes Rawalpindi	12.8	95.6	478.6 ^{a}	101.1 ^{a}	18.7	136.0	265.7
SEM	0.25	0.17	10.46	2.74	1.06	5.00	11.66
***p*-value**							
Species (Family)	0.001	0.001	0.002	0.001	0.001	0.001	0.005
Family*Location	0.001	0.345	0.001	0.010	0.576	0.292	0.039
Family	0.001	0.444	0.001	0.001	0.001	0.001	0.001
Location	0.001	0.617	0.001	0.515	0.764	0.755	0.007

DMfresh = dry matter reported at 60 °C for 48 h; DMdry = dry matter reported at 105 °C for 16 h; aNDF = amylase-treated neutral detergent fiber; CP = crude protein; EE = ether extract; NFC = non-fiber carbohydrates. Different lower-case superscripts in a column indicate a significant difference ($p < 0.01$).

3.2. Degradation Parameters and Effective CPD by Forage Species and Family

The in situ CP degradation kinetics and effective CPD for cereal and legume fodders are presented in Table 4. Significant differences in degradation kinetics and CPD were observed across cereal and legume species ($p < 0.001$). Wheat exhibited the highest washable fraction (<0.001) and showed the highest CPD value at 48 h. By contrast, sorghum demonstrated the lowest fraction a and CPD values and the highest potentially degradable fraction (fraction b). Among the legume species, mustard showed the highest fraction a value, and the highest fraction b value was found in lucerne. The CPD values were significantly higher ($p < 0.001$) in jantar and mustard than in lucerne and berseem. The fraction c and K_d values tended to differ across forage species ($p < 0.10$), and the highest K_d value was found in jantar and the lowest was found in sorghum.

Table 4. Effect of forage family, forage species, and location of growth on in situ crude protein degradation kinetics and effective degradability of cereal and legume fodders sown in Lahore and Rawalpindi. The p-values are presented in this table, where $p \leq 0.05$ shows a significant effect. The values are presented as the least square mean (forage species) with the standard error of the mean unless otherwise stated.

		In situ Parameters [1]				
Factor		a	b	c	K_d	CPD [2]
No. of Samples = 2, No. of Statistical Replicates = 2, Total No. of Observations/Feed/Incubation Interval = 4						
Species (Family)						
Cereal fodders						
Barley		0.403 ^{b}	0.437 ^{b}	0.160 ^{abc}	0.061 ^{cd}	0.617 ^{bc}
Maize		0.293 ^{c}	0.572 ^{a}	0.135 ^{b}	0.063 ^{cd}	0.599 ^{bc}
Millet		0.276 ^{c}	0.556 ^{a}	0.167 ^{ab}	0.065 ^{cd}	0.577 ^{c}
Oats		0.319 ^{c}	0.570 ^{a}	0.110 ^{c}	0.068 ^{bcd}	0.620 ^{bc}
Sorghum		0.183 ^{d}	0.615 ^{a}	0.201 ^{a}	0.045 ^{d}	0.443 ^{d}
Wheat		0.545 ^{a}	0.349 ^{b}	0.106 ^{c}	0.073 ^{bc}	0.745 ^{a}
Legume fodders						
Berseem		0.435 ^{b}	0.392 ^{b}	0.173 ^{ab}	0.099 ^{b}	0.691 ^{ab}
Jantar		0.440 ^{b}	0.432 ^{b}	0.128 ^{b}	0.140 ^{a}	0.763 ^{a}
Lucerne		0.271 ^{c}	0.583 ^{a}	0.147 ^{abc}	0.081 ^{b}	0.670 ^{ab}
Mustard		0.574 ^{a}	0.253 ^{c}	0.173 ^{ab}	0.120 ^{a}	0.755 ^{a}
SEM		0.029	0.031	0.016	0.009	0.027
Family						
Cereals		0.336 ^{b}	0.516 ^{a}	0.146	0.062 ^{b}	0.600 ^{b}
Legumes		0.429 ^{a}	0.414 ^{b}	0.155	0.110 ^{a}	0.719 ^{a}
SEM		0.014	0.015	0.008	0.004	0.013
Location						
Lahore		0.323 ^{b}	0.537 ^{a}	0.139	0.086	0.630 ^{b}
Rawalpindi		0.444 ^{a}	0.394 ^{b}	0.162	0.087	0.689 ^{a}
SEM		0.013	0.014	0.008	0.004	0.012
Family*Location						
Cereals	Lahore	0.259	0.586	0.156 ^{ab}	0.062	0.555
Cereals	Rawalpindi	0.415	0.447	0.138 ^{b}	0.063	0.645
Legumes	Lahore	0.387	0.489	0.124 ^{b}	0.110	0.705
Legumes	Rawalpindi	0.473	0.341	0.186 ^{a}	0.110	0.734
SEM		0.019	0.020	0.011	0.006	0.017
***p*-value**						
Species (Family)		0.001	0.001	0.004	0.010	0.001
Family		0.001	0.001	0.441	0.001	0.001
Location		0.001	0.001	0.044	0.895	0.002
Family*Location		0.073	0.807	0.001	0.971	0.095

a = washable fraction representing the portion of CP that had disappeared at time 0; b = potentially degradable CP fraction; c = indigestible fraction (1—a–b). The estimate of K_d from the in situ method represents the fractional rate of degradation of fraction b; CPD = crude protein degradability at 48 h of incubation. [1] Degradation parameters described according to the model by Ørskov and McDonald [16]. [2] Effective CPD was calculated from data assuming the fractional rate of passage (K_p) to be 0.05/h for forage, as used in the protein evaluation system of Hvelplund and Weisbjerg [17]. Different lower-case superscripts in a column indicate a significant difference ($p < 0.01$).

The forage family showed a significant effect on the CP degradation kinetics and CPD ($p < 0.001$). The fraction a, K_d, and CPD mean values of the legume fodders were significantly higher than those of the cereal fodders. The legumes were characterized by lower fraction b values compared to the cereal fodders.

The decreases in fraction b values (degradation curves) of the cereal and legume fodders (averaged against respective forage species) with passage of time at both locations are presented at Figure 3a,b. At both locations, legumes were degraded at faster rates than cereals up to 24 h of incubation, pointing to faster disappearance of fraction b. It is evident that the two curves are closer to each other at Location 2, whereas they are apart from each other at Location 1. It is evident that the two curves move apart at earlier hours of incubation (up to 24 h); however, the two curves move closer to each other at later incubation intervals.

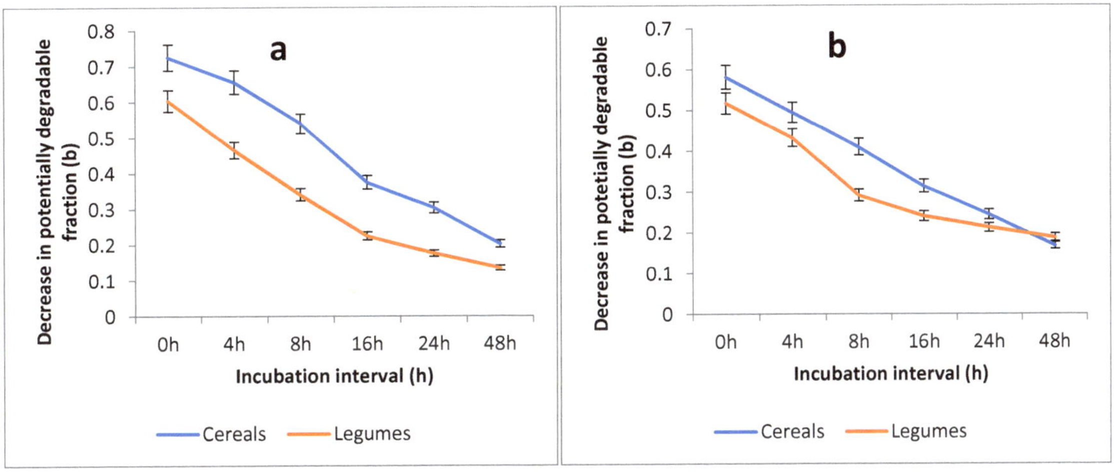

Figure 3. (**a**,**b**) The decrease in potentially degradable degradation fraction b of various fodder species averaged against cereal and legume fodders collected from Location 1 (Lahore; (**a**)) and Location 2 (Rawalpindi; (**b**)).

3.3. Degradation Parameters and Effective CPD by Location

The effect of the location of growth on the degradation parameters and effective CPD of cereals and legumes are summarized in Table 4. The location of growth significantly affected all degradation fractions and effective CPD in both cereal and legume forages ($p < 0.05$).

A family by growing location interaction tended to change fraction a and CPD for all forages ($p < 0.10$), but it significantly affected fraction c ($p < 0.001$). It tended to increase fraction a and CPD values in cereals ($p < 0.10$) but had no effect on fraction a and CPD values in legumes ($p > 0.10$). This interaction also caused an increase in the value of fraction c of legumes and a decrease in the same fraction of cereals at Location 2 compared to Location 1. This interaction did not change the values of fraction b and K_d across forage families and/or location of growth ($p > 0.10$).

The marginal means with 95% confidence interval bars of species by location of growth interactions for forage species are presented in Figure 4. The location effects were evident with different species responses. The proportion of fraction a was increased whereas that of fraction b dropped significantly for all species except maize and mustard at Location 2 compared to Location 1. The value of K_d increased for all forage species, except for jantar, millet, and wheat where it dropped at Location 2 compared to Location 1, and it remained unchanged for mustard and millet. The CPD value increased for all forage species, except for millet where it declined, and it remained unchanged for berseem and jantar at Location 2.

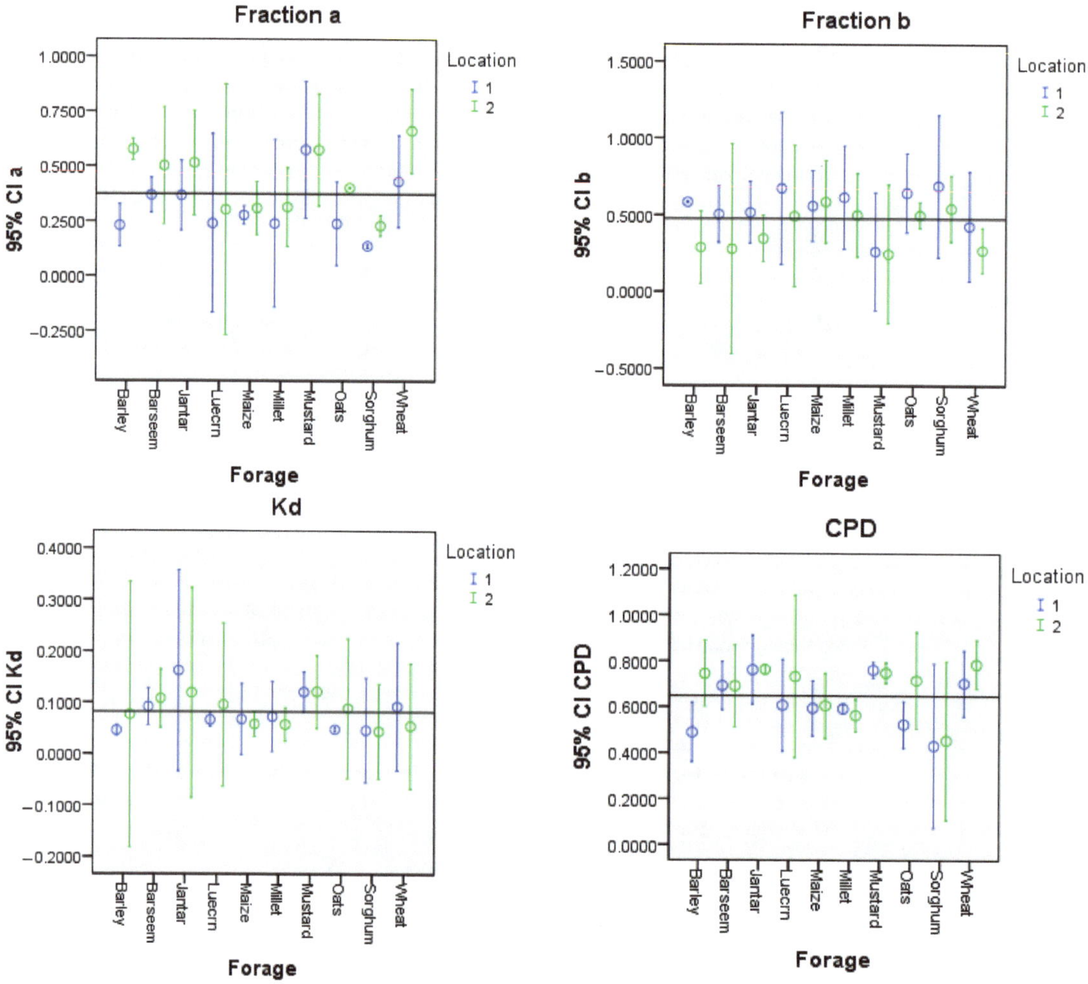

Figure 4. Marginal means with 95% confidence interval (CI) bars of various crude protein fractions of cereal and legume fodders collected from two locations (Location 1 = Lahore, Location 2 = Rawalpindi).

3.4. Relationship Between CPD and Rate of Degradation

Figure 5a,b show the relationships between K_d and CPD for the cereal and legume fodders considered summed across two locations, described using various curve fit models. The data on the regression models are presented in Table 5. The relationship between K_d and CPD was moderately positive ($R^2 = 0.44$, SE of estimate = 0.088; linear models) for cereal fodders and it did not change when quadratic or cubic models were applied. Figure 5b shows better values for legume fodders ($R^2 = 0.60$, SE of estimate = 0.035; linear models) between K_d and CPD, and the fit was improved when quadratic or cubic models were applied.

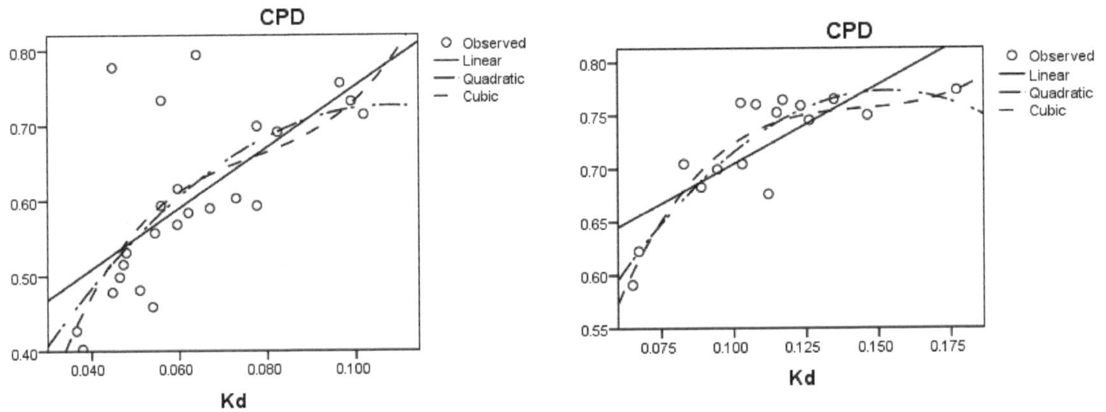

(**a**) (**b**)

Figure 5. (**a**,**b**) The relationship between rate of degradation (K_d/h) and crude protein degradation (CPD) of cereal fodders (**a**) and legume fodders (**b**) collected from two locations.

Table 5. Fitting regression models to describe the relationship between crude protein degradability and rate of degradation for cereal and legume fodders collected from Lahore and Rawalpindi.

Regression Model	*p*-Value	R^2	SE	r
Cereal fodders				
Linear	0.001	0.44	0.088	0.66
Quadratic	0.01	0.46	0.088	
Cubic	0.04	0.47	0.089	
Legume fodders				
Linear	0.001	0.60	0.035	0.78
Quadratic	0.001	0.79	0.027	
Cubic	0.001	0.81	0.027	

R^2 = coefficient of determination; SE = standard error of estimate; r = correlation.

4. Discussion

4.1. Chemical Composition of Forage Family

The CP content exhibited significant variations among forage families and species. The CP values for cereals were comparable to those reported by Sarwar et al. [18], while legumes generally demonstrated lower CP content compared to the values presented by other researchers for tropical legume forages e.g., Singh et al. [19], and higher values in comparison to those of Rehman et al. [20]. Studies have demonstrated that dietary CP content influences the in situ degradation kinetics of crude protein by modulating the solubility and degradability of protein fractions, rumen microbial activity, and passage rate of feed, all of which collectively determine CP degradation kinetics [21,22].

4.2. Degradation Parameters and Effective CPD by Forage Species and Family

The development of a comprehensive database on the ruminal protein degradation kinetics of commonly utilized feed sources is essential for optimizing ruminant nutrition. This study presents a dataset on the protein degradation characteristics of forage species prevalent in tropical and subtropical regions, thereby addressing the increasing demand for detailed information on ruminal degradation kinetics of protein.

The results of the CP degradation characteristics demonstrated significant differences in the CP degradation kinetics and CPD between cereal and legume fodders. Among the cereals, wheat exhibited the highest CPD value with a high washable fraction. However, sorghum demonstrated the lowest CPD value and the highest potentially degradable fraction. The legumes, on the other hand, showed more pronounced variability in the degradation parameters. Mustard had a high fraction *a* value, jantar showed a high K_d value, and lucerne displayed a low CPD value with the lowest K_d value. The findings of this study regarding protein fractions *a* and *b* are consistent with the results of Das et al. [23]. The CP degradability results of the cereals were also comparable to those reported by Kaithwas et al. [24]. Singh et al. [25] also identified slow degradation in sorghum. However, the results of wheat degradation in the present study are comparable to those of Islas and Soto-Navarro [26]. Conversely, the degradation parameters observed for the legumes in this study differed from those reported for lucerne by Aufrère et al. [27]. These findings highlight the potential influence of species-specific characteristics on the degradation kinetics of proteins in ruminants.

Legume forages demonstrated superiority over cereals in terms of degradation kinetics and effective CPD, confirming their enhanced protein availability. The steeper degradation curves for legumes, indicative of rapid initial degradation, suggested that they facilitate more expeditious nutrient release. By contrast, cereals exhibited slower CP degradation rates. These results align with previous studies that have reported higher levels of rapidly degradable fractions and CPD in legumes compared to cereals [24,28]. Specific leguminous plants, such as lucerne and berseem, have been shown to yield significant crude protein levels [19,29]. Valderrama et al. [22] also found that legumes demonstrated the highest fraction *a* and CPD values with quite variable degradation rates. This distinction is significant for ruminants, as higher CP degradability in the rumen leads to increased ammonia nitrogen availability [28]. The rapid degradation of crude protein in these legumes is attributed to their structural composition, which supports accelerated microbial activity within the rumen [30]. By contrast, cereal fodders generally exhibit lower protein contents but are frequently characterized by higher carbohydrate levels (aNDF) [30], which can influence the fermentation process within the rumen [24,31]. The slower protein degradation rate of cereals can limit their effectiveness as protein sources when used in isolation; however, grasses exhibit different degradation kinetics from grains due to their fiber-bound proteins and structural composition [32]. Therefore, the balance in CP degradation kinetics is crucial for ruminant nutrition.

4.3. Degradation Parameters and Effective CPD by Location

In the present study, the significant effects of the forage growing location on the CP degradation fractions and CPD of cereals and legumes highlight the influence of growing location on forage quality. The least square means of the cereals demonstrated higher fraction *a* values (49% increase) and lower fraction *b* values (13% decrease) at Location 2. Almost similar trends in the location of growth effects were found for various legume fodders for the same parameters (13% increase in fraction *a* value and 23% decrease in fraction *b* value). These results demonstrated that both rapidly soluble and slowly degradable fractions of CP are affected by the climatic conditions, regardless of the origin of the plant family. Cooler climatic conditions increased the rapidly degradable protein fraction and overall protein degradability, whereas warmer climatic conditions enhanced the slowly degradable protein fractions, thereby reducing the overall protein degradability in the tropical forages in the present study. It is worth noting that the climatic conditions had almost no effect on the rate of degradation of fraction *b*, regardless of forage family. It is quite evident that some fodder species, such as maize and sorghum among the cereal fodders and mustard among the legume fodders, remained quite non-responsive to the effects of the climatic conditions and showed quite comparable CPD values at the two locations (Figure 4). These variations in digestibility and degradation behavior can be attributed to the genetic and compositional traits of the crops, which enable adaptation to environmental conditions. Maize and sorghum are contrasting species among the C4 cereals that show adaptations to abiotic stress by reducing the leaf area, transpiration restriction, water use efficiency, germ plasm variability, drought and heat tolerance, and efficient nutrient intake, thus preserving the nutritional value in arid, semi-arid, and tropical regions [33–35]. Similarly, a study on the environment effects on mustard showed higher adaptability of mustard to environment-associated stress, such as high temperature, low rainfall, and late sowing [36]. It is further added that the magnitude of the response to location varied among forage species, with certain species exhibiting greater responsiveness, while others demonstrated increased resistance to location effects (Figure 4). This location-based variability aligns with previous studies that have emphasized geographical influences on CP chemical composition and degradation characteristics [37–39]. Gruber et al. [38] further elaborated that fractions *a* and *b* were significantly influenced by the location of growth. This agrees with the results of a previous study [39] that indicated the predominant role of genotype × environment interactions in determining the chemical composition and digestibility of grain. The study's results further demonstrated that the location of growth potentially induces variations in nutrient profiles, affecting both carbohydrate and protein fractions. These variations underscore the importance of site-specific assessment of forage quality, especially in species with wide environmental adaptability.

The geographic location determines the temperature, precipitation, and soil quality, which subsequently affect forage composition and nutritional value. Altitude has been demonstrated to significantly impact soil properties, species composition, and forage production, thereby influencing variability in CP degradability [40]. Forages cultivated at higher altitudes, such as Location 2, experience cooler temperatures and slower plant maturation, which promotes an accumulation of digestible carbohydrates and higher crude protein concentrations, thus elucidating the superior CP degradability values observed at Location 2 compared to Location 1, as reported by Dongdong et al. [41]. Altitude also influences forage quality by altering soil organic carbon and environmental factors such as temperature and precipitation. Tahir et al. [2] reported that variable climatic conditions differing in temperature and precipitation significantly affected fiber concentrations and degradability, consequently affecting forage quality. Similarly, another study showed that climate change, characterized by elevated temperatures and altered precipitation patterns, can negatively impact forage production, reducing CP availability [42].

Temperature variations across the location of growth can affect fodder digestibility by modulating the nutritional composition of forages. Buxton [43] reported that low temperatures slow down plant development, which promotes the accumulation of soluble sugars and enhances digestibility. Cooler climates also reduce the deposition of structural carbohydrates, maintaining higher CP solubility. Conversely, elevated temperatures, as at Location 1, accelerate lignification and increase the quantity of structural carbohydrates, which diminishes fodder digestibility even when the same plant species are involved [44]. Furthermore, the leaf-to-stem ratio, a critical factor in forage quality, is reduced by higher temperatures, as leaves typically possess a higher protein content than stems [41]. These findings are consistent with those of Niklas and Fernandes et al. [37,45], who observed that, at elevated temperatures, tissues with physiological activity exhibited lower levels of CP and increased concentrations of crude fiber. The implications of these structural and biochemical alterations manifested in reduced degradation kinetics under the high-temperature conditions of Location 1.

At Location 2, moderate precipitation facilitated higher CP degradation fractions, presumably due to optimal water availability for plant growth and nutrient uptake. These observations align with those of Li et al. [46], who reported that legumes exhibited higher

CP levels under moderate precipitation conditions, while cereals demonstrated increased susceptibility to reduced digestibility under water stress. Consequently, at Location 1, reduced precipitation intensified lignification and increased the accumulation of structural carbohydrates, thereby reducing CP solubility and degradation kinetics [44]. Nevertheless, species-specific responses to precipitation were observed, with legumes demonstrating insensitivity compared to cereals. These plant species exhibited resilience to low or erratic rainfall in such environments.

The degradation kinetics of CP in the rumen varied depending on the forage family in conjunction with the effect of location of growth. Usually, legumes exhibited higher variability in protein degradable fractions (fraction *a*, fraction *c*, and CPD; Table 4 and Figure 3a,b) compared to cereals [47]. In legumes, more soluble and rapidly degradable protein fractions are due to structural differences in their cell walls and nitrogen composition; however, these fractions are vulnerable to changes in climatic conditions [48]. The legume forages showed more sensitivity to environmental factors and changes in chemical composition may occur. Higher temperature and low precipitation result in more proteins complexed to lignin, complexes formed between tannins and proteins, and Maillard reaction products [49]. These components are typically neither degradable in the rumen nor digestible in the post-ruminal tract [19]. Species-based seasonal variations also occur in the amino acid composition, which may influence ruminal digestibility [50]. In addition, the molecular structure characteristics of proteins exhibit significant correlations with protein solubility, rumen degradation, and intestinal digestibility of proteins [51]. Overall, these changes cause modifications in fodder quality and changes in available CP content and ruminal CP degradability.

4.4. Relationship Between CPD and Rate of Degradation

This study revealed a strong and statistically significant relationship between K_d and CPD, indicating that increased protein degradation rates are closely associated with higher protein degradability in various fodders. This strong relationship supports the hypothesis that rapid degradation of dietary CP enhances protein solubilization, absorption, and synthesis of microbial protein within the rumen. The marginally higher R^2 values observed for legumes suggested that the greater variability in the K_d values of cereals resulted in a weaker association between the degradation rate and protein degradability. The strong K_d and CPD relationship observed for legumes in the present study aligns with previous studies highlighting the crucial role of K_d in determining degradability [22]. However, it should be noted that this relationship is quadratic and the marginal response of CPD to an increase in K_d diminishes as the value of K_d reaches beyond a certain level. The study also showed that the CPD of forage crops increased with an increase in fraction *a* and decrease in fraction *b* (observation only, analyses not performed).

5. Conclusions

Crude protein fractioning from tropical forages is significantly affected by the forage species, family, and the location of growth. The results showed considerable variability in degradation kinetics and CPD among cereal and legume fodders, with wheat and jantar ranked at the top. Legume forages had larger soluble fractions, smaller potentially degradable fractions, and rapid rates and extent of degradation of dietary proteins than cereal forages in the rumen of buffalo. Cooler climatic conditions increase the rapidly degradable protein fraction and overall protein degradability, whereas warmer climatic conditions enhance the slowly degradable protein fraction, thereby reducing the overall protein degradability in tropical forages. A strong positive relationship between K_d and CPD suggests that K_d can be a significant determinant in predicting the degradation of forage plants within the rumen. The CP degradation kinetics of cereal and legume fodders reported in this study can be utilized for optimal diet formulation.

Author Contributions: Conceptualization, M.N.T. and R.R.; Methodology, R.R. and M.N.T.; Formal analysis, M.N.T. and R.M.B.; Investigation, R.R.; Data curation, M.N.T.; Writing—original draft, R.R., F.I. and M.N.T.; Writing—review & editing, M.U.H., M.T. and R.G.; Project administration, M.N.T. and M.N.U.H.; Funding acquisition, M.N.T., M.N.U.H. and M.T. All authors have read and agreed to the published version of the manuscript.

Funding: This work was financially supported by a grant from the Punjab Agricultural Research Board (PARB) of Pakistan through a research project no. PARB-CGS-20-116. The publication charges were provided by the University of Palermo, Department SAAF (Prof. M. Todaro).

Institutional Review Board Statement: The experimental animals used in the present experiment were housed at the Livestock Farm of the Islamia University of Bahawalpur (IUB) and were maintained in accordance with the guidelines established by the Animal Care and Management Committee (The IUB Animal Use and Bioethics Committee 2015 later named as Institutional Animal Ethics Committee). The permission for the current research was obtained from the said body through ORIC no. 5765 dated 12-03-2018.

Informed Consent Statement: The buffalo used in the study were maintained at the Livestock Farm of the Faculty of Veterinary and Animals of the same university.

Data Availability Statement: The data can be made available from the corresponding author upon reasonable request.

Acknowledgments: The authors are thankful to the partners at the University of Veterinary and Animal Sciences, Lahore, Pakistan, for their meaningful collaborations and the Director of Livestock Farms, Musarrat Abbas Khan, for his technical support.

Conflicts of Interest: The authors declare no conflicts of interest.

References

1. Singh, T.; Ramakrishnan, S.; Mahanta, S.K.; Tyagi, V.C.; Roy, A.K. Tropical forage legumes in India: Status and scope for sustaining livestock production. In *Forage Groups*; Edvan, R.L.E., Santos, M., Eds.; Intechopen: London, UK, 2018; pp. 123–143.
2. Tahir, M.N.; Ihsan, M.Z.; Siddiqui, M.H.; Haque, M.N.U.; Zahra, N.; Chattha, W.S.; Bajwa, A.A. Appraisal of spatial distribution and fibre degradability of cereal-legume fodders to enhance the sustainability of livestock feed supply in sub-tropics. *Sustainability* **2024**, *16*, 4070. [CrossRef]
3. Mudau, H.S.; Mokoboki, H.K.; Ravhuhali, K.E.; Mkhize, Z. Nutrients profile of 52 browse species found in semi-arid areas of South Africa for livestock production: Effect of harvesting site. *Plants* **2021**, *10*, 2127. [CrossRef]
4. Gallardo, I.; Bárcena, R.; Pinos-Rodríguez, J.M.; Cobos, M.; Carreón, L.; Ortega, M.E. Influence of exogenous fibrolytic enzymes on in vitro and in sacco degradation of forages for ruminants. *Ital. J. Anim. Sci.* **2010**, *9*, e8. [CrossRef]
5. Miranda, L.F.; Rodriguez, N.M.; Pereira, E.S.; Queiroz, A.C.d.; Sainz, R.D.; Pimentel, P.G.; Gontijo Neto, M.M. Chemical composition and ruminal degradation kinetics of crude protein and amino acids, and intestinal digestibility of amino acids from tropical forages. *Rev. Bras. Zootec.* **2012**, *41*, 717–725. [CrossRef]
6. Veloso, C.M.; Rodriguez, N.M.; de Carvalho, G.G.P.; Pires, A.J.V.; Mourao, G.B.; Gonçalves, L.C.; Sampaio, I.B.M. Ruminal degradabilities of dry matter and crude protein of tropical forages. *Rev. Bras. Zootec.* **2006**, *35*, 613–617. [CrossRef]
7. Tahir, M.N.; Khan, Z.; Khan, N.A.; Saima, S.; Kamran, Z.; Inal, F. Dry matter degradation kinetics of selected tropical forage in Nili-Ravi buffalo and Cholistani cows at heifer and lactating stages using NorFor in situ standards. *J. Zool. Res.* **2019**, *1*, 10–19. [CrossRef]
8. Zhang, J.; Yin, B.; Xie, Y.; Li, J.; Yang, Z.; Zhang, G. Legume-cereal intercropping improves forage yield, quality and degradability. *PLoS ONE* **2015**, *10*, e0144813. [CrossRef] [PubMed]
9. Navarro Ortiz, C.A.; Roa Vega, M.L. Determination of in vitro digestibility of forage species used in ruminant feeding. *Trop. Anim. Health Prod.* **2020**, *52*, 3045–3059. [CrossRef]
10. Sarwar, M.; Khan, M.; Nisa, M.; Bhatti, S.; Shahzad, M. Nutritional management for buffalo production. *Asian-Australas. J. Anim. Sci.* **2009**, *22*, 1060–1068. [CrossRef]

11. Wanapat, M.; Kang SungChhang, K.S.; Phesatcha, K. Enhancing buffalo production efficiency through rumen manipulation and nutrition. *Buffalo Bull.* **2013**, *32*, 258–275.
12. Åkerlind, M.; Weisbjerg, M.; Eriksson, T.; Tøgersen, R.; Udén, P.; Ólafsson, B.; Harstad, O.; Volden, H. Feed analyses and digestion methods. In *NorFor-The Nordic Feed Evaluation System*; Wageningen Academic: Wageningen, The Netherlands, 2011; pp. 41–54.
13. Dumas, J. Procédés de l'analyse organique. *Ann. Chim. Phys.* **1831**, *47*, 198–205.
14. Van Soest, P.J.; Robertson, J.B.; Lewis, B.A. Methods for dietary fiber, neutral detergent fiber, and nonstarch polysaccharides in relation to animal nutrition. *J. Dairy Sci.* **1991**, *74*, 3583–3597. [CrossRef]
15. Mertens, D.R. Gravimetric determination of amylase-treated neutral detergent fiber in feeds with refluxing in beakers or crucibles: Collaborative study. *J. AOAC Int.* **2002**, *85*, 1217–1240. [PubMed]
16. Ørskov, E.-R.; McDonald, I. The estimation of protein degradability in the rumen from incubation measurements weighted according to rate of passage. *J. Agric. Sci.* **1979**, *92*, 499–503. [CrossRef]
17. Hvelplund, T.; Weisbjerg, M.R. In situ techniques for the estimation of protein degradability and postrumen availability. In *Forage Evaluation in Ruminant Nutrition*; Givens, D.I., Owen, E., Axford, R.F.E., Omed, H.M., Eds.; CABI Publishing: Oxon, UK, 2000; pp. 233–258.
18. Sarwar, M.; Mahmood, S.; Abbas, W.; Ali, C. In situ ruminal degradation kinetics of forages and feed byproducts in male Nili-Ravi buffalo calves. *Asian-Australas. J. Anim. Sci.* **1996**, *9*, 533–538. [CrossRef]
19. Singh, S.; Kushwaha, B.; Nag, S.; Mishra, A.; Singh, A.; Anele, U. In vitro ruminal fermentation, protein and carbohydrate fractionation, methane production and prediction of twelve commonly used Indian green forages. *Anim. Feed Sci. Technol.* **2012**, *178*, 2–11. [CrossRef]
20. Rehman, S.; Nizam, S.; Saima Rubab, S.B.; Wei, X. Evaluation of protein content in some fodder crops. *Hamdard Med.* **2023**, *66*, 1–14.
21. Habib, G.; Khan, N.; Ali, M.; Bezabih, M. In situ ruminal crude protein degradability of by-products from cereals, oilseeds and animal origin. *Livest. Sci.* **2013**, *153*, 81–87. [CrossRef]
22. Valderrama, X.L.; Anrique, R.G. In situ rumen degradation kinetics of high-protein forage crops in temperate climates. *Chil. J. Agric. Res.* **2011**, *71*, 572. [CrossRef]
23. Das, L.K.; Kundu, S.; Kumar, D.; Datt, C. Fractionation of carbohydrate and protein content of some forage feeds of ruminants for nutritive evaluation. *Vet. World* **2015**, *8*, 197. [CrossRef] [PubMed]
24. Kaithwas, M.; Singh, S.; Prusty, S.; Mondal, G.; Kundu, S. Evaluation of legume and cereal fodders for carbohydrate and protein fractions, nutrient digestibility, energy and forage quality. *Range Manag. Agrofor.* **2020**, *41*, 126–132.
25. Singh, S.; Joshi, D.; Kumar, R. Carbohydrate and protein fractions, nutritive value and energetic efficiency in different sorghum accessions. *Range Manag. Agrofor.* **2018**, *39*, 251–259.
26. Islas, A.; Soto-Navarro, S. Effect of supplementation of dried distillers grains with solubles on forage intake and characteristics of digestion of beef heifers grazing small-grain pasture. *J. Anim. Sci.* **2011**, *89*, 1229–1237. [CrossRef]
27. Aufrère, J.; Graviou, D.; Baumont, R.; Detour, A.; Demarquilly, C. Degradation in the rumen of proteins from fresh lucerne forage in various stages of growth and conserved as silage or hay. *Anim. Res.* **2000**, *49*, 461–474. [CrossRef]
28. Singh, K.; Das, M.; Samanta, A.; Kundu, S.; Sharma, S. Evaluation of certain feed resources for carbohydrate and protein fractions and in situ digestion characteristics. *Indian J. Anim. Sci.* **2002**, *72*, 794–797.
29. Gerdes, L.; Barbosa, C.M.P.; Giacomini, A.A.; Mattos, W.T.d.; Gimenes, F.M.d.A.; Batista, K.; Uzan, B.Z. Introduction of forage legumes into Aruana Guineagrass pasture. *Bol. De Indústria Anim.* **2020**, *77*, 1–10. [CrossRef]
30. Iommelli, P.; Zicarelli, F.; Musco, N.; Sarubbi, F.; Grossi, M.; Lotito, D.; Lombardi, P.; Infascelli, F.; Tudisco, R. Effect of cereals and legumes processing on in situ rumen protein degradability: A review. *Fermentation* **2022**, *8*, 363. [CrossRef]
31. Musco, N.; Koura, I.B.; Tudisco, R.; Awadjihè, G.; Adjolohoun, S.; Cutrignelli, M.I.; Mollica, M.P.; Houinato, M.; Infascelli, F.; Calabrò, S. Nutritional characteristics of forage grown in south of Benin. *Asian-Australas. J. Anim. Sci.* **2016**, *29*, 51–61. [CrossRef] [PubMed]
32. Foster, J.; Carter, J.; Sollenberger, L.; Blount, A.; Myer, R.; Maddox, M.; Phatak, S.; Adesogan, A. Nutritive value, fermentation characteristics, and in situ disappearance kinetics of ensiled warm-season legumes and bahiagrass. *J. Dairy Sci.* **2011**, *94*, 2042–2050. [CrossRef] [PubMed]
33. Choudhary, S.; Guha, A.; Kholova, J.; Pandravada, A.; Messina, C.D.; Cooper, M.; Vadez, V. Maize, sorghum, and pearl millet have highly contrasting species strategies to adapt to water stress and climate change-like conditions. *Plant Sci.* **2020**, *295*, 110297. [CrossRef]
34. Hadebe, S.; Modi, A.; Mabhaudhi, T. Drought tolerance and water use of cereal crops: A focus on sorghum as a food security crop in sub—Saharan Africa. *J. Agron. Crop Sci.* **2017**, *203*, 177–191. [CrossRef]
35. Rao, I.M.; Ayarza, M.A.; García, R. Adaptive attributes of tropical forage species to acid soils I. Differences in plant growth, nutrient acquisition and nutrient utilization among C4 grasses and C3 legumes. *J. Plant Nutr.* **1995**, *18*, 2135–2155. [CrossRef]

36. Gunasekera, C.; Martin, L.; Siddique, K.; Walton, G. Genotype by environment interactions of Indian mustard (*Brassica juncea* L.) and canola (*B. napus* L.) in Mediterranean-type environments: 1. Crop growth and seed yield. *Eur. J. Agron.* **2006**, *25*, 1–12. [CrossRef]
37. Fernandes, G.A.; de Assis, J.R.; da Silva, E.B.; Morales, R.L.; Costa, F.M.; Olini, L.M.G.; da Conceição Padilha, V.H.T.; Nunes, D.; Dias, M.R. Influence of the time of the year on the nutritional value of forage consumed by beef cattle raised on grassland. *Sci. Electron. Arch.* **2022**, *15*. [CrossRef]
38. Gruber, L.; Taferner, K.; Haberl, L.; Maierhofer, G.; Gasteiner, J.; Urdl, M. Ruminal in situ degradation as well as crude protein and carbohydrate fractions of CNCPS of forage maize depending on vegetative stage, variety, conservation and site. In Proceedings of the 15th International Science Symposium on Nutrition of Domestic Animals, Zadravec-Erjavec Days', Radenci, Slovenia, 9–10 November 2006.
39. Kaufman, R.; Wilson, J.; Bean, S.; Galant, A.; Perumal, R.; Tesso, T.; Herald, T.; Shi, Y.C. Influence of genotype × location interaction on grain sorghum grain chemistry and digestibility. *Agron. J.* **2018**, *110*, 1681–1688. [CrossRef]
40. Roukos, C.; Koutsoukis, C.; Akrida-Demertzi, K.; Karatassiou, M.; Demertzis, G.; Kandrelis, S. The effect of altitudinal zone on soil properties, species composition and forage production in a subalpine grassland in northwest Greece. *Appl. Ecol. Environ. Res.* **2017**, *15*, 609–626. [CrossRef]
41. Dongdong, C.; Qi, L.; Zhe, L.; Fuquan, H.; Xin, C.; Shixiao, X.; Xinquan, Z.; Liang, Z. Variations of forage yield and nutrients with altitude gradients and their influencing factors in alpine meadow of Sanjiangyuan, China. *J. Soil Sci. Plant Nutr.* **2020**, *20*, 2164–2174. [CrossRef]
42. Pembleton, K.; Cullen, B.; Rawnsley, R.; Ramilan, T. Climate change effects on pasture-based dairy systems in south-eastern Australia. *Crop. Pasture Sci.* **2020**, *72*, 666–677. [CrossRef]
43. Buxton, D.R. Quality-related characteristics of forages as influenced by plant environment and agronomic factors. *Anim. Feed Sci. Technol.* **1996**, *59*, 37–49. [CrossRef]
44. Vázquez-de-Aldana, B.R.; García-Ciudad, A.; García-Criado, B. Interannual variations of above-ground biomass and nutritional quality of Mediterranean grasslands in Western Spain over a 20-year period. *Aust. J. Agric. Res.* **2008**, *59*, 769–779. [CrossRef]
45. Niklas, K.J. Plant allometry, leaf nitrogen and phosphorus stoichiometry, and interspecific trends in annual growth rates. *Ann. Bot.* **2006**, *97*, 155–163. [CrossRef] [PubMed]
46. Li, X.; An, J.; Hou, X. Effects of six consecutive years of irrigation and phosphorus fertilization on alfalfa yield. *Plants* **2023**, *12*, 2227. [CrossRef] [PubMed]
47. Enriquez-Hidalgo, D.; Barrera, K.; Rivero, M.; Merino, V.; Teixeira, D.; Vargas-Bello-Pérez, E. Effect of Feeding Lucerne and a Mixed Diet of Oats and Berseem Clover as a Source of Fresh Forage on Ruminal Characteristics and Nitrogen Use Efficiency in Dairy Cows during the Winter Period. *Ruminants* **2022**, *2*, 212–225. [CrossRef]
48. Seibt, D.C.; Olivo, C.J.; Alessio, V.; Sauter, C.P.; Bratz, V.F.; Aguirre, P.F. Forage mass and nutritional value of elephant grass intercropped with forage legumes1. *Rev. Ceres* **2021**, *68*, 429–440. [CrossRef]
49. Moore, K.J.; Jung, H.-J.G. Lignin and fiber digestion. *J. Range Manag.* **2001**, *54*, 420–430. [CrossRef]
50. Gaillard, C.; Bhatti, H.S.; Novoa-Garrido, M.; Lind, V.; Roleda, M.Y.; Weisbjerg, M.R. Amino acid profiles of nine seaweed species and their in situ degradability in dairy cows. *Anim. Feed Sci. Technol.* **2018**, *241*, 210–222. [CrossRef]
51. Peng, Q.; Khan, N.A.; Wang, Z.; Yu, P. Relationship of feeds protein structural makeup in common Prairie feeds with protein solubility, in situ ruminal degradation and intestinal digestibility. *Anim. Feed Sci. Technol.* **2014**, *194*, 58–70. [CrossRef]

animals

MDPI

Article

Exploring the In Vitro Effects of Cassava Diets and Enterococcus Strains on Rumen Fermentation, Gas Production, and Cyanide Concentrations

Saowalak Lukbun [1], Chanon Suntara [1], Rittikeard Prachumchai [2], Waroon Khota [3] and Anusorn Cherdthong [1,*]

[1] Tropical Feed Resources Research and Development Center (TROFREC), Department of Animal Science, Faculty of Agriculture, Khon Kaen University, Khon Kaen 40002, Thailand; saowalaklukbun@kkumail.com (S.L.); chansun@kku.ac.th (C.S.)

[2] Department of Animal Science, Faculty of Agricultural Technology, Rajamangala, University of Technology Thanyaburi, Pathum Thani 12130, Thailand; rittikeard_p@rmutt.ac.th

[3] Department of Animal Science, Faculty of Natural Resources, Rajamangala University of Technology Isan, Sakon Nakhon Campus, Sakon Nakhon 47160, Thailand; waroon.kh@rmuti.ac.th

* Correspondence: anusornc@kku.ac.th; Tel.: +66-4320-2362

Simple Summary: Cassava roots or leaves serve as a cost-effective source of energy and protein for ruminants but contain high levels of hydrocyanic acid (HCN), which pose toxicity risks. Cyanide-utilizing bacteria (CUB), specifically *Enterococcus faecium* KKU-BF7 and *Enterococcus gallinarum* KKU-BC15, found in rumen fluid, have demonstrated the capability to reduce HCN levels. This study investigated the effects of adding CUB to cassava-based diets on HCN reduction, gas production, and in vitro digestibility. The results highlighted that supplementing cassava diets with *E. faecium* KKU-BF7, particularly at a HCN concentration of 600 mg/kg, significantly improved HCN degradation, cumulative gas production, in vitro digestibility, and volatile fatty acid (VFA) concentration. These findings underscore the potential of CUB in mitigating HCN toxicity and enhancing feed quality, thereby promoting sustainable livestock production and animal health.

Abstract: This study examined the effects of adding CUB alongside HCN sources from fresh cassava diets on HCN reduction, gas production, and in vitro digestibility. A completely randomized design (CRD) with a $2 \times 2 \times 3 + 1$ factorial approach was used, where Factor A was the HCN source [fresh cassava root (FCR) or leaf (FCL)], Factor B was the HCN concentration (300 and 600 mg/kg dry matter (DM)), and Factor C was the bacterial supplement [no-CUB, *E. faecium* KKU-BF7 (CUB1), and *E. gallinarum* KKU-BC15 (CUB2)]. Statistical analysis was performed using the PROC GLM procedure in SAS. No interaction was observed among the main factors on gas kinetics and cumulative gas ($p > 0.05$). The addition of CUB1 or CUB2 enhanced cumulative gas production compared to the no-CUB group ($p = 0.04$). Cyanide degradation efficiency was high when FCR was included at a high HCN level. At 12 h post-incubation, HCN degradation efficiency was higher in the CUB2 and CUB1 groups, reaching 98.44–99.07% compared to the no-CUB group. The higher HCN level increased in vitro acid detergent fiber digestibility (IVADFD) ($p = 0.01$) by 7.20% compared to the low HCN level, and CUB2 further improved IVADFD. Compared to the FCL-fed group, FCR supplementation increased total VFA concentration ($p = 0.03$) and propionic acid (C3) concentration ($p = 0.04$). The addition of CUB2 further enhanced propionic acid concentration by 8.97% compared to no-CUB supplementation ($p = 0.04$). These results indicate that supplementing *E. gallinarum* KKU-BC15 at the highest HCN levels in FCR boosts HCN degradation efficiency, fiber digestibility, total VFA, and C3 concentration.

Keywords: cattle nutrition; cyanide-utilizing bacteria; degradation efficiency; feeding regime; nutritional components

check for updates

Citation: Lukbun, S.; Suntara, C.; Prachumchai, R.; Khota, W.; Cherdthong, A. Exploring the In Vitro Effects of Cassava Diets and Enterococcus Strains on Rumen Fermentation, Gas Production, and Cyanide Concentrations. *Animals* **2024**, *14*, 3269. https://doi.org/10.3390/ani14223269

Academic Editor: Lizhi Wang

Received: 15 October 2024
Revised: 2 November 2024
Accepted: 6 November 2024
Published: 13 November 2024

1. Introduction

In tropical regions, fresh cassava roots and leaves provide an affordable source of energy and protein for ruminants, positioning them as a viable alternative to cassava chips. This accessibility, along with the ease of local cultivation, makes cassava a popular feed choice among farmers [1,2]. Introducing fresh cassava diets as a feed alternative may effectively improve the use of cassava feeds without requiring preprocessing [3,4]. Fresh cassava can be a valuable energy source in ruminant diets, but its use is limited by the presence of hydrocyanic acid (HCN), which poses a risk of prussic acid toxicity or hydrocyanic poisoning in animals [5,6]. Toxicity risks from cassava depend on both the amount consumed and the animal's physiological condition [7]. Poisoning is typically triggered when animals consume excessive quantities, particularly in cases where cassava constitutes a large part of the diet or is fed without processing. Ingesting feeds with over 500 mg HCN/kg on a dry-weight basis is considered dangerous, as noted by Aminlari et al. [8]. Fresh cassava roots have been found to contain approximately 114 mg/kg of HCN on a fresh weight basis, which, if unprocessed, can contribute to significant HCN exposure in livestock diets [9]. Prolonged exposure can impair animal health, manifesting as respiratory distress, reduced growth performance, or even mortality in severe cases.

Reducing HCN concentrations in cassava feed is critical for ensuring its safe inclusion in ruminant diets, as high levels of HCN pose toxicity risks to these animals. However, ruminants possess a natural capacity to detoxify HCN through the action of enzymes present in both rumen microorganisms and animal tissues [10]. This detoxification mechanism mitigates the potential toxicity of HCN, allowing ruminants to tolerate cassava feed when consumed under controlled levels. Ingested HCN is rapidly absorbed, with toxicity symptoms appearing within 15 min when large quantities are ingested, suggesting that absorption primarily occurs in the rumen [8,11]. The degradation rate of 60% after 4 h still allows significant HCN absorption, highlighting the importance of rapid detoxification mechanisms. Rhodanese and β-mercaptopyruvate sulfurtransferase are key enzymes involved in this process [10,12]. Rhodanese catalyzes the transfer of sulfur to cyanide (HCN), converting it into thiocyanate (SCN), a compound significantly less toxic than HCN, which is subsequently excreted in the urine. This reaction is crucial as it reduces the systemic toxicity of HCN, enabling ruminants to handle cyanogenic feeds better. β-mercaptopyruvate sulfurtransferase works similarly by transferring sulfur from sulfur-containing molecules to HCN, facilitating its conversion to thiocyanate [11]. These enzymes are not only active in rumen microorganisms but are also present in the animal's liver and kidneys, where rhodanese activity is particularly high, ensuring detoxification occurs both in the digestive system and within body tissues [9,10]. Through the combined activity of these enzymes, ruminants convert approximately 65–80% of ingested HCN into thiocyanate, thereby reducing HCN levels to non-toxic concentrations [13].

The process requires adequate sulfur for optimal function, and the addition of sulfur to the diet has been shown to enhance HCN detoxification. For instance, the inclusion of elemental sulfur with cassava-based feeds can neutralize HCN, as demonstrated by Cherdthong et al. [3]. However, excessive sulfur intake can lead to adverse effects, such as reduced feed consumption and polioencephalomalacia [10,11]. Furthermore, the rumen pH and forage-to-concentrate ratio also influence HCN degradation. A rumen pH below 6.0, typically associated with high-concentrate diets, can impair microbial activity and slow HCN degradation [14]. Therefore, understanding the balance of dietary components is crucial in optimizing HCN detoxification and minimizing toxicity risks in ruminants [15,16].

Biotechnology could be employed to incorporate innovative feeding techniques that reduce the toxicity of HCN in fresh cassava roots. In support of this, Sumadong et al. [17] found that rumen bacteria activate rhodanese and β-mercaptopyruvate sulfurtransferases, enabling ruminants to detoxify small amounts of HCN efficiently. The utilization of enzymes and microorganisms in technology is intriguing. According to Bhalla et al. [18], five enzymes, namely cyanide hydratase, cyanide dehydratase, cyanide monooxygenase, and rhodanese, have been identified for the purpose of reducing HCN poisoning. The thiosulfate receptor binds with HCN at an atomic level. This results in the formation of a thiocyanate compound, which has the ability to decrease its level of toxicity [7,19]. Rhodanese can be synthesized by various bacteria, including *Thiobacillus denitrificans* [20], *Pseudomonas aeruginosa*, *Pseudomonas fluorescens*, *Escherichia coli* [21], *Thiobacillus neopolitanus* [22], *Bacillus subtilis* [23], and *Fusarium solani* [24]. A recent study conducted by Khota et al. [4] has revealed the presence of cyanide-utilizing bacteria (CUB), i.e., *E. faecium* KKU-BF7 and *E. gallinarum* KKU-BC15, in rumen fluid. A reduction in HCN levels is the result of the efficient utilization of HCN as a nitrogen source for microbial growth by these bacteria. The CUB group exhibits a significant resistance to a diverse array of HCN molecules, potentially surpassing other bacteria in its ability to metabolize HCN [10].

The hypothesis was that supplementing fresh cassava root or leaf with cyanide-utilizing ruminal bacteria (*E. faecium* KKU-BF7 and *E. gallinarum* KKU-BC15) would reduce HCN levels while improving nutrient digestibility and fermentation characteristics in the rumen. The aim of this study was to investigate the effects of these bacterial supplements on HCN reduction, gas kinetics, and in vitro digestibility using gas production techniques.

2. Materials and Methods

2.1. Treatments and Experimental Design

The experiment followed a complete randomized design (CRD) and had a 2 × 2 × 3 (+1) factorial structure. Factor A was the sources of HCN, which included fresh cassava root and fresh cassava leaves. Factor B represented the concentrations of HCN at 300 and 600 parts per million (ppm) in fresh samples. The HCN levels of 300 and 600 mg/kg DM were selected based on the findings of Supapong and Cherdtong [14], where these levels, combined with rhodanese enzyme at 1 mg per 10^4 ppm HCN, demonstrated safe HCN detoxification and improved feed utilization in Thai native beef cattle. Factor C involved the introduction of different types of cyanide-utilizing bacteria (CUB) inoculants, including no CUB, *E. faecium* KKU-BF7, and *E. gallinarum* KKU-BC15, at a concentration of 10^8 FU/mL. The (+1) treatment in the test bottle did not contain any additional elements.

2.2. Cyanide-Utilizing Bacteria in Rumen Culture

The CUB was derived through the process of screening and isolation in the rumen of swamp buffalo and Thai native beef cattle, as documented in a study conducted by Khota et al. [4]. The isolated strains, identified as KKU-BF7, were found to be closely related to the species *E. faecium*, as labeled by KKU-BF7. Similarly, KKU-BC15 was identified as *E. gallinarum*, as labeled by KKU-BC15. The bacteria of interest were selected by enriching them in nutrient broth (Difco Laboratories, Detroit, MI, USA) using the method described by Moradkhani et al. [21]. Following a series of culture enrichments in a mineral medium incubation, the growing culture was diluted in a solution containing 8.5 g per liter of NaCl at dilutions ranging from 10^{-1} to 10^{-5}. Lactobacilli MRS broth medium (Difco Laboratories, Detroit, MI, USA) was used to cultivate CUB for 24 h. A concentration of approximately 10^8 CFU/mL was determined by measuring the absorbance at 660 nm [21]. The bacteria were then prepared for use in an in vitro investigation.

2.3. The Preparation of Rumen Inoculums and Animals

Two male Thai cattle, each weighing around 450 ± 30 kg, were used to collect rumen fluid. The bulls were kept in separate cages. The concentrate diet had a crude protein (CP) content of 18% and a total digestible nutrient (TDN) content of 75%. The meal was divided into two equal portions and fed to the cattle at around 08:00 h and 17:00 h, with each portion amounting to 0.5% of body weight (BW). Rice straw was constantly available. Mineral blocks and water were freely available. In the early morning, approximately one hour prior to feeding, the cattle's ruminal fluid was collected. Multiple layers of cheesecloth were employed to filter the fluid, which was subsequently transferred to heated thermos flasks for transport to the laboratory. Artificial saliva was synthesized utilizing the methodology proposed by Menke and Steingass [25]. By combining synthetic saliva and rumen fluid in a 2:1 ratio, a composite rumen inoculum was produced. To remove the oxygen, CO_2 was added to the serum bottles before submerging them in a water bath heated to 39 °C. Forty milliliters of the rumen solution was injected into the bottles by force one hour before being used as inoculum.

2.4. Substrates

The Kasetsart 50 variety of fresh cassava, which included both the root and the leaf, was gathered from a local market in Khon Kaen, Thailand. One-year-old cassava was freshly picked, cleaned, and cut into pieces that were passed through a 1 mm sieve. The cassava was prepared to achieve HCN concentrations of either 300 or 600 mg/kg DM, corresponding to fresh weights of 1.08 and 2.16 g. A substrate consisting of a mixture of rice straw and concentrate (in a ratio of 60:40) was utilized, with a total weight of 0.5 g placed in the bottles. Laboratory analytical procedures were performed on the concentrate, straw, and HCN sources to ensure accurate composition and appropriate concentration levels. In order to achieve a particle size that was able to pass through a filter with a pore size of one millimeter, the fresh cassava, rice straw, and concentrate diet samples were precisely pulverized. These ground samples were then prepared for the investigation of their nutritional composition and for the gas production study. The CUB cultures were each put into the bottom at a concentration of 10^8 CFU/mL. Table 1 provides an inventory of the diets that were used as substrates in the experiment, including their contents and their chemical composition.

Table 1. Chemical composition of the substrates used in the in vitro experiment (DM basis, %).

Item	Concentrate, %DM	FCR	FCL	Rice Straw
Ingredients, %DM				
Corn meal	50			
Soybean meal	14			
Fine rice bran	17			
Palm kernel meal	12.5			
Urea	0.5			
Di-calcium phosphate	1.5			
Molasses	3			
Mineral premix	1.5			
Chemical composition				
Dry matter, %	89.61	38.50	26.80	92.50
Organic matter, %DM	94.12	96.87	91.77	89.50
Crude protein, %DM	18.71	3.00	21.70	2.30
Ash, %DM	5.88	3.13	8.23	10.50
Neutral detergent fiber, %DM	52.19	8.00	48.80	71.20
Acid detergent fiber, %DM	22.01	6.90	32.50	44.20
Hydrocyanic acid, ppm	-	100.60	221.37	-

FCR, fresh cassava root; FCL, fresh cassava leaf.

The temperature in the incubator was set to 39 °C so that the bottles could be used for an in vitro gas test. The caps were sealed with rubber and metal caps. Every 2 h, the bottles were stirred during each sampling step. The cumulative assessment of gas output was conducted through three distinct runs, each consisting of 3 replicates [(3 bottles per treatment × 13 treatments) + 5 bottles of blank)] × 3 runs. To establish baseline measurements, a total of five control bottles containing only ruminal inoculum were used as blanks. These blanks were prepared in the same manner as the experimental bottles but without the addition of any substrates. The mean gas output values from these control bottles served as a reference point. The net gas output for each experimental bottle was determined by subtracting the mean values of the blanks from the corresponding measured values. A total of 78 bottles were used for the experiment, with 3 bottles per treatment and 13 treatments. Two sampling durations, 4 and 8 h of incubation, were considered. Ruminal pH, ammonia–nitrogen (NH_3-N), and VFA were measured independently for each bottle. An assessment of digestibility was conducted using a distinct set of 78 bottles (3 bottles per treatment × 13 treatments × 2 sample times at 12 and 24 h incubation). A total of 130 bottles were used to prepare the HCN concentration analysis. Each treatment consisted of 2 bottles, and there were 13 treatments in total. The sampling was performed at 0, 4, 8, 12 and 24 h of incubation, resulting in 5 sample times.

2.5. Analysis of Chemical Composition in Diets

Table 1 displays the samples of rice straw, concentrate, fresh cassava root, and fresh cassava leaves that underwent chemical analysis. The DM, ash, CP, and organic matter (OM) measurements were carried out in compliance with the recommendations provided by the AOAC [26]. The methodology described by Van Soest et al. [27] was used to analyze the neutral detergent fiber (NDF) and acid detergent fiber (ADF). The cumulative assessment of gas output was monitored at various incubation times, specifically at 0, 1, 2, 4, 6, 8, 10, 12, 18, 24, 36, 48, 72 and 96 h. The gas production data were analyzed using the Ørskov and McDonald [28] model, which can be expressed as follows:

$$Y = a + b\,(1 - e^{-ct})$$

where Y = gas produced at time "t" (mL), a = gas production from the immediately soluble fraction (mL), b = gas production from the insoluble fraction (mL), c = gas production rate constant for the insoluble fraction (mL/h), t = incubation time (h), and a + b = the potential extent of gas production (mL).

After the incubation period, fermentation fluid was collected at 4 and 8 h. The purpose of this was to measure the ruminal pH. The collected liquor was then filtered using four layers of cheesecloth. The fermentation liquors were subjected to centrifugation at a force of 16,000 times the acceleration due to gravity for a duration of 15 min at a temperature of 4 °C. This process was carried out in order to collect the liquid portion above the sediment, which was then used for the determination of NH_3-N [26]. Additionally, the VFA content was analyzed using gas chromatography. The gas chromatograph used for this analysis was the Wilmington, DE 5890A Series II model (Agilent Technologies, Inc, Santa Clara, CA, United States), and a glass column measuring 180 cm in length and 4 mm in diameter was employed. A combination of 100 g/L SP-1200 and 10 g/L H_3PO_4 were packed into the column using 80/100 mesh Chromosorb WAW, which was provided by Supelco, a company based in Bellefonte, PA, USA. Following 12 and 24 h of incubation, the in vitro digestibility of DM (IVDMD) and OM (IVOMD) was assessed. The resulting content was then used to determine the in vitro NDF degradability (IVNDFD) and in vitro ADF degradability (IVADFD) using an Ankom filter bag [4].

A modified version of Fisher and Brown's [29] procedure, notably the picric acid method, was used to quantify the amounts of HCN in the fermentation liquid at various time points (0, 4, 8 and 12 h of incubation). Standard potassium cyanide (KCN) solutions were used to create a linear calibration curve. A solution that contained 0.5% (*w*/*v*) picric acid and 0.25 M Na_2CO_3 was added to 0.05 mL parts of KCN solutions. This resulted in the achievement of the desired result. The KCN solutions were centrifuged at 15,000× *g* for 10 min at 4 °C before the aliquots were added. The resulting solutions were heated to boiling for a duration of 5 min, then diluted to a volume of 1 mL with 0.85 mL of distilled water and cooled for a period of 30 min in tap water. Absorbance at 520 nm was measured using a spectrophotometer, with distilled water and picric acid reagent serving as the blank. For the purpose of calculating the degrading efficiency (*DE*) of HCN, the following formula was utilized:

$$DE\,(\%) = \left(\frac{Ic - Rc}{Ic}\right) \times 100$$

where *Ic* = initial concentration of HCN (ppm) and *Rc* = residual concentration of HCN (ppm).

2.6. Statistical Analysis

Statistical analysis was conducted using the PROC GLM procedure in the SAS program version no 9.1.3 [30] to assess the main effects and interactions among these factors. Tukey's multiple comparison test was employed to identify significant differences among treatment means, with $p < 0.05$ considered statistically significant.

To ensure the validity of the statistical analysis, the assumptions of normality and homoscedasticity were assessed. Residuals were examined for normality using the Shapiro–Wilk test, and homoscedasticity was evaluated by plotting residuals against fitted values. Where necessary, data transformations were applied to meet these assumptions.

The model utilized in the analysis was as follows:

$$Yijkl = \mu + ai + bj + ck + abij + acik + bcjk + abcijk + \varepsilon ijkl$$

where Yijkl represents the response variable, μ is the overall mean, ai is the effect of HCN source (fresh cassava root or fresh cassava leaves), bj is the effect of HCN level (300 or 600 mg/kg DM), ck is the effect of bacterial treatment (no CUB, *E. faecium* KKU-BF7, or *E. gallinarum* KKU-BC15), and εijkl is the residual error term. Interaction terms were specified as follows: abij: the interaction between HCN source and HCN level, examining how different concentrations of HCN affect responses for each HCN source; acik: the interaction between HCN source and bacterial treatment, assessing how the bacterial treatments interact with HCN sources to influence key outcomes; bcjk: the interaction between HCN level and bacterial treatment, determining how different bacterial treatments interact with HCN concentrations; abcijk: the three-way interaction among HCN source, HCN level, and bacterial treatment, capturing the combined effects of these factors on responses.

3. Results

3.1. Gas Dynamics and Total Gas Output

The effects of CUB supplementation, along with various HCN sources and levels, on gas kinetics and accumulation are summarized in Table 2, and the cumulative gas production for each substrate treatment is illustrated as gas production curves in Figure 1. No interaction was observed between the HCN source, HCN level, and CUB type on the gas production kinetics, including gas from immediately soluble fractions (a), insoluble fractions (b), the rate of gas production from insoluble fractions (c), the potential total gas production (a + b), and cumulative gas after 96 h ($p > 0.05$). For the main effect of the HCN source, FCR led to significantly higher gas production compared to FCL, with cumulative gas (96 h) values of 203.92 mL and 179.25 mL, respectively ($p = 0.02$). Higher HCN levels (600 mg/kg) resulted in increased cumulative gas production ($p = 0.02$) and gas production from the insoluble fraction ($p = 0.01$) compared to the 300 mg/kg level, with increases of 18% and 16%, respectively. The addition of *E. faecium* KKU-BF7 or *E. gallinarum* KKU-BC15 enhanced cumulative gas production significantly compared to the no-CUB group ($p = 0.04$), with increases ranging from 10.7% to 11.8%.

Table 2. Effect of cyanide-utilizing bacteria (CUB) supplementation with different sources and levels of cyanide on gas kinetics and cumulative gas at 96 h after incubation.

Treatment	Cyanide Source	Level of Cyanide	Type of CUB	Gas Kinetics				Cumulative Gas (mL), 96 h
				a	b	c	|a| + b	
T1		Control		−15.06	136.68	0.03	151.75	113.23
T2	FCR	300	No CUB	−27.37	179.86	0.07	207.23	150.70
T3			*E. faecium* KKU-BF7	−29.82	239.45	0.05	269.27	205.03
T4			*E. gallinarum* KKU-BC15	−34.51	254.92	0.06	289.43	216.20
T5		600	No CUB	−29.76	249.37	0.05	279.14	208.70
T6			*E. faecium* KKU-BF7	−26.62	234.01	0.06	260.63	208.30
T7			*E. gallinarum* KKU-BC15	−29.50	272.11	0.05	301.61	234.57
T8	FCL	300	No CUB	−19.20	170.77	0.04	189.97	144.30
T9			*E. faecium* KKU-BF7	−16.76	183.31	0.04	200.07	161.50
T10			*E. gallinarum* KKU-BC15	−19.73	182.28	0.05	202.02	158.00
T11		600	No CUB	−23.24	235.93	0.04	259.17	209.03
T12			*E. faecium* KKU-BF7	−20.04	226.22	0.05	246.26	203.45
T13			*E. gallinarum* KKU-BC15	−21.97	223.08	0.05	245.05	199.23
SEM				2.21	23.93	0.01	25.60	20.93

Table 2. *Cont.*

Treatment	Cyanide Source	Level of Cyanide	Type of CUB	Gas Kinetics				Cumulative Gas (mL), 96 h
				a	b	c	|a| + b	
Main effect								
HCN source		FCR		−29.60	238.29 ^{d}	0.05	267.89 ^{d}	203.92 ^{d}
		FCL		−20.16	203.60 ^{e}	0.04	223.76 ^{e}	179.25 ^{e}
p-value				0.09	0.01	0.71	0.01	0.02
HCN dose		300		−24.57	201.77 ^{e}	0.05	226.33 ^{d}	172.62 ^{e}
		600		−25.49	240.12 ^{d}	0.05	265.31 ^{e}	210.55 ^{d}
p-value				0.11	0.01	0.66	<0.01	0.02
CUB specie		No CUB		−24.89	208.98	0.05	233.88	178.18 ^{e}
		E. faecium KKU-BF7		−23.61	220.75	0.05	244.06	199.57 ^{d}
		E. gallinarum KKU-BC15		−26.43	233.10	0.05	259.53	202.00 ^{d}
p-value				0.34	0.09	0.66	0.10	0.04
Interaction								
Control vs. other				0.01	0.01	0.05	0.01	0.01
A × B				0.01	0.01	0.01	0.01	0.01
A × C				0.23	0.08	0.19	0.08	0.08
B × C				0.23	0.08	0.19	0.08	0.08
A × B × C				0.34	0.09	0.66	0.10	0.06

SEM = standard error of the mean. ^{d,e} means in the same column with different superscripts differ ($p < 0.05$). FCR = fresh cassava root; FCL = fresh cassava leaf; a = gas production from the immediately soluble fraction (mL); b = gas production from the insoluble fraction (mL); c = gas production rate constant for the insoluble fraction (mL/h); and a + b = the potential extent of gas production (mL). A = cyanide source (FCR and FCL); B = level of cyanide (300, 600 mg/kg DM); C = type of cyanide-utilizing ruminal bacteria (no CUB, *E. faecium* KKU-BF7, and *E. gallinarum* KKU-BC15). Study interaction effect: A × B is the interaction effect between cyanide source and level of cyanide, A × C is the interaction effect between cyanide source and type of cyanide-utilizing ruminal bacteria, B × C is the interaction effect between level of cyanide 300 and 600 mg/kg DM and type of cyanide-utilizing ruminal bacteria, A × B × C is the interaction effect among cyanide source level of cyanide type of CUB. Significance levels are represented by *p*-value < 0.05.

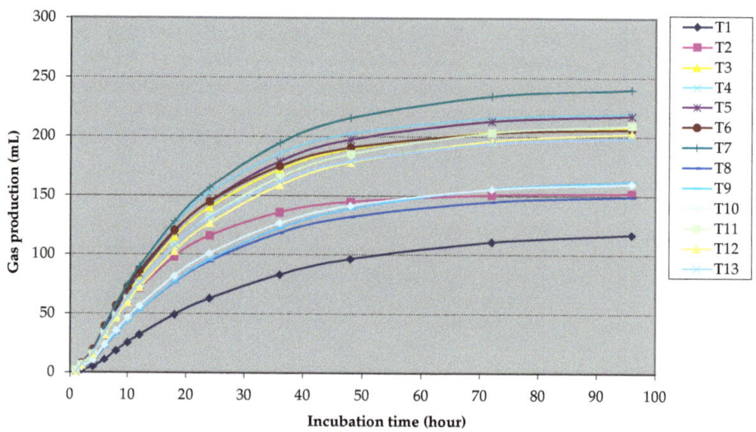

Figure 1. The effect of cyanide-utilizing bacteria (CUB) supplementation with different sources and levels of cyanide on cumulative gas production at different times of incubation. [T1 = control treatment; T2 = fresh cassava root + 300 mg/kg DM of HCN + no CUB; T3 = fresh cassava root + 300 mg/kg DM of HCN + *E. faecium* KKU-BF7; T4 = fresh cassava root + 300 mg/kg DM of HCN + *E. gallinarum* KKU-BC15; T5 = fresh cassava root + 600 mg/kg DM of HCN + no CUB; T6 = fresh cassava root + 600 mg/kg DM of HCN + *E. faecium* KKU-BF7; T7 = fresh cassava root + 300 mg/kg DM of HCN + *E. gallinarum* KKU-BC15; T8 = fresh cassava leaf + 300 mg/kg DM of HCN + no CUB; T9 = fresh cassava leaf + 300 mg/kg DM of HCN + *E. faecium* KKU-BF7; T10 = fresh cassava leaf + 300 mg/kg DM of HCN + *E. gallinarum* KKU-BC15; T11 = fresh cassava leaf + 600 mg/kg DM of HCN + no CUB; T12 = fresh cassava leaf + 600 mg/kg DM of HCN + *E. faecium* KKU-BF7; T13 = fresh cassava leaf + 600 mg/kg DM of HCN + *E. gallinarum* KKU-BC15].

3.2. Ruminal pH and Ammonia-Nitrogen (NH_3-N) and Cyanide Degradation Efficiency In Vitro

The effects of CUB supplementation with varying sources and levels of HCN on ruminal pH, NH_3-N concentration, and HCN degradation efficiency in vitro are presented in Table 3. No interactions were observed between HCN sources, levels, and CUB species on ruminal pH and NH_3-N ($p > 0.05$). However, HCN degradation efficiency showed a significant interaction among these factors. At 4 h of incubation, in vitro rumen pH in the FCR group was lower than in the FCL group (6.65 vs. 6.75; $p = 0.04$). The NH_3-N concentration at 8 h of incubation was highest in the FCL group at 1.03 mg/dL, compared to FCR ($p = 0.01$). CUB supplementation did not change ruminal pH or NH_3-N concentration ($p > 0.05$). Cyanide degradation efficiency was high when FCR was included at 600 mg/kg DM HCN. At 12 h after incubation, HCN degradation efficiency was significantly higher in the *E. gallinarum* KKU-BC15 and *E. faecium* KKU-BF7 groups, reaching 98.44–99.07%. This was significantly greater than in the no-CUB group, which showed a degradation efficiency of 83.20%.

Table 3. Effect of cyanide-utilizing bacteria (CUB) supplementation with different sources and levels of pH, ammonia–nitrogen, and cyanide degradation efficiencies.

Treatment	Cyanide Source	Level of Cyanide	Type of CUB	pH		Ammonia–Nitrogen, mg/dL		Cyanide Degradation Efficiency, %		
				4 h	8 h	4 h	8 h	4 h	8 h	12 h
T1	Control			6.92	7.05	22.78	22.22	0^{g}	0^{g}	0^{g}
T2	FCR	300	No CUB	6.72	7.09	22.71	20.39	63.2^{e}	80.83^{d}	86.83^{d}
T3			*E. faecium* KKU-BF7	6.85	7.05	23.04	23.02	75.72^{c}	90.00^{b}	100^{a}
T4			*E. gallinarum* KKU-BC15	6.59	7.17	22.85	19.32	76.44^{c}	89.81^{b}	100^{a}
T5		600	No CUB	6.59	6.97	23.04	19.73	76.14^{c}	83.52^{c}	92.69^{c}
T6			*E. faecium* KKU-BF7	6.58	6.83	23.06	23.21	$87.11^{a,b}$	93.85^{a}	$98.70^{a,b}$
T7			*E. gallinarum* KKU-BC15	6.58	6.80	22.05	19.87	87.56^{a}	93.80^{a}	100^{a}
T8	FCL	300	No CUB	6.3	6.71	22.49	22.30	51.20^{f}	66.27^{e}	85.61^{d}
T9			*E. faecium* KKU-BF7	6.71	6.93	22.94	23.13	74.93^{c}	88.49^{b}	99.79^{a}
T10			*E. gallinarum* KKU-BC15	6.68	6.74	22.69	23.43	75.48^{c}	88.30^{b}	$95.85^{b,c}$
T11		600	No CUB	7.09	7.00	21.53	21.85	67.02^{d}	78.87^{d}	92.21^{c}
T12			*E. faecium* KKU-BF7	6.9	7.08	22.11	19.35	$86.45^{a,b}$	93.34^{a}	$97.77^{a,b}$
T13			*E. gallinarum* KKU-BC15	7.02	6.98	22.07	21.64	85.92^{b}	94.12^{a}	$97.91^{a,b}$
SEM				0.09	0.05	0.29	1.00	0.49	0.80	1.28
Main effect										
HCN source	FCR			6.65^{b}	6.98^{a}	22.79^{a}	20.92^{b}	77.70^{a}	88.64^{a}	95.20^{a}
	FCL			6.75^{a}	6.90^{b}	22.31^{b}	21.95^{a}	73.50^{b}	84.95^{b}	91.94^{b}
p-value				0.04	0.01	0.01	0.01	<0.01	<0.01	0.01
HCN dose	300			6.64^{b}	6.95	22.79^{a}	21.93^{a}	69.50^{b}	83.95^{b}	92.18^{b}
	600			6.76^{a}	6.94	22.31^{b}	20.94^{b}	81.70^{a}	89.63^{a}	94.95^{a}
p-value				0.04	0.75	0.01	0.01	<0.01	<0.01	0.01
CUB specie	No CUB			6.61	6.94	22.44	21.07	64.39^{b}	77.38^{b}	83.20^{b}
	E. faecium KKU-BF7			6.76	6.97	22.79	22.17	81.05^{a}	91.50^{a}	99.07^{a}
	E. gallinarum KKU-BC15			6.71	6.92	22.42	21.06	81.36^{a}	91.5^{a}	98.44^{a}
p-value				0.57	0.06	0.91	1.00	<0.01	<0.01	<0.01
Interaction										
Control vs. other				0.06	0.06	0.47	0.46	<0.01	<0.01	<0.01
A × B				0.06	0.06	0.47	0.09	<0.01	<0.01	<0.01
A × C				0.13	0.51	0.08	0.52	<0.01	<0.01	0.12
B × C				0.13	0.51	0.08	0.52	<0.01	<0.01	0.12
A × B × C				0.57	0.60	0.91	0.99	<0.01	<0.01	<0.01

SEM = standard error of the mean. $^{a–g}$ means in the same column with different superscripts differ ($p < 0.05$). FCR = fresh cassava root; FCL = fresh cassava leaf. A = cyanide source (FCR and FCL); B = level of cyanide (300 and 600 mg/kg DM); C = type of cyanide-utilizing ruminal bacteria (no CUB, *E. faecium* KKU-BF7, and *E. gallinarum* KKU-BC15). Study interaction effect: A × B is the interaction effect between cyanide source and level of cyanide, A × C is the interaction effect between cyanide source and type of cyanide-utilizing ruminal bacteria, B × C is the interaction effect between level of cyanide 300 and 600 mg/kg DM and type of cyanide-utilizing ruminal bacteria, A × B × C is the interaction effect among cyanide source level of cyanide type of CUB. Significance levels are represented by *p*-value < 0.05.

3.3. In Vitro Digestibility

The effects of CUB supplementation, along with different HCN sources and concentrations (300 and 600 mg/kg DM), on IVOMD, IVDMD, IVNDFD, and IVADFD are presented in Table 4. There was no significant interaction between HCN source, HCN level, and CUB type for IVOMD, IVDMD, and IVNDFD ($p > 0.05$). However, the higher HCN level of 600 mg/kg DM significantly increased IVOMD ($p < 0.01$) and IVNDFD ($p = 0.01$) by 4.99% and 7.20%, respectively, compared to the 300 mg/kg DM level. The addition of CUB had no effect on IVOMD, IVDMD, or IVNDFD, except for IVADFD, where a significant interaction among the factors was observed ($p = 0.02$). Specifically, the addition of *E. gallinarum* KKU-BC15 or the high HCN level of 600 mg/kg DM increased IVADFD by 1.42% and 5.53%, respectively, compared to no CUB supplementation or the lower HCN level of 300 mg/kg DM.

Table 4. Effect of cyanide-utilizing bacteria (CUB) supplementation with different sources and level of cyanide on in vitro digestibility.

Treatment	Cyanide Source	Level of Cyanide	Type of CUB	IVOMD, %	IVDMD, %	IVNDFD, %	IVADFD, %
T1		Control		82.59	77.97	47.37	22.38 ^{f}
T2	FCR	300	No CUB	83.54	74.26	59.80	23.27 ^{f}
T3			*E. faecium* KKU-BF7	83.35	73.75	61.45	21.66 ^{f}
T4			*E. gallinarum* KKU-BC15	83.24	72.16	63.43	31.67 ^{c,d}
T5		600	No CUB	83.88	77.90	67.16	33.13 ^{b,c,d}
T6			*E. faecium* KKU-BF7	84.47	78.80	69.67	27.47 ^{e}
T7			*E. gallinarum* KKU-BC15	84.49	79.34	69.10	29.00 ^{d,e}
T8	FCL	300	No CUB	86.39	78.62	62.97	30.03 ^{d,e}
T9			*E. faecium* KKU-BF7	85.48	76.31	58.80	34.73 ^{a,b,c}
T10			*E. gallinarum* KKU-BC15	84.82	75.81	58.56	37.36 ^{a}
T11		600	No CUB	87.14	81.75	67.47	33.97 ^{b,c,d}
T12			*E. faecium* KKU-BF7	87.09	81.56	67.30	34.75 ^{a,b,c}
T13			*E. gallinarum* KKU-BC15	87.05	81.46	67.50	36.31 ^{a,b}
SEM				0.49	0.65	0.81	1.27
Main effect							
A		FCR		83.82 ^{b}	76.04 ^{a}	65.10 ^{a}	26.83 ^{a}
		FCL		86.33 ^{a}	79.25 ^{b}	63.77 ^{b}	35.77 ^{b}
p-value				<0.01	<0.01	0.01	<0.01
B		300		84.47 ^{b}	75.15 ^{b}	60.83 ^{b}	28.54 ^{b}
		600		85.68 ^{a}	80.14 ^{a}	68.03 ^{a}	34.07 ^{a}
p-value				<0.01	<0.01	0.01	<0.01
C		No CUB		85.24	78.13	64.35	30.80 ^{b}
		E. faecium KKU-BF7		85.10	77.6	64.30	30.90 ^{b}
		E. gallinarum KKU-BC15		84.90	77.19	64.65	32.22 ^{a}
p-value				0.34	0.06	0.61	0.02
Interaction							
Control vs. other				0.01	<0.01	<0.01	<0.01
A × B				0.01	0.63	<0.01	<0.01
A × C				0.18	0.21	0.01	0.92
B × C				0.18	0.21	0.01	0.92
A × B × C				0.34	0.06	0.61	0.02

SEM = standard error of the mean. $^{a–f}$ means in the same column with different superscripts differ ($p < 0.05$). FCR = fresh cassava root; FCL = fresh cassava leaf. IVDMD, in vitro digestibility of dry matter; IVOMD, in vitro digestibility of organic matter; IVNDFD, in vitro digestibility of neutral detergent fiber degradability; IVADFD, in vitro digestibility of acid detergent fiber degradability. A = cyanide source (FCR and FCL); B = level of cyanide (300 and 600 mg/kg DM); C = type of cyanide-utilizing ruminal bacteria (no CUB, *E. faecium* KKU-BF7, and *E. gallinarum* KKU-BC15). Study interaction effect: A × B is the interaction effect between cyanide source and level of cyanide, A × C is the interaction effect between cyanide source and type of cyanide-utilizing ruminal bacteria, B × C is the interaction effect between level of cyanide 300 and 600 mg/kg DM and type of cyanide-utilizing ruminal bacteria, A × B × C is the interaction effect among cyanide source level of cyanide type of CUB. Significance levels are represented by *p*-value < 0.05.

3.4. Concentration of Volatile Fatty Acids (VFAs)

The effects of cyanide-utilizing bacteria (CUB) supplementation with different HCN sources and levels on VFA concentrations are shown in Table 5. There was no significant interaction between HCN source, HCN level, and CUB type on overall VFA concentrations or profiles ($p > 0.05$). However, compared to the FCL-fed group, FCR supplementation significantly increased total VFA concentration ($p = 0.03$) and propionic acid concentration ($p = 0.04$), raising them from 43.10 to 48.21 mmol/L and 23.48% to 25.06%, respectively. Additionally, a higher HCN level (600 mg/kg DM) increased the propionic acid concentration to 25.17% ($p = 0.02$). The addition of *E. gallinarum* KKU-BC15 further enhanced the propionic acid concentration by 8.97% compared to the group without CUB supplementation ($p = 0.04$).

Table 5. Effect of cyanide-utilizing bacteria (CUB) supplementation with different sources and levels of cyanide on volatile fatty acids (VFAs) concentrations.

Treatment	Cyanide Source	Level of Cyanide	Type of CUB	Total VFA, mmol/L	VFA Profiles, %		
					Acetic Acid	Propionic Acid	Butyric Acid
T1		Control		39.22	64.80	25.08	10.12
T2	FCR	300	No CUB	43.04	67.55	23.34	9.11
T3			*E. faecium* KKU-BF7	50.36	65.36	25.41	9.23
T4			*E. gallinarum* KKU-BC15	51.98	66.68	23.21	10.11
T5		600	No CUB	45.84	67.00	23.77	9.23
T6			*E. faecium* KKU-BF7	48.16	64.22	25.96	9.82
T7			*E. gallinarum* KKU-BC15	49.90	62.56	28.67	8.77
T8	FCL	300	No CUB	44.90	67.00	19.88	13.12
T9			*E. faecium* KKU-BF7	46.80	63.77	24.69	11.54
T10			*E. gallinarum* KKU-BC15	40.17	63.45	23.67	12.88
T11		600	No CUB	42.71	65.73	21.82	12.45
T12			*E. faecium* KKU-BF7	44.07	66.14	22.44	11.42
T13			*E. gallinarum* KKU-BC15	39.97	63.63	28.36	8.01
SEM				1.79	1.26	1.60	0.48
Main effect							
HCN source		FCR		48.21 [a]	65.56	25.06 [a]	9.38
		FCL		43.10 [b]	64.95	23.48 [b]	11.57
p-value				0.03	0.08	0.04	0.59
HCN dose		300		46.07	65.54	23.59 [a]	10.87
		600		45.11	64.88	25.17 [b]	9.95
p-value				0.42	0.08	0.02	0.59
CUB species		No CUB		44.61	66.28	22.93 [a]	10.79
		E. faecium KKU-BF7		46.80	65.15	24.29 [a,b]	10.56
		E. gallinarum KKU-BC15		45.57	64.63	25.19 [b]	10.19
p-value				0.92	0.89	0.04	0.17
Interaction							
Control vs. other				0.01	0.01	<0.01	0.03
A × B				0.01	0.01	<0.01	0.03
A × C				0.02	0.80	0.87	0.75
B × C				0.02	0.80	0.87	0.75
A × B × C				0.92	0.89	0.48	0.17

SEM = standard error of the mean. [a,b] means in the same column with different superscripts differ ($p < 0.05$). FCR = fresh cassava root; FCL = fresh cassava leaf. A = cyanide source (FCR and FCL); B = level of cyanide (300 and 600 mg/kg DM); C = type of cyanide-utilizing ruminal bacteria (no CUB, *E. faecium* KKU-BF7, and *E. gallinarum* KKU-BC15). Study interaction effect: A × B is the interaction effect between cyanide source and level of cyanide, A × C is the interaction effect between cyanide source and type of cyanide-utilizing ruminal bacteria, and B × C is the interaction effect between level of cyanide 300 and 600 mg/kg DM and type of cyanide-utilizing ruminal bacteria. Significance levels are represented by *p*-value < 0.05.

4. Discussion

In this study, experimental concentrate diets were formulated to support optimal growth, health, and milk or meat production in cattle, with crude protein levels between 14 and 18%, as recommended by the NRC [31], to meet these production goals. The rice straw in this experiment contained 2.30% crude protein, and fresh cassava roots provided only 3% crude protein on a dry matter basis, compared to 21.70% in fresh cassava leaves. This variation highlights the limited protein contribution from rice straw and fresh cassava roots, emphasizing the need for protein-rich supplements in the diet to achieve adequate overall crude protein levels for cattle. Cyanide concentrations in fresh cassava leaves (221.37 mg/kg fresh weight) were higher than those in fresh cassava roots (100.60 mg/kg fresh weight). The HCN concentration of Manihot esculenta Kasetsart 50 was found to be 103.5 mg/kg, according to Dagaew et al. [32]. This particular plant was acquired from a local farmer in the Khon Kaen district of Thailand. Prachumchai et al. [11] similarly reported a HCN level of 106.00 mg/kg in fresh cassava roots. Nguyen et al. [9] found that fresh cassava roots had a HCN content of 114 mg/kg. Vetter [33] also reported that HCN levels varied by cassava type, with bitter and sweet varieties containing between 310 and 468 mg/kg. It is important to note that the HCN levels found in our study were slightly lower than those reported in previous studies. This discrepancy may be attributed to various factors, including differences in cassava variety, geographic conditions, and growing practices [9,16]. The variability among cassava varieties and their growing conditions should be acknowledged as potential limitations of this study, as these factors could significantly influence both nutritional content and cyanide levels. Further research is warranted to explore these variabilities and their impact on the nutritional safety of cassava in cattle diets.

The inclusion of concentrate, roughage, and HCN sources (fresh cassava leaves and roots) in the diets led to increased gas production, particularly when CUB were added. This may be due to the availability of starch in fresh cassava roots, which provides a substrate for rumen microbes to generate gas [11]. On the other hand, fresh cassava leaves reduced gas accumulation at 96 h, which may be attributed to their high fiber content (NDF and ADF), potentially limiting digestion and gas production. However, it is important to consider that other factors, such as the presence of anti-nutritional compounds or differences in microbial activity, could also contribute to this effect [34]. Prachumchai et al. [11] observed that increasing the inclusion of fresh cassava root in the diet positively influenced gas production from both the immediately soluble and insoluble fractions, as well as cumulative gas production over a 96 h incubation period. This increase can be attributed to the high starch content in fresh cassava root, which enhances nutrient availability and promotes fermentation processes within the rumen. The starch serves as a readily fermentable carbohydrate, providing energy sources for the ruminal microbes, leading to increased gas production as a by-product of microbial metabolism [32]. Furthermore, the supplementation of CUB is hypothesized to enhance feed digestion and mitigate the toxic effects of HCN by converting HCN into less toxic compounds. This interaction allows for more efficient microbial function and improved fermentation dynamics. Supporting this notion, Sumadong et al. [17] demonstrated that elevated HCN levels negatively impacted gas production from the insoluble fractions and cumulative gas production after 96 h of incubation. However, this detrimental effect was significantly alleviated by CUB supplementation, indicating that these bacteria may aid in maintaining microbial viability and fermentation efficiency in the presence of HCN, ultimately enhancing nutrient utilization and gas production.

In this study, the ruminal pH ranged from 6.30 to 7.17, a level conducive to optimal microbial activity and fiber degradation. This pH range supports bacterial growth, especially for cellulolytic bacteria, which thrive in mildly acidic to neutral conditions, thereby maintaining a balance that allows for the efficient breakdown of fiber [35]. Similar to the findings of Bach et al. [35], ruminal pH levels between 6.5 and 7.0 facilitate fiber digestion without hindering microbial activity. Furthermore, Sombuddee et al. [36] observed that adding CUB to a feed containing HCN helped stabilize pH by enhancing microbial resilience to cyanide stress, thereby creating a favorable environment for fermentation.

The NH_3-N concentrations observed in this study (19.32–23.43 mg%) indicate adequate nitrogen availability for microbial growth. NH_3-N is primarily generated through protein hydrolysis and deamination by rumen microbes, serving as an essential nitrogen source for microbial protein synthesis. Adequate NH_3-N levels ensure the growth of ammonia-utilizing bacteria, which convert NH_3-N into microbial protein, contributing to the animal's protein intake [36]. Prachumchai et al. [11] reported that fresh cassava root supplementation increases NH_3-N concentrations, likely due to the additional fermentable carbohydrates that provide energy for proteolytic bacteria, enhancing NH_3-N production. Similarly, Cherdthong et al. [3] observed that NH_3-N levels rose with fresh cassava root supplementation, underscoring its role in balancing energy and nitrogen for microbial growth.

The efficiency of HCN degradation in the rumen was significantly influenced by CUB supplementation. Cyanide-utilizing bacteria metabolize HCN as a nitrogen source, using the rhodanese enzyme to convert cyanide into thiocyanate, a less toxic form that can be safely utilized by rumen microbes [6,11]. This enzymatic degradation of HCN not only detoxifies the rumen environment but also provides nitrogen that can be assimilated into microbial biomass, supporting overall microbial growth and activity.

Regarding in vitro digestibility, fresh cassava roots—high in nonstructural carbohydrates—facilitate rapid microbial fermentation, as these carbohydrates are easily accessible for degradation, thereby promoting bacterial population growth and nutrient utilization [11]. However, increasing fresh cassava leaf levels tended to reduce digestibility. This is likely due to the high fiber content of fresh cassava leaves, which could inhibit microbial accessibility to substrates and slow down the digestion rate [37]. The high fiber in leaves may require more cellulolytic activity, which can be challenging for bacteria in the presence of limited fermentable carbohydrate sources.

The inclusion of CUB appears to support microbial digestion indirectly by reducing feed toxins, such as HCN, thus promoting ammonia production and a favorable microbial environment [4]. Dagaew et al. [32] noted that an increased ratio of fresh cassava roots to rice straw improved fiber digestibility, likely because the starch in cassava root provides additional energy for the microbes involved in fiber degradation. Sumadong et al. [17] observed that sulfur combined with fresh cassava roots enhanced gas production, rumen characteristics, and degradability. This effect is attributed to the synergistic action of nonstructural carbohydrates in cassava roots, which enhance microbial proliferation and activity, leading to improved digestion efficiency. Sombuddee et al. [36] further demonstrated that CUB supplementation increases in vitro dry matter digestibility by 11.16%, likely due to the ability of CUB to reduce toxins and support a microbial environment conducive to optimal fermentation.

Total volatile fatty acids are critical for ruminant energy, providing over 70% of the animal's energy [14,32]. In this study, supplementation with FCR resulted in significantly greater total VFA and propionic acid concentrations compared to FCL, likely due to its higher energy density, beneficial fermentation characteristics, and reduced toxicity risks from cyanogenic compounds [17]. Cassava roots are particularly rich in carbohydrates, especially starch, which provides an easily fermentable energy source for rumen microorganisms [11]. This high energy availability enhances fermentation activity, leading to an increased VFA production, particularly propionic acid, which is more prominent in starch fermentation. Although cassava leaves have a higher protein content, they also contain more fiber, which can reduce their digestibility and fermentation efficiency in the rumen [19]. Sumadong et al. [17] found that fresh cassava root supplementation significantly affected acetate and butyrate concentrations, with higher total VFA and propionate concentrations but lower acetate and butyrate levels observed. Similarly, Sombuddee et al. [36] reported that potassium HCN and HCN-utilizing bacteria impacted VFA concentrations, with the lowest VFA concentrations occurring at higher HCN levels. This suggests that while HCN levels can initially stimulate certain microbial populations, excessive HCN ultimately hampers VFA production, emphasizing the need for balanced cassava supplementation in ruminant diets [38]. This study additionally revealed that the inclusion of *E. gallinarum* KKU-BC15 increases propionic acid levels by 8.97% compared to the groups lacking CUB supplementation, a result of several interconnected factors. *E. gallinarum* KKU-BC15 is recognized for its capacity to enhance the fermentation process within the rumen. This particular strain effectively utilizes dietary substrates, especially those sourced from cassava, which boosts the production of VFAs, particularly propionic acid [4]. The introduction of CUBs like *E. gallinarum* KKU-BC15 can enhance microbial activity and diversity, which are essential for optimizing the fermentation pathways that favor propionate synthesis [36]. Certain strains of *E. gallinarum* have metabolic functions that facilitate the conversion of substrates into propionic acid; for instance, they can use fumarate as an electron acceptor, transforming it into succinate and then into propionate [4]. This specific metabolic pathway is less active in groups without CUB supplementation, leading to lower concentrations of propionic acid.

5. Conclusions

This study highlights the potential of using *E. gallinarum* KKU-BC15 as a valuable feed supplement in diets containing high levels of HCN from fresh cassava root, specifically with 600 mg/kg HCN. The significant increases in gas production, VFA concentration, propionic acid, and in vitro digestibility associated with this supplementation not only suggest improved fermentation efficiency but also point to a promising strategy for mitigating HCN toxicity in ruminants. By optimizing the utilization of cassava as a feed resource, this approach could enhance overall feed efficiency and promote healthier livestock, thereby contributing to sustainable livestock production in regions where cassava is a primary feed source. Further in vivo studies are recommended to explore the potential of *E. gallinarum* KKU-BC15 as a HCN mitigation agent and its application in cattle management.

Author Contributions: Planning and design of this study, S.L., C.S. and A.C.; conducting and sampling, S.L. and A.C.; sample analysis, S.L.; statistical analysis, S.L., C.S., R.P., W.K. and A.C.; manuscript drafting, S.L., C.S., R.P. and A.C.; manuscript editing and finalizing, S.L., C.S., R.P., W.K. and A.C. All authors have read and agreed to the published version of the manuscript.

Funding: The authors express their most sincere gratitude to the Khon Kaen University Fundamental Fund through the National Science, Research, and Innovation Fund (NSRF) and the National Research Council of Thailand (NRCT) (Grant No. NRCT5-RSA63003-01) for providing financial support. Miss. Saowalak Lukbun is grateful for receiving a scholarship from the National Research Council of Thailand (NRCT) and Animal Feed Inter Trade Co., Ltd., via Research and Researchers for Industries-RRI (contract grant N41A650396).

Institutional Review Board Statement: The Institutional Animal Care and Use Committee of Khon Kaen University (KKU) approved all of the procedures using animals in the metabolism experiments (Record No. IACUC-KKU-45/64).

Informed Consent Statement: Not applicable.

Data Availability Statement: The data that support the findings of this study are available from the corresponding author upon reasonable request.

Acknowledgments: The authors would like to express their sincere thanks to the Tropical Feed Resources Research and Development Center (TROFREC), Department of Animal Science, Faculty of Agriculture, Khon Kaen University (KKU), for the use of the animals and research facilities.

Conflicts of Interest: The authors declare no conflicts of interest.

References

1. Galvão, L.T.O.; Vargas, J.A.C.; Mezzomo, R.; Oliveira, L.R.S.; Dos Santos, P.M.; Maciel, R.P.; Soares, E.A.; Farias, M.L.C.; Caldas, C.S.; Faciola, A.P.; et al. Effects of supplementing cassava root silage to grazing dairy cows on nutrient utilization, milk production and composition in the tropics. *Trop. Anim. Health Prod.* **2024**, *56*, 120. [CrossRef] [PubMed]
2. Li, M.; Zhou, H.; Zi, X.; Lv, R.; Tang, J.; Ou, W.; Chen, S. Feeding value assessment of five varieties whole-plant cassava in tropical China. *Fermentation* **2024**, *10*, 45. [CrossRef]

3. Cherdthong, A.; Khonkhaeng, B.; Seankamsorn, A.; Supapong, C.; Wanapat, M.; Gunun, N.; Gunun, P.; Chanjula, P.; Polyorach, S. Effects of feeding fresh cassava root with high-sulfur feed block on feed utilization, rumen fermentation, and blood metabolites in Thai native cattle. *Trop. Anim. Health Prod.* **2018**, *50*, 1365–1371. [CrossRef]
4. Khota, W.; Kaewpila, C.; Kimprasit, T.; Seemakram, W.; Kakaisorn, S.; Wanapat, M.; Cherdthong, A. The isolation of rumen enterococci strains along with high potential utilizing cyanide. *Sci. Rep.* **2023**, *13*, 13176. [CrossRef] [PubMed]
5. Jumare, F.I.; Salleh, M.M.; Ihsan, N.; Hussin, H. Cassava waste as an animal feed treatment: Past and future. *Rev. Environ. Sci. Biotechnol.* **2024**, *23*, 839–868. [CrossRef]
6. Arsov, A.; Tsigoriyna, L.; Batovska, D.; Armenova, N.; Mu, W.; Zhang, W.; Petrov, K.; Petrova, P. Bacterial degradation of antinutrients in foods: The genomic insight. *Foods* **2024**, *13*, 2408. [CrossRef]
7. Giantin, S.; Franzin, A.; Brusa, F.; Montemurro, V.; Bozzetta, E.; Caprai, E.; Fedrizzi, G.; Girolami, F.; Nebbia, C. Overview of cyanide poisoning in cattle from *Sorghum halepense* and *S. bicolor* cultivars in Northwest Italy. *Animals* **2024**, *14*, 743. [CrossRef]
8. Aminlari, M.; Gilanpour, H. Comparative studies on the distribution of rhodanese in different tissues of domestic animals. *Comp. Biochem. Physiol. B* **1991**, *99*, 673–677. [CrossRef]
9. Nguyen, T.L.; Ogle, R.B.; Preston, T.R. Cassava root silage for crossbred pigs under village conditions in Central Vietnam. *Livest. Res. Rural Dev.* **1997**, *9*, 2.
10. Molina-Botero, I.C.; Gaviria-Uribe, X.; Rios-Betancur, J.P.; Medina-Campuzano, M.; Toro-Trujillo, M.; González-Quintero, R.; Ospina, B.; Arango, J. Methane emission, carbon footprint, and productivity of specialized dairy cows supplemented with bitter cassava (*Manihot esculenta* Crantz). *Animals* **2024**, *14*, 19. [CrossRef]
11. Prachumchai, R.; Cherdthong, A.; Wanapat, M. Screening of cyanide-utilizing bacteria from rumen and in vitro evaluation of fresh cassava root utilization with a pellet containing high sulfur diet. *Vet. Sci.* **2021**, *8*, 10. [CrossRef] [PubMed]
12. Kaleta, K.; Janik, K.; Rydz, L.; Wróbel, M.; Jurkowska, H. Bridging the gap in cancer research: Sulfur metabolism of leukemic cells with a focus on L-cysteine metabolism and hydrogen sulfide-producing enzymes. *Biomolecules* **2024**, *14*, 746. [CrossRef]
13. Khajali, F.; Rafiei, F. A Review of plant anti-nutritional factors in animal health and production: The classification, biological properties, and the passivation strategy. *J. Agric. Food Res.* **2024**, *18*, 101290. [CrossRef]
14. Supapong, C.; Sommai, S.; Khonkhaeng, B.; Suntara, C.; Prachumchai, R.; Phesatcha, K.; Chanjula, P.; Cherdthong, A. Effect of rhodanese enzyme addition on rumen fermentation, cyanide concentration, and feed utilization in beef cattle receiving various levels of fresh cassava root. *Fermentation* **2022**, *8*, 146. [CrossRef]
15. Wang, W.; Wang, Y.; Li, Z.; An, R.; Ren, Q.; Li, J. Effects of feeding different Sorghum-Sudangrass hybrids silages on growth performance, fatty acid composition of longissimus dorsi muscle, ruminal bacteria, and volatile fatty acid formation of weaned Small-Tailed Han lambs. Anim. *Feed Sci. Technol.* **2024**, *307*, 115851. [CrossRef]
16. Promkot, C.; Wanapat, M. Effect of elemental sulfur supplementation on rumen environment parameters and utilization efficiency of fresh cassava foliage and cassava hay in dairy cattle. *Asian-Australas. J. Anim. Sci.* **2009**, *22*, 1366–1376. [CrossRef]
17. Sumadong, P.; Cherdthong, A.; So, S.; Wanapat, M. Sulfur, fresh cassava root, and urea independently enhanced gas production, ruminal characteristics, and in vitro degradability. *BMC Vet. Res.* **2021**, *17*, 304. [CrossRef]
18. Bhalla, T.C.; Kumar, V.; Kumar, V. Microbial remediation of cyanides. In *Bioremediation: Current Research and Application*; Rathoure, A.K., Ed.; International Publishing House: New Delhi, India, 2017; pp. 88–110.
19. Prachumchai, R.; Cherdthong, A.; Wanapat, M.; So, S.; Polyorach, S. Fresh cassava root replacing cassava chip could enhance milk production of lactating dairy cows fed diets based on high sulfur-containing pellet. *Sci. Rep.* **2022**, *12*, 3809. [CrossRef]
20. Bowen, T.J.; Butler, P.J.; Happold, F.C. Some properties of the rhodanese system of *Thiobacillus denitrificans*. *Biochem. J.* **1965**, *97*, 651–657. [CrossRef]
21. Moradkhani, M.; Yaghmaei, S.; Nejad, Z.G. Biodegradation of cyanide under alkaline conditions by a strain of *Pseudomonas putida* isolated from gold mine soil and optimization of process variables through response surface methodology (RSM). *Period. Polytech. Chem. Eng.* **2018**, *62*, 265–273. [CrossRef]
22. Taylor, B.F. Oxidation of elemental sulfur by an enzyme system from *Thiobacillus neopolitanus*. *Biochim. Biophys. Acta* **1968**, *170*, 112–122. [CrossRef] [PubMed]
23. Villarejo, M.; Westley, J. Sulfur metabolism of *Bacillus subtilis*. *Biochim. Biophys. Acta* **1966**, *117*, 209–216. [CrossRef] [PubMed]
24. Dumestre, A.; Chone, T.H.; Portal, J.; Gerard, M.; Berthelin, J. Cyanide degradation under alkaline conditions by a strain of *Fusarium solani* isolated from contaminated soils. *Appl. Environ. Microbiol.* **1997**, *63*, 2729–2734. [CrossRef] [PubMed]
25. Menke, H.H.; Steingass, H. Estimation of the energetic feed value obtained from chemical analysis and in vitro gas production using rumen fluid. *Anim. Res. Dev.* **1988**, *28*, 7–55.
26. AOAC. *Official Methods of Analysis*, 19th ed.; Association of Official Analytical Chemists: Gaithersburg, MD, USA, 2012.
27. Van Soest, P.V.; Robertson, J.B.; Lewis, B.A. Methods for dietary fiber, neutral detergent fiber, and non-starch polysaccharides in relation to animal nutrition. *J. Dairy Sci.* **1991**, *74*, 3583–3597. [CrossRef]
28. Ørskov, E.R.; McDonald, I. The estimation of protein degradability in the rumen from incubation measurements weighted according to rate of passage. *J. Agric. Sci.* **1979**, *92*, 499–503. [CrossRef]
29. Fisher, F.; Brown, J. Colorimetric determination of cyanide in stack gas and waste water. *Anal. Chem.* **1952**, *24*, 1440–1444. [CrossRef]
30. SAS Institute Inc. *SAS User's Guide: Statistics, Version 9.1.3*; SAS Institute Inc.: Cary, NC, USA, 2003.
31. National Research Council (NRC). *Nutrient Requirements of Dairy Cattle*, 7th ed.; National Academic Press: Washington, DC, USA, 2001.

32. Dagaew, G.; Cherdthong, A.; Wanapat, M.; So, S.; Polyorach, S. Ruminal fermentation, milk production efficiency, and nutrient digestibility of lactating dairy cows receiving fresh cassava root and solid feed-block containing high sulfur. *Fermentation* **2021**, *7*, 114. [CrossRef]
33. Vetter, J. Plant cyanogenic glycosides. *Toxicon* **2000**, *38*, 11–36. [CrossRef]
34. Terranova, M.; Kreuzer, M.; Li, Y.; Schwarm, A.; Clauss, M. Indications for a lower methane yield from digested fibre in ruminants digesting fibre more efficiently. *Anim. Feed Sci. Technol.* **2024**, *307*, 115834. [CrossRef]
35. Bach, A.; Baudon, M.; Elcoso, G.; Viejo, J.; Courillon, A. Effects on rumen pH and feed intake of a dietary concentrate challenge in cows fed rations containing pH modulators with different neutralizing capacity. *J. Dairy Sci.* **2023**, *106*, 4580–4598. [CrossRef] [PubMed]
36. Sombuddee, N.; Suntara, C.; Khota, W.; Boontiam, W.; Chanjula, P.; Cherdthong, A. Effect of cyanide-utilizing bacteria and sulfur supplementation on reducing cyanide concentration and in vitro degradability using in vitro gas production technique. *Fermentation* **2022**, *8*, 436. [CrossRef]
37. Sawanon, S.; Kobayashi, Y. Synergistic fibrolysis in the rumen by cellulolytic *Ruminococcus flavefaciens* and non-cellulolytic *Selenomonas ruminantium*: Evidence in defined cultures. *Anim. Sci. J.* **2006**, *77*, 208–214. [CrossRef]
38. Zheng, Y.; Zhao, Y.; Xue, S.; Wang, W.; Wang, Y.; Cao, Z.; Yang, H.; Li, S. Feeding value assessment of substituting cassava (*Manihot esculenta*) residue for concentrate of dairy cows using an in vitro gas test. *Animals* **2021**, *11*, 307. [CrossRef]

animals

Article

Impacts of Protein and Energy Levels on Rumen Fermentation and Microbial Activity Under Different Incubation Temperatures

Yong-Ho Jo [1,2,†], Won-Seob Kim [1,†], Yoo-Rae Kim [1], Mun-Su Ju [1], Jalil Ghassemi Nejad [1] and Hong-Gu Lee [1,*]

[1] Department of Animal Science and Technology, Sanghuh College of Life Sciences, Konkuk University, Seoul 05029, Republic of Korea; jyh3977@naver.com (Y.-H.J.); kws9285@konkuk.ac.kr (W.-S.K.); yooraekim95@gmail.com (Y.-R.K.); ju2139@naver.com (M.-S.J.); jalilghasseminejad@gmail.com (J.G.N.)

[2] IANS Co., Ltd., Cheonan-si 31090, Republic of Korea

[*] Correspondence: hglee66@konkuk.ac.kr; Tel.: +82-02-450-0523

[†] These authors contributed equally to this work.

Simple Summary: This study explored the effects of protein and energy levels on rumen fermentation under different incubation temperatures using an in vitro system. We found that higher incubation temperatures increased NH_3-N and total volatile fatty acids (TVFA) but decreased liquid-associated bacteria (LAB). Conversely, higher protein levels elevated NH_3-N and acetate levels but reduced propionate, while higher energy levels had the opposite effect on rumen fermentation properties. In addition, incubation temperatures and energy levels were affected on rumen fermentation characteristics and LAB protein amounts. However, there were no significant interactions between energy or protein levels and incubation temperatures for TVFA and LAB. The findings suggested that the adjustment of single nutrient levels of protein or energy would not be sufficient to enhance microbial protein synthesis under different incubation temperatures.

Abstract: This study aimed to explore the effects of different incubation temperatures on ruminal fermentation and rumen microorganisms and determine the appropriate protein and energy levels to enhance microbial protein synthesis using an in vitro system. Rumen inoculum was collected from two fistulated Holstein heifers (trial 1: BW: 652.3 kg ± 25.2; trial 2: BW: 683.3 kg ± 30.2) and assessed using a closed-batch culture system. The experimental model employed a 2 × 5 factorial arrangement using incubation temperatures set to 39 and 41 °C, with protein levels set to 12.0, 13.5, 15.0, 16.5, and 18.0% of DM in trial 1 or with energy levels set to 2.4, 2.5, 2.6, 2.7, and 2.8 Mcal/kg of DM in trial 2. The data were analyzed using the MIXED procedure. The results showed increased ($p < 0.05$) NH_3-N concentrations and total volatile fatty acids (TVFAs) with higher incubation temperatures, while the liquid-associated bacterial (LAB) amounts decreased ($p < 0.05$) in trials 1 and 2. The interaction between the energy level and incubation temperature affected ($p < 0.05$) the LAB protein levels in trial 2. Higher protein levels led to increased ($p < 0.05$) NH_3-N and acetate concentrations, but it decreased ($p < 0.05$) the propionate percentage. Conversely, higher energy levels decreased ($p < 0.05$) the amount of acetate and increased the propionate concentration, altering the acetate-to-propionate ratio. However, no interaction involving TVFA and LAB was observed between the incubation temperature and the protein or energy levels. Changes in the NH_3-N, TVFAs, and LAB protein amounts were observed under different incubation temperatures and energy levels. In conclusion, these findings provide insight into the metabolic adaptation under different ruminal temperatures and the impacts of dietary adjustments on rumen fermentation and microbial activity. However, there are limitations to replicating the complex physiological responses that occur within the whole body solely through in vitro experiments.

Keywords: energy; incubation temperatures; protein; rumen fermentation properties

Citation: Jo, Y.-H.; Kim, W.-S.; Kim, Y.-R.; Ju, M.-S.; Nejad, J.G.; Lee, H.-G. Impacts of Protein and Energy Levels on Rumen Fermentation and Microbial Activity Under Different Incubation Temperatures. *Animals* **2024**, *14*, 3093. https://doi.org/10.3390/ani14213093

Academic Editors: Lizhi Wang and Alejandro Plascencia

Received: 11 September 2024
Revised: 25 October 2024
Accepted: 25 October 2024
Published: 26 October 2024

1. Introduction

External environmental temperature is closely related to the ruminal microbial ecosystem in cattle [1–3]. Cows reduce their feed intake and choose feeds with less fermentation heat to prevent an increase in body temperature under high-ambient-temperature conditions [2,4,5]. Ruminal and core temperature are also influenced by the external environment and type of diet [6,7]. This results in a decrease in ruminal fluid pH, leading to abnormal fermentation characteristics. Consequently, ruminal bacteria are reduced [1,2]; this process is closely correlated to decreasing amounts of microbial proteins. Microbial proteins make up an average of 59% of the total metabolic protein [8]. For microbial protein synthesis, the amount of digestible organic matter (OM) and the balance of energy and protein are important factors [8,9]. As previously mentioned, cows decrease their OM intake under high-ambient-temperature conditions, a mechanism that is not easy to prevent through nutritional manipulation. To accomplish an appropriate balance between energy and protein, it is necessary to consider not only the energy or protein content but also the rate of degradation. The Cornell Net Carbohydrate and Protein System is used for classifying proteins into fractions A, B1, B2, B3, and C [10]. In addition, energy values can be calculated according to the NRC [11] equation, and neutral detergent fiber (NDF), non-fiber carbohydrate (NFC), and crude fat are adjusted when the energy is increased. Since rumen fermentation heat promotes the intake of certain feed ingredients, during the summer, the consumption of concentrated feed becomes more crucial than that of forage-based feed [12]. The intake of concentrated feed increases the fermentation rate, leading to a drop in pH, especially during summer, and it results in a lower acetate-to-propionate ratio. During the summer season, both intake and rumination times are reduced. Additionally, the reduction in chewing stimulation decreases the saliva immersion ratio.

Numerous studies demonstrate the impacts of varying energy and protein levels on microbial protein synthesis under optimal fermentation temperatures [13,14]. However, these previous studies have limitations because using animals to validate the impacts of external environmental temperature on rumen fermentation properties overlooks the decrease in DMI and the physiological adaptations these animals make to maintain their balance. Thus, in vitro models for evaluating different incubation temperatures should incorporate varied nutrient levels. To the best of our knowledge, there is no existing study investigating the relationship between the quantities of microbial proteins and the nutrient compositions under different incubation temperatures. To address this knowledge gap, we hypothesize that the compositions of protein and energy levels in a diet influence rumen fermentation properties, and optimizing these diet levels can improve rumen fermentation under different incubation temperatures in ruminants.

We conducted two distinct trials to investigate these hypotheses. The objective of trial 1 was to determine the appropriate protein level while maintaining consistent levels of energy components such as NFC and NDF, subsequently increasing the protein content based on its degradation rate. In trial 2, the objectives were to identify the suitable energy level while maintaining the protein composition constant and to increase the energy levels by raising the NFC and reducing the NDF. Additionally, we aimed to confirm the effects of different incubation temperatures on ruminant microorganisms independent of the physiological response of the host.

2. Materials and Methods

All the procedures involving animals were approved by the Institutional Animal Care and Use Committee (IACUC) of Konkuk University (Approval No: KU21011) (Seoul, Republic of Korea).

2.1. Experiment Design, Collecting Ruminal Inoculum and In Vitro Incubation Procedures

Two rumen-fistulated Holstein heifers (trial 1 BW: 652.3 kg, SD: 25.2; trial 2: BW: 683.3 kg, SD: 30.2) were utilized to acquire rumen fluid samples. These cows were fed a 6:4 ratio of forage to concentrate for 2 weeks prior to sampling, according to the maintenance requirements outlined by the NRC [11], with water and minerals available ad libitum. The experimental model employed a 2 × 5 factorial arrangement using incubation temperatures of 39 and 41 °C with 5 protein levels (12.0, 13.5, 15, 16.5, and 18%) or 5 energy levels (2.4, 2.5, 2.6, 2.7, and 2.8 Mcal/kg) for each trial. Rumen fluid was collected from the ventral and dorsal sacs of two heifers 2 h prior to morning feeding during a thermoneutral period. The temperature–humidity index remained consistently below 70, which is recognized as the upper critical threshold for cattle [15]. The collected rumen fluid was then promptly transported to a laboratory within 20 min of collecting using preheated thermos bottles. Subsequently, the rumen fluid was filtered through a nylon filter with a pore size of 50 μm. Finally, the rumen fluid collected from the two heifers was mixed in equal proportions.

The mixed rumen fluid was prepared by combining Menke's buffer and rumen fluid in a ratio of 1:3 (*v*/*v*) [16]. The inoculum was heated at 39 °C and continuously purged with CO_2. Menke's buffer consisted of 9.52 mM Na_2PO_4, 10.8 mM KH_2PO_4, 0.58 mM $MgSO_4 \cdot 7H_2O$, 1.08 μmol $CaCl_2 \cdot 2H_2O$, 0.61 μmol $MnCl_2 \cdot 4H_2O$, 0.05 μmol $CoCl_2 \cdot 6H_2O$, 0.05 μmol $FeCl_2 \cdot 4H_2O$, 8.66 mM NH_4HCO_3, 71.3 mM $NaHCO_3$, 0.05 μmol Resazurin sodium salt, 147 μmol NaOH, and 2.97 μmol $Na_2S \cdot 9H_2O$.

The substrates were ground through a 2-mm screen and used as a substrate for incubation. Then, 30 mL of buffered rumen fluid was added to 50 mL serum bottles containing F57 filter bags (pore size 25 μm, Ankom Technology Corp., Macedon, NY, USA); each bag contained 0.3 g of substrate [17]. Ultra-high purity (99.99%) CO_2 was flushed into the headspaces of the bottles, which were incubated for 3, 6, 12, 24, and 48 h during each treatment. The experiments were conducted three times, with two replications per experiment. Thus, a total of 2 (39 and 41 °C) × 5 (concentrations) × 5 (incubation time) × 3 (incubation replication) × 2 (bottle replication) × 2 (trials; protein or energy) + 60 bottles (blank bottle per each treatment) = 660 bottles were used.

Table 1 presents the results of the chemical composition based on the feed ingredient used in trial 1. In trial 1, five substrates were formulated with varying protein concentrations of 12.0, 13.5, 15, 16.5, and 18% while maintaining a constant metabolizable energy (ME) level of 2.7 (Table 2). The chemical composition of the feed ingredients used in trial 2 is presented in Table 3. In trial 2, five substrates were formulated with varying metabolizable energy concentrations of 2.4, 2.5, 2.6, 2.7, and 2.8 Mcal/kg while maintaining a constant protein content of 17.5% dry matter (DM) and considering the protein fraction (Table 4). Dietary protein and energy levels were selected based on recommendations from the eighth beef cattle NRC (2016) guidelines and findings from previous studies [5,18].

Table 1. The chemical compositions of the feed ingredients used in trial 1.

Items (% of DM) [1]	Corn Grain, Ground, Dry	Wheat	Wheat Bran	Soybean Meal	Distillers Grains with Soluble, Dry	Rapeseed Meal	Sesame Meal	Corn Gluten Feed, Dry	Corn Gluten Meal, Dry	Rice Straw, Mature	Tall-Fescue Hay, Mature	Alfalfa, Hay
CP	7.84	12.46	16.48	48.48	29.10	37.80	42.50	19.40	66.40	3.73	5.67	15.10
A	0.30	1.37	1.79	3.01	1.93	5.75	3.64	8.07	2.04	0.00	0.00	1.07
B1	0.86	1.53	2.43	1.12	1.39	6.78	2.47	2.95	2.01	0.98	0.92	0.22
B2	5.55	6.47	7.73	40.76	16.91	17.65	21.01	5.85	53.73	0.63	2.92	10.82
B3	0.99	2.98	4.12	3.43	8.10	0.00	1.90	2.05	8.62	1.43	0.49	2.42
C	0.14	0.11	0.41	0.16	0.77	7.62	13.48	0.48	0.00	0.69	1.34	0.57
NDIP	1.13	3.09	4.52	3.59	8.87	7.43	15.38	2.53	8.62	2.12	1.83	2.99
ADIP	0.14	0.11	0.41	0.16	0.77	7.62	13.48	0.48	0.00	0.69	1.34	0.57
NDF	11.15	15.87	41.23	15.30	48.13	32.14	42.87	43.19	12.10	69.51	71.07	35.42
ADF	2.53	3.65	12.22	6.75	13.72	28.99	36.22	21.30	2.68	36.74	41.01	24.94
ADL	0.41	0.39	3.04	0.59	1.38	16.93	20.12	2.97	1.05	4.58	5.06	4.70
EE	3.21	0.84	3.25	0.77	7.68	0.61	13.40	0.99	1.49	1.03	0.62	1.33
NFC	77.68	72.16	38.55	32.30	18.43	27.52	8.51	33.25	24.69	16.26	14.27	41.94
Ash	1.25	1.77	5.02	6.75	5.51	9.38	8.13	5.72	3.95	11.6	10.2	9.20
TDN [2]	88.23	87.33	67.56	80.27	80.27	55.99	67.35	68.10	85.69	51.27	51.60	63.86
tdNFC	76.13	73.55	37.78	31.66	18.06	26.97	8.34	32.59	24.20	15.93	13.98	41.11
tdCP	7.79	12.41	16.31	48.41	28.82	34.73	37.08	19.18	66.39	2.99	4.27	14.43
tdFA	2.21	0.00	2.25	0.00	6.68	0.00	12.40	0.00	0.49	0.03	0.00	0.33
tdNDF	6.35	8.38	20.46	7.20	25.36	1.30	1.04	23.33	1.00	39.27	40.35	14.59
DE [3]	3.81	3.84	3.27	4.04	3.77	2.83	3.34	3.12	4.52	2.19	2.22	2.88
ME [4]	3.40	3.42	2.85	3.63	3.37	2.41	2.97	2.70	4.12	1.76	1.79	2.46

[1] CP: Crude protein; A: protein A fraction represents the soluble nonprotein N. Degradable rate in rumen is direct; B1: protein B1 fraction is the soluble true protein. Degradable rate in rumen is 120–400%/h; B2: protein B2 fraction represents protein with intermediate rates of degradation. Degradable rate in rumen is 3–16%/h; B3: protein B3 fraction is the CP insoluble in neutral detergent solution but soluble in acid detergent solution. Degradable rate in rumen is 0.06–0.55%/h; C: protein C fraction is the unavailable nitrogen [10,19]; NDIP: CP insoluble in neutral detergent solution; ADIP: CP insoluble in acid detergent solution; NDF: neutral detergent fiber; ADF: acid detergent fiber; ADL: acid detergent lignin; EE: ether extract; NFC: non-fiber carbohydrates. [2] True digestibility nutrition (%) = true digestibility non-fiber carbohydrate (tdNFC) + true digestibility crude protein (tdCP) + true digestibility fatty acid (tdFA) × 2.25 + true digestibility neutral detergent fiber (tdNDF) − 7 [11]. [3] Digestible energy (DE) = (tdNFC/100) × 4.2 + (tdCP/100) × 5.6 + (tdFA/100) × 9.4 − 0.3) [11]. [4] If EE was lower than 3%, metabolizable energy (ME) was (1.01 × DE) − 0.45, if EE was higher than 3%, ME was ((1.01 × DE) − 0.45) + 0.0046 × (EE − 3) [11].

Table 2. The feed ingredients and chemical compositions used in trial 1.

Ingredient (%)	CP 12.0%	CP 13.5%	CP 15.0%	CP 16.5%	CP 18.0%
Corn grain, ground, dry	33.81	32.10	22.33		
Wheat			7.29	29.45	26.31
Soybean Meal, solvent, 44% CP	3.86				
Distiller grains with soluble, dried	7.22	2.78	4.16		
Rapeseed meal			1.11	2.37	4.66
Sesame meal		1.59	1.48	0.07	
Corn gluten feed, dry	1.10	1.98		0.01	0.17
Corn gluten meal, dry		6.53	7.55	8.42	10.40
Rice straw, mature	10.65	31.03	29.90	9.19	14.77
Tall-fescue hay, mature	18.93			17.51	11.93
Alfalfa, hay	24.43	24.02	26.20	33.00	31.75
Chemical Composition, % of DM [1]	**CP 12.0%**	**CP 13.5%**	**CP 15.0%**	**CP 16.5%**	**CP 18.0%**
CP	12.00	13.50	15.00	16.50	18.00
A	0.72	0.85	0.98	1.21	1.37
B1	0.80	0.90	0.98	1.10	1.26
B2	8.00	9.00	10.00	11.00	12.00
B3	1.91	2.20	2.51	2.62	2.72
C	0.58	0.64	0.71	0.71	0.83
NDIP	2.49	2.84	3.22	3.32	3.54
ADIP	0.58	0.64	0.71	0.71	0.83
NDF	37.80	37.30	37.60	37.00	37.00
ADF	20.10	19.80	20.00	20.80	20.90
ADL	2.91	3.25	3.44	3.53	3.85
EE	2.23	2.21	2.07	1.04	1.05
NFC	43.90	43.00	41.50	41.80	40.30
Ash	6.55	6.86	7.03	6.97	7.17
Ca	0.48	0.52	0.55	0.60	0.60
P	0.28	0.26	0.26	0.23	0.23
TDN, % [2]	70.30	69.80	69.50	69.10	68.60
tdNFC	43.00	42.10	40.90	41.80	40.30
tdCP	11.40	13.00	14.50	15.90	17.40
tdFA	1.31	1.21	1.09	0.16	0.16

Table 2. *Cont.*

Chemical Composition, % of DM [1]	CP 12.0%	CP 13.5%	CP 15.0%	CP 16.5%	CP 18.0%
tdNDF	19.90	19.00	18.80	18.10	17.70
DE (Mcal/kg) [3]	3.11	3.11	3.12	3.12	3.12
ME (Mcal/kg) [4]	2.69	2.69	2.70	2.70	2.70

[1] CP: Crude protein; A: protein A fraction represents the soluble nonprotein N. Degradable rate in rumen is direct; B1: protein B1 fraction is the soluble true protein. Degradable rate in rumen is 120–400%/h; B2: protein B2 fraction represents protein with intermediate rates of degradation. Degradable rate in rumen is 3–16%/h; B3: protein B3 fraction is the CP insoluble in neutral detergent solution but soluble in acid detergent solution. Degradable rate in rumen is 0.06–0.55%/h; C: protein C fraction is the unavailable nitrogen [10,19]; NDIP: CP insoluble in neutral detergent solution; ADIP: CP insoluble in acid detergent solution; NDF: neutral detergent fiber; ADF: acid detergent fiber; ADL: acid detergent lignin; EE: ether extract; NFC: non-fiber carbohydrates. [2] True digestibility nutrition (%) = true digestibility non-fiber carbohydrate (tdNFC) + true digestibility crude protein (tdCP) + true digestibility fatty acid (tdFA) × 2.25 + true digestibility neutral detergent fiber (tdNDF) − 7 [11]. [3] Digestible energy (DE) = (tdNFC/100) × 4.2 + (tdCP/100) × 5.6 + (tdFA/100) × 9.4 − 0.3) [11]. [4] If EE was lower than 3%, metabolizable energy (ME) was (1.01 × DE) − 0.45, if EE was higher than 3%, ME was ((1.01 × DE) − 0.45) + 0.0046 × (EE − 3) [11].

Table 3. The chemical compositions of the feed ingredients used in trial 2.

Items [1] (% of DM)	Corn Grain, Ground, Dry	Wheat	Soybean Meal	Distillers Grains with Soluble, Dry	Rapeseed Meal	Corn Gluten Feed, Dry	Corn Gluten Meal, Dry	Tall-Fescue Hay, Mature	Alfalfa, Hay
CP	7.34	11.7	44.8	29.0	37.7	22.3	66.0	7.97	13.3
A	0.28	1.28	2.78	1.92	5.74	3.25	2.03	0.00	0.95
B1	0.81	1.43	1.03	1.39	6.77	8.88	2.00	1.29	0.19
B2	5.20	6.05	37.65	16.84	17.58	3.25	53.37	4.10	9.54
B3	0.93	2.79	3.16	8.07	0.00	6.42	8.57	0.69	2.14
C	0.13	0.11	0.15	0.76	7.62	0.53	0.00	1.89	0.50
NDIP	1.06	2.89	3.31	8.84	7.62	6.95	8.57	2.58	2.64
ADIP	0.13	0.11	0.15	0.76	7.62	0.53	0.00	1.89	0.50
NDF	11.09	15.87	12.61	41.01	28.58	40.10	21.62	65.27	40.53
ADF	3.03	3.65	7.60	16.1	19.8	10.4	2.68	41.4	25.2
ADL	0.41	0.39	0.59	1.38	16.9	2.97	1.05	4.70	5.06
EE	3.11	1.39	1.47	7.72	1.08	1.73	2.20	0.22	1.60
NFC	78.60	72.57	38.31	26.44	32.36	36.36	16.79	19.66	40.95
Ash	0.92	1.40	6.15	4.68	7.88	6.44	2.00	9.46	6.23
TDN [2]	80.63	81.30	78.12	87.12	65.50	77.02	83.49	48.62	63.25
tdNFC	77.02	71.12	37.55	25.91	31.71	35.63	16.45	19.26	40.13
tdCP	7.29	11.62	44.71	28.68	34.67	22.11	65.96	6.00	12.74

Table 3. *Cont.*

Items [1] (% of DM)	Corn Grain, Ground, Dry	Wheat	Soybean Meal	Distillers Grains with Soluble, Dry	Rapeseed Meal	Corn Gluten Feed, Dry	Corn Gluten Meal, Dry	Tall-Fescue Hay, Mature	Alfalfa, Hay
tdFA	2.11	0.39	0.47	6.72	0.08	0.73	1.20	0.00	0.60
tdNDF	6.36	8.52	5.49	20.3	0.40	18.11	7.32	35.77	18.20
DE(Mcal/kg) [3]	3.81	3.73	4.06	3.88	3.00	3.26	4.50	2.35	2.92
ME(Mcal/kg) [4]	3.14	3.17	3.03	3.43	2.47	2.98	3.27	1.66	2.31

[1] CP: Crude protein; A: protein A fraction represents the soluble nonprotein N. Degradable rate in rumen is direct; B1: protein B1 fraction is the soluble true protein. Degradable rate in rumen is 120–400%/h; B2: protein B2 fraction represents protein with intermediate rates of degradation. Degradable rate in rumen is 3–16%/h; B3: protein B3 fraction is the CP insoluble in neutral detergent solution but soluble in acid detergent solution. Degradable rate in rumen is 0.06–0.55%/h; C: protein C fraction is the unavailable nitrogen [10,19]; NDIP: CP insoluble in neutral detergent solution; ADIP: CP insoluble in acid detergent solution; NDF: neutral detergent fiber; ADF: acid detergent fiber; ADL: acid detergent lignin; EE: ether extract; NFC: non-fiber carbohydrates. [2] True digestibility nutrition (%) = true digestibility non-fiber carbohydrate (tdNFC) + true digestibility crude protein (tdCP) + true digestibility fatty acid (tdFA) × 2.25 + true digestibility neutral detergent fiber (tdNDF) − 7 [11]. [3] Digestible energy (DE) = (tdNFC/100) × 4.2 + (tdCP/100) × 5.6 + (tdFA/100) × 9.4 − 0.3) [11]. [4] If EE was lower than 3%, metabolizable energy (ME) was (1.01 × DE) − 0.45, if EE was higher than 3%, ME was ((1.01 × DE) − 0.45) + 0.0046 × (EE − 3) [11].

Table 4. The feed ingredients and chemical compositions used in trial 2.

Ingredient (%)	ME 2.4	ME 2.5	ME 2.6	ME 2.7	ME 2.8
Corn grain, ground, dry	10.00	15.18	20.18	25.74	31.29
Wheat	10.00	13.00	16.00	19.25	22.50
Soybean Meal, solvent, 44% CP	8.24	8.80	9.60	10.30	10.99
Distiller grains with soluble, dried		0.30	0.60	0.94	1.28
Rapeseed meal	5.42	5.50	6.00	6.79	7.59
Corn gluten feed, dry		0.70	1.43	0.71	
Corn gluten meal, dry	5.00	4.77	4.14	3.72	3.31
Tall-fescue hay, mature	30.00	26.75	23.05	19.27	15.48
Alfalfa, hay	31.34	25.00	19.00	13.28	7.56
Chemical Composition, % of DM [1]	**ME 2.4**	**ME 2.5**	**ME 2.6**	**ME 2.7**	**ME 2.8**
CP	17.50	17.50	17.50	17.50	17.50
A	1.10	1.13	1.20	1.23	1.28
B1	1.22	1.33	1.45	1.48	1.50
B2	12.07	11.96	11.80	11.80	11.80

Table 4. *Cont.*

Chemical Composition, % of DM [1]	ME 2.4	ME 2.5	ME 2.6	ME 2.7	ME 2.8
B3	1.94	1.98	2.00	1.96	1.92
C	1.17	1.10	1.06	1.03	1.00
NDIP	3.11	3.08	3.05	2.99	2.92
ADIP	1.17	1.10	1.06	1.03	1.00
NDF	38.65	35.45	32.16	28.60	25.01
ADF	22.81	20.30	17.79	15.24	12.70
ADL	4.09	3.69	3.36	3.04	2.73
EE	1.31	1.44	1.57	1.71	1.86
NFC	39.60	43.15	46.78	50.70	54.62
Ash	6.06	5.54	5.04	4.48	3.94
Ca	0.21	0.19	0.18	0.17	0.16
P	0.66	0.68	0.72	0.78	0.84
TDN [2]	68.33	70.61	72.84	75.22	77.56
tdNFC	38.81	42.28	45.84	49.68	53.53
tdCP	16.54	16.64	16.72	16.80	16.88
tdFA	0.54	0.65	0.75	0.87	0.98
tdNDF	18.76	17.23	15.58	13.79	11.95
DE (Mcal/kg) [3]	3.10	3.19	3.29	3.39	3.49
ME (Mcal/kg) [4]	2.40	2.50	2.60	2.70	2.80

[1] CP: Crude protein; A: protein A fraction represents the soluble nonprotein N. Degradable rate in rumen is direct; B1: protein B1 fraction is the soluble true protein. Degradable rate in rumen is 120–400%/h; B2: protein B2 fraction represents protein with intermediate rates of degradation. Degradable rate in rumen is 3–16%/h; B3: protein B3 fraction is the CP insoluble in neutral detergent solution but soluble in acid detergent solution. Degradable rate in rumen is 0.06–0.55%/h; C: protein C fraction is the unavailable nitrogen [10,19]; NDIP: CP insoluble in neutral detergent solution; ADIP: CP insoluble in acid detergent solution; NDF: neutral detergent fiber; ADF: acid detergent fiber; ADL: acid detergent lignin; EE: ether extract; NFC: non-fiber carbohydrates. [2] True digestibility nutrition (%) = true digestibility non-fiber carbohydrate (tdNFC) + true digestibility crude protein (tdCP) + true digestibility fatty acid (tdFA) × 2.25 + true digestibility neutral detergent fiber (tdNDF) − 7 [11]. [3] Digestible energy (DE) = (tdNFC/100) × 4.2 + (tdCP/100) × 5.6 + (tdFA/100) × 9.4 − 0.3) [11]. [4] If EE was lower than 3%, metabolizable energy (ME) was (1.01 × DE) − 0.45, if EE was higher than 3%, ME was ((1.01 × DE) − 0.45) + 0.0046 × (EE − 3) [11].

2.2. Chemical Analysis

For the in vitro batch culture, twelve feed ingredients were used. These ingredients were mixed to adjust the protein or energy level of the experimental substrate (Tables 1 and 3). Each feed ingredient was dried completely for three days at 55 °C, and the crude protein (CP) was measured using the Kjeldahl automatic distiller (Kjeltec 8400 Analyzer, Hillerød, Denmark), as described by AOAC, with N × 6.25 (method 976.05) [20]. To differentiate proteins based on their digestion rate, nonprotein nitrogen (fraction A), true soluble protein (fraction B1), insoluble protein-neutral soluble detergent protein (fraction B2), and neutral insoluble detergent protein that was soluble in acid detergent (fraction B3) and insoluble in acid detergent (fraction C) were measured using the method outlined by Licitra, Hernandez and Van Soest [10]. Ether extract (EE) was measured using the Ankom XT15 extractor (Ankom Technology Corp., Macedon, NY, USA) with method 920.39, while ash (method 942.05; 550 °C) was measured using a furnace (FHX-63, Daihan Scientific Co., Wonju, Republic of Korea), as described by AOAC [20]. Neutral detergent fiber (NDF) and acid detergent fiber (ADF) were analyzed using the method described by Van Soest et al. [21] with the Ankom 200 fiber analyzer (Ankom Technology Corp., NY, USA); heat-stable α-amylase (Ankom Technology Corp., NY, USA) was used to analyze the NDF. Acid detergent lignin (ADL) was measured using the method outlined by Van Soest and Robertson [22]. Non-fiber carbohydrate (NFC) was calculated using the following equation with the chemical analysis parameters: NFC = 100 − ash − EE − (CP − NDIP) − NDF. Additionally, the equations provided by the NRC [11] were used to evaluate the energy values of the feed ingredients, expressed as total digestible nutrients (TDN) and metabolizable energy (ME).

The experimental feed of trial 1 increased the protein content while keeping the energy constant, reducing the non-fiber carbohydrates from 43.9 to 40.3%. However, as the energy value was set higher in feeds with increased protein contents, the ether extract was reduced from 2.23 to 1.04%. Consequently, the five-step protein concentration feeds used in this experiment had an ME of 2.7 Mcal/kg (Table 2).

The experimental feeds of trial 2 were designed to maintain a constant protein fraction ratio with a fixed protein content of 17.5%. As the energy levels increased, the NFC gradually rose from 39.60 to 54.62%. Conversely, with higher ME levels, the NDF progressively decreased from 38.65 to 25.0%. The EE experienced a slight increase from 1.31 to 1.86% since excessive concentrations could affect the fermentation in the rumen. As a result, the five-step ME concentration feeds used in this experiment had a CP of 17.5% (Table 4).

2.3. Analyses of Fermentation Properties

Following the incubation of the vials for 3, 6, 12, 24, and 48 h, the pH values were measured using a digital pH meter (S20 SevenEasy pH; Mettler Toledo Co., Ltd., Greifensee, Switzerland). Residual rumen fluid samples were then immediately stored at −20 °C for the subsequent analysis of volatile fatty acids (VFAs) and NH_3-N. Upon thawing, 10 mL of the sample was mixed with 1 mL of $HgCl_2$ 2% (*w*/*v*) and briefly centrifuged at 2000× *g* for 10 min at 4 °C to remove the feed particles. The resulting supernatants were used for VFA and NH_3-N analysis. To prepare the samples for the VFA analysis, 1.4 mL of the supernatants was first mixed with 0.28 mL of 25% (*w*/*v*) meta-phosphoric acid and then centrifuged again at 20,000× *g* for 20 min at 4 °C. Next, 1 mL of the supernatant was mixed with 50 μL of 2% (*w*/*v*) pivalic acid as an internal standard. The VFA profile was measured using a gas chromatograph (HP 6890 series GC system; Agilent Technologies Inc., Santa Clara, CA, USA) equipped with a flame ionization detector and a capillary column (DB-FFAP; Agilent Technologies Inc., Santa Clara, CA, USA). The inlet, oven, and detector temperatures were set to 220, 100, and 250 °C, respectively. Each sample for VFA analysis was duplicated three times.

To determine the NH_3-N concentration, the centrifuged samples were subjected to catalyzed indophenol reactions [23] and analyzed using spectrophotometry (Synergy2; Biotek Instruments, Inc., Winooski, VT, USA). In brief, NH_3-N standards and samples

(2 μL) were mixed with 147 μL of phenol color reagent (531 mM phenol, 0.95 mM sodium nitroferricyanide) and 125 μL of alkali-hypochlorite (625 mM sodium hydroxide, 28.2 mM sodium hypochlorite) in a 96-well cell plate. The reaction was carried out in a 55 °C dry oven for 10 min, followed by the measurement of absorbance at 630 nm using spectrophotometry. Each sample was triplicated and analyzed three times (3 × 3).

2.4. Protein Analysis of Liquid-Associated Bacteria (LAB)

The residual rumen fluid intended for LAB analysis was stored at −4 °C and examined using a bicinchoninic acid (BCA) protein assay kit (Thermo Fisher, Waltham, MA, USA). Prior to measuring the microbial protein content, the pretreatment method outlined by Makkar et al. [24] was followed. The stored rumen fluid was thawed and vortexed at 3000 rpm for 10 s to gently detach the microbes from the substrate or the samples. Subsequently, it was centrifuged at 400× *g* for 5 min to eliminate feed and protozoa. Next, 1 mL of supernatant was transferred and centrifuged at 25,000× *g* for 20 min. The supernatant was removed, leaving only the pellet at the bottom of the tubes. Then, 1 mL of deionized water was added, and the mixture was vortexed at 3000 rpm for 10 s, followed by another centrifugation at 25,000× *g* for 20 min. After discarding the supernatant, 1 mL of 0.25 N NaOH was added and vortexed at 3000 rpm for 10 s. The mixture was digested using a heating machine at 100 °C for 10 min before being centrifuged at 25,000× *g* for 30 min. Lastly, 25 μL of the supernatant was transferred onto a 96-well plate. The total protein content in the supernatant was determined using the bicinchoninic acid (BCA) assay (Thermo Fisher Scientific, Waltham, MA, USA) with spectrophotometry (Synergy 2, Biotek Instruments Inc., Santa Clara, CA, USA), measuring the absorbance at 562 nm.

2.5. Statistical Analysis

The fermentation properties and LAB data were analyzed using the MIXED procedure in SAS version 9.4 (SAS Institute Inc., Cary, NC, USA). The model used was as follows:

$$Y_{ijk} = \mu + \alpha_i + \beta_j + \alpha\beta_{ij} + \gamma_k + \varepsilon_{ijk}, \quad (1)$$

where Y_{ijk} represents the observation of incubation run *k* for the given treatments *i* and *j*, μ is the overall mean, α_i is the fixed effect of incubation temperature *i* (39 and 41 °C), β_j is the fixed effect of the protein or energy level of substrate *j* (CP 12.0, 13.5, 15.0, 16.5, and 18.0% of DM or ME 2.4, 2.5, 2.6, 2.7, and 2.8 Mcal/kg), $\alpha\beta_{ij}$ is the interaction between the incubation temperature and the protein or energy level of the substrate, γ_k is the random effect of incubation run *k* nested in treatments *i* and *j*, and ε_{ij} is the residual effect. The data were presented as least squares means with the standard error of the mean. When a treatment indicated a significant effect, the least squares means were compared using Tukey's post hoc comparison. Differences were considered statistically significant if the *p*-value was less than 0.05. If the *p*-value was greater than 0.05, it indicated that there was no statistically significant difference.

3. Results

3.1. In Vitro Fermentation Properties

In trial 1, the results indicated that there were significant changes in the pH and NH_3-N concentrations as well as in the TVFA compositions with increasing fermentation temperatures. The pH of the inoculum was significantly higher at 41 °C after 3 h of fermentation ($p < 0.001$, Table 5), but there was no significant difference in the pH after that time. On the other hand, the NH_3-N concentration was found to be significantly higher in the 41 °C fermentation environment compared to the 39 °C fermentation environment after 3, 12, 24, and 48 h. In terms of VFA composition, it was found that the concentration of *iso*-butyrate and *iso*-valerate gradually increased at a fermentation temperature of 41 °C. The concentration of *iso*-butyrate increased to about 6, 7, 11, and 13% after 6, 12, 24, and 48 h, respectively, while the concentration of *iso*-valerate gradually increased to about 6, 14, and 16% after 12, 24, and 48 h compared to the 39 °C fermentation environment. It was confirmed that, as the incubation temperature increased to 41 °C, the TVFAs also increased after 3, 6, and 12 h ($p < 0.05$, Table 5). In particular, the TVFA levels rose sharply by approximately 30% within the first 3 h. However, after 48 h incubated at 41 °C, the TVFA levels decreased by about 19.14% from 104.6 to 84.6 mM ($p = 0.01$, Table 5). Additionally, analyzing the change in VFA composition under 41 °C incubation temperatures for 48 h revealed a slight reduction in the acetate concentration, while the butyrate, *iso*-butyric acid, valerate, and *iso*-valerate concentrations all increased ($p < 0.05$).

Table 5. The results of in vitro ruminal fermentation after 3, 6, 12, 24, and 48 h of incubations according to five protein levels (CP 12.0, 13.5, 15.0, 16.5, 18.0% of DM basis) and two incubation temperatures (39 and 41 °C).

Items [1]	**39 °C**					**41 °C**					**SEM**	***p*-Value** [2]		
CP %	**12.0**	**13.5**	**15.0**	**16.5**	**18.0**	**12.0**	**13.5**	**15.0**	**16.5**	**18.0**		**CP**	**Tem**	**CP × Tem**
Ruminal fermentation after 3 h														
pH	6.73	6.72	6.72	6.73	6.73	6.75	6.74	6.74	6.74	6.74	0.04	0.445	<0.001	0.937
NH_3-N, mg/dL	43.17	41.96	43.35	49.74	47.82	47.57	44.57	45.68	49.07	53.45	2.46	0.001	0.011	0.377
TVFA, mM	30.74	30.67	31.73	33.54	32.76	38.29	38.43	41.68	43.98	45.10	5.46	0.777	0.004	0.983
C2, %	68.56	68.50	69.12	68.03	68.00	69.35	69.67	69.25	69.33	69.22	0.67	0.812	0.019	0.825
C3, %	18.29	18.29	17.16	18.93	18.85	18.06	17.81	18.11	18.25	18.42	0.71	0.168	0.537	0.377
iC4, %	0.88	0.82	0.90	0.95	0.92	0.88	0.84	0.84	0.87	0.88	0.16	0.875	0.476	0.959
C4, %	10.18	10.37	10.66	9.86	9.99	9.70	9.80	9.89	9.45	9.39	1.17	0.169	0.004	0.972
iC5, %	0.80	0.79	0.82	0.86	0.86	0.74	0.69	0.69	0.73	0.75	0.09	0.780	0.009	0.967
C5, %	1.29	1.24	1.34	1.37	1.38	1.28	1.18	1.23	1.36	1.35	0.09	0.281	0.412	0.965
BCFA, %	1.68	1.60	1.72	1.81	1.79	1.61	1.53	1.52	1.60	1.62	0.25	0.832	0.108	0.974
A:P	3.76	3.76	5.79	3.60	3.62	3.85	3.92	3.83	3.80	3.77	0.65	0.401	0.517	0.406
LAB, μg/mL	241.4	231.3	343.6	253.2	231.9	246.0	210.3	229.0	238.6	241.8	39.09	0.768	0.590	0.927
Ruminal fermentation after 6 h														
pH	6.69	6.69	6.69	6.68	6.68	6.69	6.70	6.69	6.69	6.68	0.05	0.498	0.418	0.847
NH_3-N, mg/dL	42.08	41.04	42.28	48.19	49.64	39.06	37.52	40.93	49.07	50.63	2.93	<0.001	0.386	0.748
TVFA, mM	38.32	37.83	38.08	41.41	42.56	41.06	40.01	42.48	42.88	44.33	5.73	0.036	0.014	0.864
C2, %	68.22	68.19	67.60	67.65	67.94	67.73	68.01	67.18	67.74	67.36	0.64	0.184	0.117	0.809
C3, %	18.50	18.34	18.52	18.59	18.47	18.77	18.43	18.69	18.31	18.62	0.63	0.914	0.679	0.914
iC4, %	0.87	0.84	0.85	0.88	0.88	0.89	0.87	0.90	0.91	0.93	0.10	0.249	0.025	0.919
C4, %	10.37	10.71	10.98	10.43	10.38	10.59	10.78	11.16	10.78	10.65	1.12	0.164	0.181	0.985
iC5, %	0.75	0.71	0.73	0.79	0.76	0.75	0.72	0.76	0.77	0.81	0.05	0.016	0.291	0.476
C5, %	1.29	1.21	1.32	1.65	1.58	1.27	1.19	1.32	1.50	1.62	0.06	<0.001	0.296	0.310
BCFA, %	1.61	1.55	1.58	1.67	1.64	1.64	1.59	1.66	1.68	1.74	0.15	0.065	0.078	0.745
A:P	3.69	3.72	3.65	3.65	3.69	3.61	3.69	3.60	3.71	3.64	0.12	0.864	0.544	0.895
LAB, μg/mL	266.8	251.6	250.7	236.6	274.5	244.0	232.7	245.0	235.1	247.2	18.93	0.474	<0.001	0.862
Ruminal fermentation after 12 h														
pH	6.55	6.56	6.58	6.60	6.62	6.55	6.59	6.60	6.61	6.62	0.06	0.002	0.334	0.948
NH_3-N, mg/dL	36.54	36.72	41.37	52.93	54.24	40.93	43.75	50.62	56.88	62.69	4.25	<0.001	<0.001	0.632
TVFA, mM	53.39	51.91	54.57	54.12	54.11	57.94	54.93	55.24	57.06	57.31	6.31	0.450	0.004	0.738
C2, %	63.96	64.02	65.06	66.53	66.45	64.10	64.20	64.73	66.20	66.54	0.48	<0.001	0.865	0.942
C3, %	21.73	21.08	20.17	18.84	18.79	21.55	20.82	20.09	18.56	18.48	1.02	<0.001	0.455	0.999
iC4, %	0.86	0.87	0.91	0.96	0.93	0.92	0.94	0.97	1.01	1.02	0.05	<0.001	<0.001	0.777
C4, %	11.52	12.12	11.85	11.31	11.45	11.40	12.05	12.08	11.92	11.54	0.98	0.380	0.551	0.871
iC5, %	0.75	0.76	0.78	0.84	0.85	0.81	0.81	0.84	0.87	0.89	0.02	<0.001	<0.001	0.739
C5, %	1.18	1.14	1.24	1.53	1.54	1.21	1.19	1.28	1.44	1.53	0.03	<0.001	0.606	0.063
BCFA, %	1.61	1.64	1.69	1.80	1.78	1.73	1.75	1.81	1.88	1.91	0.07	<0.001	<0.001	0.930
A:P	2.95	3.05	3.24	3.57	3.57	2.98	3.09	3.23	3.61	3.64	0.20	<0.001	0.618	0.997
LAB, μg/mL	315.1	307.7	323.4	298.4	285.4	216.7	220.5	210.2	191.3	196.5	71.49	0.952	0.339	1.000

Table 5. *Cont.*

Items [1]	39 °C					41 °C					SEM	*p*-Value [2]		
CP %	12.0	13.5	15.0	16.5	18.0	12.0	13.5	15.0	16.5	18.0		CP	Tem	CP × Tem
Ruminal fermentation after 24 h														
pH	6.48	6.48	6.52	6.54	6.55	6.50	6.48	6.50	6.54	6.57	0.06	<0.001	0.606	0.472
NH_4, mg/dL	70.07	67.34	73.07	81.56	83.02	88.42	81.99	85.36	92.27	96.40	9.65	<0.001	<0.001	0.704
TVFA, mM	70.19	68.05	72.52	74.64	79.99	74.03	70.50	69.65	68.01	72.13	6.66	0.565	0.401	0.524
C2, %	62.85	63.20	64.92	66.58	67.05	63.64	62.78	63.62	65.11	65.32	0.55	<0.001	0.026	0.163
C3, %	21.22	20.62	19.23	17.45	17.35	20.53	20.63	19.69	17.82	17.91	0.80	<0.001	0.574	0.514
iC4, %	1.20	1.18	1.14	1.16	1.12	1.29	1.27	1.27	1.28	1.33	0.05	0.736	<0.001	0.355
C4, %	12.21	12.57	12.35	12.27	12.03	11.84	12.67	12.76	13.03	12.58	0.96	0.481	0.258	0.653
iC5, %	1.13	1.10	1.04	1.03	0.99	1.22	1.22	1.19	1.17	1.21	0.04	0.114	<0.001	0.418
C5, %	1.39	1.33	1.33	1.50	1.45	1.47	1.44	1.47	1.58	1.65	0.06	0.001	<0.001	0.564
BCFA, %	2.33	2.28	2.17	2.19	2.11	2.51	2.49	2.46	2.46	2.54	0.08	0.390	<0.001	0.379
A:P	2.97	3.07	3.38	3.84	3.88	3.10	3.05	3.24	3.69	3.67	0.16	<0.001	0.215	0.470
LAB, µg/mL	225.3	202.2	178.2	185.4	192.2	149.5	152.6	153.2	141.0	150.8	68.18	0.991	<0.001	1.000
Ruminal fermentation after 48 h														
pH	6.48	6.50	6.48	6.52	6.52	6.49	6.48	6.49	6.56	6.55	0.07	0.003	0.197	0.463
NH3-N, mg/dL	110.28	114.58	116.44	127.20	137.57	137.22	129.18	130.78	135.89	145.89	15.29	0.026	0.002	0.579
TVFA, mM	104.81	95.69	111.33	105.71	105.39	84.66	84.78	83.99	84.07	85.38	11.39	0.974	0.010	0.964
C2, %	64.23	63.77	65.20	66.69	66.67	62.81	63.05	63.19	64.84	64.79	0.79	0.004	0.002	0.870
C3, %	20.07	19.73	18.85	17.05	17.05	20.32	19.89	19.33	17.43	17.37	0.62	<0.001	0.194	0.993
iC4, %	1.41	1.44	1.33	1.33	1.36	1.60	1.54	1.54	1.52	1.54	0.08	0.770	0.002	0.957
C4, %	11.34	12.05	11.87	12.09	12.05	11.86	12.27	12.69	12.97	12.97	0.98	0.179	0.016	0.890
iC5, %	1.41	1.47	1.30	1.27	1.29	1.65	1.59	1.56	1.50	1.52	0.10	0.340	0.001	0.948
C5, %	1.54	1.54	1.45	1.57	1.58	1.76	1.65	1.69	1.73	1.81	0.10	0.546	0.001	0.912
BCFA, %	2.82	2.91	2.63	2.60	2.65	3.25	3.13	3.09	3.02	3.06	0.18	0.516	0.001	0.948
A:P	3.20	3.23	3.46	3.93	3.92	3.10	3.17	3.27	3.75	3.75	0.13	<0.001	0.043	0.961
LAB, µg/mL	169.8	143.0	147.5	145.1	138.0	166.2	162.1	159.8	147.1	136.8	24.12	0.634	<0.001	0.996

[1] C2, acetate (mol/100 mol); C3, propionate (mol/100 mol); iC4, *iso*-butyrate (mol/100 mol); C4, butyrate (mol/100 mol); iC5, *iso*-valerate (mol/100 mol); C5, valerate (mol/100 mol); BCFA, branched chain fatty acid (added iC4 and iC5, mol/100 mol); A:P, acetate to propionate ratio; TVFA, total volatile fatty acids; LAB, rumen liquid-associated bacteria. [2] Data were analyzed using the MIXED procedure of SAS as mixed model, including crude protein (CP), incubation temperature (Tem) and CP × Tem as fixed effects and incubation run as a random effect.

In trial 2, upon increasing the incubation temperature from 39 to 41 °C, significant differences in the NH_3-N, TVFA, and LAB were observed ($p < 0.01$, Table 6). The NH_3-N levels were 19.7, 10.6, 12.4, and 30.9% higher at fermentation times of 3, 6, 12, and 24 h ($p < 0.001$). Compared to an incubation temperature of 39 °C, the TVFA levels increased by 13.5, 10.6, and 11.1% after 3, 6, and 12 h when the temperature was 41 °C ($p < 0.01$), with no significant differences after 24 and 48 h of fermentation.

Table 6. The results of in vitro ruminal fermentation after 3, 6, 12, 24, and 48 h of incubations according to five metabolizable energy levels (2.4, 2.5, 2.6, 2.7, and 2.8 Mcal/kg of DM basis) and two incubation temperatures (39 and 41 °C).

Items [1]	39 °C					41 °C					SEM	p-Value [2]		
ME, Mcal/kg	2.4	2.5	2.6	2.7	2.8	2.4	2.5	2.6	2.7	2.8		Energy	Tem	Energy × Tem
						Ruminal fermentation after 3 h								
pH	6.90 [abc]	6.89 [bc]	6.89 [bc]	6.90 [bc]	6.88 [c]	6.91 [abc]	6.89 [bc]	6.92 [ab]	6.91 [abc]	6.93 [a]	0.01	0.140	<0.001	0.011
NH_3-N, mg/dL	62.63 [cd]	59.75 [d]	66.22 [bcd]	68.50 [bc]	66.71 [bcd]	80.18 [a]	77.87[a]	77.92 [a]	77.53 [a]	74.13 [ab]	1.61	0.133	<0.001	0.009
TVFA, mM	25.77	25.85	25.61	27.71	24.17	28.70	29.19	29.83	29.41	29.36	1.41	0.670	<0.001	0.678
C2, %	69.80	69.64	69.28	69.27	69.16	69.55	69.99	69.69	68.90	69.14	0.41	0.141	0.898	0.666
C3, %	17.15	16.73	17.14	16.91	16.80	17.17	17.03	16.98	16.94	16.78	0.45	0.353	0.793	0.827
iC4, %	0.88	0.92	0.90	0.92	0.93	0.92	0.88	0.90	1.00	0.96	0.02	0.032	0.153	0.190
C4, %	10.12	10.66	10.71	10.78	10.98	10.21	10.06	10.35	10.82	10.88	0.23	0.003	0.110	0.291
iC5, %	0.88	0.90	0.87	0.92	0.91	0.91	0.86	0.90	1.01	0.96	0.03	0.019	0.105	0.163
C5, %	1.17	1.15	1.09	1.20	1.21	1.24	1.18	1.19	1.34	1.28	0.03	0.001	<0.001	0.525
BCFA, %	1.76	1.82	1.77	1.84	1.84	1.82	1.74	1.80	2.01	1.92	0.05	0.023	0.122	0.173
A:P	4.07	4.16	4.05	4.10	4.13	4.05	4.12	4.11	4.07	4.12	0.13	0.688	0.855	0.921
LAB, μg/mL	48.73 [bc]	34.73 [c]	47.59 [bc]	49.21 [bc]	75.05 [a]	63.29 [ab]	61.48 [ab]	52.91 [bc]	60.42 [ab]	51.43 [bc]	3.86	0.009	0.011	<0.001
						Ruminal fermentation after 6 h								
pH	6.85	6.86	6.87	6.86	6.86	6.86	6.87	6.85	6.85	6.85	0.01	0.215	0.261	0.067
NH_3-N, mg/dL	70.90	67.49	66.85	65.11	63.44	72.23	77.06	72.05	78.23	75.61	2.49	0.696	<0.001	0.137
TVFA, mM	29.94 [d]	30.81 [cd]	30.61 [cd]	31.00 [cd]	29.69 [d]	33.22 [bc]	31.28 [cd]	35.95 [a]	35.59 [ab]	32.17 [cd]	0.60	<0.001	<0.001	0.002
C2, %	68.67 [ab]	68.59 [ab]	68.27 [abc]	68.14 [abc]	67.18 [bcd]	68.89 [a]	68.54 [ab]	66.91 [cd]	65.72 [d]	66.03 [d]	0.32	<0.001	<0.001	0.003
C3, %	17.42 [bc]	17.27 [c]	17.42 [bc]	17.74 [bc]	18.00 [abc]	17.43 [bc]	17.32 [c]	18.61 [ab]	19.09 [a]	18.35 [abc]	0.33	0.001	0.002	0.036
iC4, %	0.93	0.95	0.94	0.89	0.97	0.92	0.95	0.90	0.93	0.96	0.03	0.213	0.632	0.715
C4, %	10.81	10.91	11.12	11.07	11.53	10.56	10.95	11.36	12.02	12.42	0.37	<0.001	0.016	0.053
iC5, %	0.89	0.92	0.92	0.86	0.95	0.87	0.92	0.87	0.91	0.93	0.03	0.263	0.736	0.592
C5, %	1.27	1.36	1.33	1.29	1.38	1.33	1.32	1.35	1.34	1.31	0.04	0.693	0.795	0.506
BCFA, %	1.82	1.87	1.86	1.75	1.92	1.79	1.87	1.77	1.84	1.89	0.07	0.251	0.684	0.651
A:P	3.94 [a]	3.97 [a]	3.92 [a]	3.85 [ab]	3.74 [abc]	3.95 [a]	3.96 [a]	3.60 [bc]	3.44 [c]	3.60 [bc]	0.07	<0.001	<0.001	0.012
LAB, μg/mL	57.29 [b]	67.55 [b]	69.70 [b]	60.99 [b]	56.36 [b]	51.97 [b]	101.60 [a]	63.90 [b]	70.15 [b]	58.72 [b]	7.25	0.001	0.091	0.025
						Ruminal fermentation after 12 h								
pH	6.76 [ab]	6.75 [b]	6.76 [b]	6.73 [b]	6.72 [b]	6.81 [a]	6.76 [b]	6.76 [b]	6.72 [b]	6.74 [ab]	0.01	<0.001	0.059	0.044
NH_3-N, mg/dL	70.14 [bcd]	67.57 [bcd]	54.66 [e]	62.76 [cde]	59.95 [de]	85.83 [a]	73.11 [bc]	75.98 [ab]	66.51 [bcd]	69.38 [bcd]	2.33	<0.001	<0.001	0.006
TVFA, mM	42.23 [cde]	40.00 [e]	41.67 [de]	42.75 [cde]	47.05 [abcd]	51.37 [a]	45.03 [bcde]	48.14 [abc]	49.43 [ab]	46.19 [abcd]	1.19	0.010	<0.001	0.006
C2, %	65.10	65.27	63.86	62.22	60.56	65.44	65.65	62.19	60.05	60.18	0.60	<0.001	0.066	0.120
C3, %	20.39	19.86	20.58	21.35	23.03	20.29	19.76	21.04	22.55	22.31	0.49	<0.001	0.595	0.274
iC4, % *	0.86	0.92	0.90	0.87	0.81	0.91	0.88	0.86	0.82	0.87	0.02	0.013	0.909	0.016
C4, %	11.53 [c]	11.75 [c]	12.40 [bc]	13.35 [ab]	13.46 [ab]	11.17 [c]	11.58 [c]	13.71 [a]	14.48 [a]	14.43 [a]	0.33	<0.001	0.002	0.010

Table 6. *Cont.*

Items [1]	39 °C					41 °C					SEM	p-Value [2]		
ME, Mcal/kg	2.4	2.5	2.6	2.7	2.8	2.4	2.5	2.6	2.7	2.8		Energy	Tem	Energy × Tem
Ruminal fermentation after 12 h														
iC5, % *	0.82	0.87	0.87	0.86	0.83	0.87	0.83	0.83	0.80	0.86	0.02	0.830	0.261	0.037
C5, %	1.31	1.33	1.40	1.35	1.31	1.31	1.31	1.36	1.31	1.35	0.04	0.359	0.632	0.760
BCFA, % *	1.68	1.79	1.77	1.73	1.64	1.79	1.71	1.69	1.61	1.72	0.04	0.221	0.509	0.024
A:P	3.20	3.29	3.10	2.92	2.64	3.23	3.33	2.96	2.67	2.70	0.09	<0.001	0.323	0.319
LAB, µg/mL	76.08	82.96	88.02	62.35	78.51	76.26	72.78	110.54	86.13	75.90	10.23	0.099	0.284	0.314
Ruminal fermentation after 24 h														
pH	6.69	6.68	6.64	6.62	6.62	6.70	6.67	6.66	6.62	6.61	0.01	<0.001	0.615	0.364
NH_3-N, mg/dL	88.34	82.91	77.29	78.03	78.18	106.45	109.57	111.89	102.15	99.69	3.02	0.039	<0.001	0.112
TVFA, mM	54.13	52.25	61.60	60.99	57.46	56.20	62.77	67.31	63.67	68.29	2.18	0.002	<0.001	0.163
C2, %	62.77	62.62	59.97	58.35	58.20	63.81	63.37	59.33	58.50	57.64	1.00	<0.001	0.820	0.884
C3, %	20.76	20.61	22.81	23.46	23.43	20.18	20.70	23.13	23.28	22.82	0.70	<0.001	0.551	0.853
iC4, %	1.05	1.02	0.98	1.00	0.91	1.04	1.10	1.12	1.01	1.08	0.07	0.798	0.060	0.534
C4, %	12.83	13.29	13.80	14.66	15.06	12.43	12.30	13.80	14.67	15.72	0.68	0.002	0.735	0.808
iC5, %	1.01	0.98	0.99	0.99	0.95	1.02	1.05	1.09	1.03	1.15	0.07	0.974	0.060	0.632
C5, %	1.58	1.48	1.44	1.54	1.45	1.52	1.48	1.55	1.51	1.59	0.08	0.920	0.541	0.684
BCFA, %	2.06	2.00	1.97	1.99	1.85	2.05	2.15	2.20	2.04	2.23	0.13	0.980	0.059	0.594
A:P	3.02	3.04	2.65	2.49	2.49	3.16	3.06	2.58	2.52	2.54	0.11	<0.001	0.579	0.850
LAB, µg/mL	102.82	85.68	82.23	87.47	77.98	78.84	85.40	99.02	76.20	97.56	6.87	0.666	0.969	0.022
Ruminal fermentation after 48 h														
pH	6.66	6.66	6.63	6.58	6.59	6.69	6.67	6.61	6.59	6.59	0.01	<0.001	0.389	0.065
NH_3-N, mg/dL	143.75	128.69	139.64	134.25	133.72	140.73	141.87	142.03	150.58	146.15	7.79	0.813	0.062	0.554
TVFA, mM	70.58	76.68	73.21	78.01	79.03	69.91	75.17	77.92	79.04	76.94	1.83	0.002	0.801	0.375
C2, %	62.47	62.31	60.76	58.62	57.78	63.78	60.97	58.90	57.49	56.39	0.64	<0.001	0.043	0.150
C3, %	20.46	19.97	20.79	22.06	22.18	19.68	20.54	21.54	22.28	22.81	0.45	<0.001	0.337	0.450
iC4, %	1.54	1.53	1.43	1.42	1.45	1.44	1.43	1.36	1.44	1.35	0.04	0.085	0.014	0.534
C4, %	12.12	12.82	13.69	14.49	15.10	11.78	13.63	14.71	15.15	15.70	0.52	<0.001	0.111	0.731
iC5, %	1.54	1.54	1.46	1.49	1.56	1.47	1.49	1.42	1.54	1.51	0.05	0.336	0.298	0.712
C5, %	1.87	1.83	1.87	1.91	1.94	1.84	1.94	2.06	2.10	2.24	0.08	0.074	0.009	0.409
BCFA, %	3.08	3.07	2.89	2.92	3.00	2.91	2.92	2.79	2.99	2.85	0.08	0.336	0.074	0.612
A:P	3.06	3.13	2.93	2.66	2.61	3.24	2.97	2.74	2.58	2.48	0.08	<0.001	0.164	0.170
LAB, µg/mL	137.98 ^{bcd}	138.38 ^{bcd}	168.73 ^{ab}	152.25 ^{abc}	186.90 ^{a}	108.45 ^{d}	104.26 ^{d}	102.84 ^{d}	109.74 ^{cd}	98.19 ^{d}	8.69	0.125	<0.001	0.014

[1] C2, acetate (mol/100 mol); C3, propionate (mol/100 mol); iC4, iso-butyrate (mol/100 mol); C4, butyrate (mol/100 mol); iC5, iso-valerate (mol/100 mol); C5, valerate (mol/100 mol); BCFA, branched chain fatty acid (added iC4 and iC5, mol/100 mol); A:P, acetate to propionate ratio; TVFA, total volatile fatty acids; LAB, rumen liquid-associated bacteria. [2] Data were analyzed using the MIXED procedure of SAS as mixed model, including energy levels, incubation temperature and energy × incubation temperature (Tem) as fixed effects and incubation run as a random effect. $^{a–e}$ Indicate that the means of the interactions of energy levels and incubation temperature are significantly different. * Indicates when there is a statistically significant difference in interaction in the MIXED procedure, but no significant difference in post-test Tukey.

3.2. In Vitro Fermentation Properties According to the Protein Levels of the Trial 1 Diet

The concentration of NH_3-N was consistent at protein levels of 12.0, 13.5, and 15.0%, but significantly increased at 16.5 and 18.0% CPs ($p < 0.05$, Table 5). After 3 h of fermentation, the concentration of generated NH_3-N increased by 8.9% and 11.6% for CPs of 16.5% and 18.5%, respectively, compared to a CP of 12.0% (Table 5, $p = 0.001$). Throughout all the fermentation times, the NH_3-N levels remained the highest for CPs of 16.5 and 18.0%, with a particularly large difference in concentration of 41.7–50.9% after 12 h of fermentation. The difference decreased to about 10.9% after 24 h of fermentation. Additionally, the increase in acetate was found to be significant ($p < 0.01$) with an increasing protein content after 12 h of fermentation, while the propionate decreased from CPs of 16.5 and 18.0% after 24 h of fermentation compared to a protein content of 12.0%. Valerate was also found to increase by approximately 24.1% over a 6 h fermentation period when using CPs of 16.5 and 18.0% (Table 5).

3.3. In Vitro Fermentation Properties According to the Energy Levels of the Trial 2 Diet

Changes in the energy level influenced the pH, acetate, propionate, butyrate, TVFA, and the ratio of acetate to propionate ($p < 0.01$, Table 6). After 12, 24, and 48 h of fermentation, the pH decreased as the energy level increased ($p < 0.001$). The acetate concentrations declined when the energy level exceeded 2.6 Mcal/kg, with higher energy levels resulting in lower acetate production rates ($p < 0.001$). Conversely, the propionate concentrations increased, displaying an inverse relationship with the production ratio pattern of acetate ($p < 0.001$). Thus, the ratio of acetate to propionate was found to decrease with increasing energy levels. The production rates of butyrate increased linearly as the energy level rose. Based on an energy level of 2.4 Mcal/kg after 24 h of fermentation, the increases in the production ratios of butyrate were 1.3, 9.2, 16.1, and 21.8% at 2.5, 2.6, 2.7, and 2.8 Mcal/kg, respectively. The TVFA concentrations did not show significant differences even when the energy level increased to 2.6, 2.7, and 2.8 Mcal/kg after 24 h of fermentation and up to 2.5 Mcal/kg after 48 h. Increasing the energy level above 2.5 Mcal/kg while maintaining a constant protein content of 17.5% did not result in further increases in the TVFA concentrations. The BCFA, valerate, and protein amounts in LAB were not affected by the energy levels (Table 6).

After 3 h of fermentation, the pH exhibited a decreasing pattern with an increase in the energy level, but, at 41 °C, the pH increased ($p = 0.011$). NH_3-N also displayed different patterns between 39 and 41 °C: the NH_3-N increased as the energy levels rose at 39 °C but decreased at 41 °C ($p = 0.009$). The TVFA concentration increased at 2.6 and 2.7 Mcal/kg at 41 °C compared to 39 °C after 6 h of fermentation ($p = 0.002$). Regarding the VFA composition, a significant difference was observed between a decrease in acetate and an increase in propionate at 2.7 Mcal/kg at 41 °C ($p < 0.05$) resulting in a reduced A:P ratio ($p = 0.025$). After 12 h of fermentation, the NH_3-N concentration was higher at 41 °C for ME 2.4 and 2.6 Mcal/kg compared to 39 °C, and in the case of TVFA, increases were shown at 41 °C for ME 2.4, 2.6, and 2.7 Mcal/kg treatments compared to 39 °C ($p = 0.006$). However, after 24 and 48 h of fermentation, no interaction was found between the energy levels and the incubation temperatures in the fermentation properties.

3.4. Changes in Protein Amount of LAB

In trial 1, during fermentation at an incubation temperature of 41 °C for 6 h, the protein of LAB decreased by 6% compared to 256 µg/mL at 39 °C, reaching 240.8 µg/mL ($p < 0.001$, Table 5). While there was no significant difference after 12 h of culture due to high variations, by 24 h, the protein level decreased by approximately 24% to 149.4 µg/mL at 41 °C compared to the value of 196.7 µg/mL at 39 °C ($p < 0.001$, Table 5).

In trial 2, the protein amounts in LAB initially increased by 13.4 and 11.0% after 3 and 6 h, respectively, at a temperature of 41 °C, but they decreased by 33.3% after 48 h of fermentation (Table 6). The protein amounts in LAB were not affected by the energy levels. An interaction between incubation temperatures and the energy level was observed after

48 h of fermentation ($p < 0.001$): the protein amounts in LAB increased as the energy level rose at an incubation temperature of 39 °C; however, under 41 °C incubation conditions, the protein amounts in LAB remained approximately 52.1 µg/mL lower than those at 39 °C despite the increase in the energy level.

4. Discussion

4.1. In Vitro Fermentation Properties

In an in vitro experiment, it should be noted that protozoa and fungi exhibit reduced activity and have lower absorption and secretion rates within rumen epithelial cells. However, in vitro experiments like the one conducted in this study maintain constant substrate levels corresponding to the feed intake and consistent incubation times, representing the passage rates. This methodology allows for the examination of the direct effects of different incubation temperatures on rumen microorganisms while excluding other confounding factors.

A previous study reported that the reticular temperature exceeded 40.5 °C in dairy cows exposed to high ambient temperatures [4]. They also observed a correlation between reticular temperature and core temperature under high-ambient-temperature conditions. In this study, the incubation temperature was set based on the positive correlation between the highest rumen reticular temperature and an incubation temperature close to 41 °C [4]. The experiment was designed to expose rumen microorganisms to high ambient temperatures.

It is commonly observed that the pH inside the rumen often decreases during heat stress experiments with cattle [25]. In our current study, in trials 1 and 2, incubation temperatures did not induce a decrease in the pH. Conversely, C King et al. [26] reported that an increase in the incubation temperature resulted in a higher pH. It is believed that the pH reduction associated with ruminal temperature does not have a direct effect on ruminal microorganisms but rather a response to the host experiencing high ambient temperatures. Factors contributing to a decreased pH include changes in DMI, a high-concentrate diet, reduced saliva secretion, and alterations in the digestive system's mobility and absorption [27–29]. However, we cultured only rumen microorganisms in vitro and found that the pH did not decrease when high incubation temperature was applied directly to these microorganisms. While the exact reason for the stability of the pH remains unclear, this suggests that the decrease in ruminal pH during the summer season in vivo may be attributed not to the microorganisms themselves but rather to the animal's homeostatic responses or behavioral changes.

The increase in TVFA during fermentation (3–12 h) at 41 °C observed in this study was consistent with the findings of trial 1, where an increase in TVFA was observed for up to 12 h of fermentation. The optimal temperature for rumen microorganisms is unknown, as the rumen-reticular temperature varies from 38 to 40 °C and consists of an ecosystem of over 200 bacterial species, more than 25 species of protozoa, and five types of fungi [30]. Each microbial species has a unique optimal temperature at which the maximum growth rate and activity are achieved. For instance, *Saccharomyces cerevisiae*'s optimal incubation temperature is 40.9 °C [31], while that of *Ruminococcus albus* is 44.0 °C [32]. Although the kinds of microbial species in this study that would be activated at the 41 °C incubation temperature are still unknown, combining the findings of trial 1 and trial 2 confirmed that the TVFA concentration in ruminants increased at 41 °C until the 12 h mark, suggesting that certain microbial species may become more active under high-ambient-temperature conditions.

Additionally, the *iso*-butyrate and *iso*-valerate concentrations increased by approximately 6–16% after 12 h of fermentation in trial 1. These two VFAs are branched-chain fatty acids (BCFAs) that form when amino acids such as valine, isoleucine, leucine, and proline are used as precursors during decomposition. Although the association between BCFAs and high incubation temperature has not been extensively studied, fibrolytic bacteria require BCFAs for growth and proliferation [33,34]. The increase in BCFAs is thought to be a result of reduced BCFA usage for synthesis due to decreased microbial protein synthesis from NH_3-N under high incubation temperature, causing a relative increase in BCFA concentrations.

The results of trial 1, trial 2, and C King et al. [26] indicate that the changes in the proportions and amounts of acetate, propionate, butyrate, and valerate observed under high-ambient-temperature conditions in animals may be due to physiological adjustments to maintain homeostasis rather than being a direct result of microbial activity.

During the initial stage of fermentation, including the period from 3 to 12 h, an upward trend in the protein amounts of LAB was observed, accompanied by an increase in TVFA at an incubation temperature of 41 °C. These findings suggest an elevation in the NH_3-N level that can be attributed to the augmented degradation rate of the substrates. However, after 12 h of fermentation, declines in both the TVFA and protein amounts of LAB were observed under high incubation temperatures. This showed that prolonged high incubation temperatures adversely affected microbial activity, resulting in reduced microbial synthesis when utilizing NH_3-N [35,36]. The observed decreases in the protein amounts in LAB under high incubation temperatures further supported the notion of reduced microbial protein synthesis. Specifically, the protein amounts of LAB decreased by 33.3% after 48 h of fermentation in trial 2, suggesting the significant impact of high incubation temperature on rumen microorganisms.

4.2. In Vitro Fermentation Properties According to the Protein Levels of the Trial 1 Diet

When the CP levels were at 16.5 and 18.0%, NH_3-N was increased after 3 h of fermentation. This occurred because, as protein levels rose, the NPN also increased to maintain the same protein fraction ratio. Additionally, after 12 h of fermentation, the acetate levels in treatments with CPs of 16.5 and 18.0% rose by approximately 21% compared to those with a CP of 12.0%. This was believed to result from a reduction in the NFC content even though the NDF content remained similar. After 24 h, propionate concentration declined by around 15.5% in treatments with CPs of 16.5 and 18.0% compared to a CP of 12.0%. This was considered to be the reverse of the acetate increase observed [37]. This delay in fermentation may be attributed to the fact that acetate production was at 2.81 moles/12 h, while propionate production was at a slower rate of 0.82 moles/12 h, as previously reported [38]

Valerate production after 6 h of fermentation in treatments with CPs of 16.5 and 18.0% increased by approximately 14.1% compared to a CP of 12.0%. Branched-chain amino acids, including leucine and isoleucine, are precursors in valerate production; thus, an increase in protein content enhances the valerate levels [39]. Moreover, when the ratio of soluble protein to a CP of 14.0% increases to 20–50% (equivalent to a CP of 2.8 to 7% of DM basis), the valerate amounts increase in conjunction with 30% or more soluble protein content [40]. In this study, since all the protein types increased with the rising CP levels, this could be attributed to the increase in soluble protein.

Significant differences were observed in the fermentation properties and LAB between the incubation temperatures and protein levels. The results indicated that while protein levels significantly influenced the ruminal fermentation properties, no interaction was observed between protein levels and incubation temperatures. Thus, it might not be necessary to consider protein levels when evaluating ruminal fermentation properties and the quantity of microbial protein synthesis under high-ambient-temperature conditions.

4.3. In Vitro Fermentation Properties According to the Energy Levels of the Trial 2 Diet

When investigating the effects of high temperatures on rumen fermentation properties using animals, the direct impact was that the pH decreased after 12 h of fermentation when using ME 2.6, 2.7, and 2.8 Mcal/kg. The increase in NFC and the decrease in NDF in the diet are potential reasons for this effect [41,42]. In the current study, a level of energy setting of at least 5 levels was required to statistically determine the adequate level of energy. When the concentration was set to increase linearly, it reached an ME of 2.8 Mcal/kg. With the ingredient used in this study, setting the energy concentration higher than this led to extreme changes in the ratio of concentrate and roughage. The objective of this study was to gradually increase the energy requirement from 2.4 Mcal/kg while maintaining constant NFC and NDF contents. For this reason, although decreasing the NDF content at the ME 2.8 Mcal/kg level could potentially cause ruminal acidosis when fed to cows, it was challenging to consider the impact on cows when determining the energy content.

Acetate is indeed one of the primary VFAs produced during ruminal fermentation, and it is generated in the highest proportion compared to other VFAs like propionate and butyrate. Acetate-producing bacteria play a vital role in the process of converting one molecule of pyruvate and one molecule of H_2O into one molecule of acetate, one molecule of CO_2, and two hydrogen atoms [43]. In this study, the production ratio of acetate decreased with increasing energy levels starting from 6 h of fermentation. This indicated that increasing the energy level in the fermentation resulted in a linear decrease in the production ratio of acetate, with the highest reduction observed at ME 2.8 Mcal/kg. This decrease in acetate concentration could be attributed to the increased availability of NFC and the decreased availability of NDF in the substrate. It is important to note that acetate is mainly synthesized from slowly digestible fiber sources, such as NDF, which has a degradation rate of 0.03–0.09/h [44]. Therefore, the linear decrease in the acetate production ratio after 48 h of fermentation with increasing energy levels could have been due to the higher availability of rapidly fermentable carbohydrates and the lower availability of slowly fermentable NDF when using higher energy levels, especially compared to ME 2.4 Mcal/kg, which contained more NDF.

Propionate-producing bacteria could convert one molecule of pyruvate and four hydrogen atoms into one molecule of propionate and one molecule of H_2O through two different pathways: the succinate and acrylate pathways [43]. Starch-rich diets have been shown to support the development of propionate-producing bacterial species and are associated with an increase in the proportion of propionate, as reported by Makkar and McSweeney [30]. The production ratio of propionate increased with increasing energy levels starting from 6 h of fermentation. After 24 h of fermentation, the production ratio of propionate increased by 0.9, 12.2, 14.1, and 12.9 when ME was increased to 2.5, 2.6, 2.7, and 2.8 Mcal/kg of DM compared to 2.4 Mcal/kg, respectively. Similarly, after 48 h of fermentation, the increase in the production ratio of propionate was 0.9, 5.5, 10.5, and 12.1% for ME levels of 2.5, 2.6, 2.7, and 2.8 Mcal/kg of DM, respectively. The rapid digestion of NFC, which had a degradation rate of 0.05 to 0.50/h, may explain the increase in propionate concentration within a 24 h period [44,45]. The decrease in acetate and increase in propionate led to a decrease in the A:P ratio after 6 h of fermentation, with reductions of 7.1, 17.0, 18.8, and 19.2% when ME was increased to 2.5, 2.6, 2.7, and 2.8 Mcal/kg of DM compared to 2.4 Mcal/kg, respectively. The levels of NFC and reduced NDF could have influenced the concentration of acetate, propionate, and the A:P ratio [41,42].

The study by Dijkstra, Forbes and France [43] explains the process of transforming two molecules of pyruvate into one molecule of butyrate and two molecules of CO_2. The findings of this study showed that the butyrate production rate increased as the energy level increased during fermentation, particularly after 6 h and continuing linearly over 24 and 48 h of fermentation. The observed changes in the C4 production rates could potentially alter the population of butyrate-producing bacteria, such as *Butyrivibrio fibrisolvens* or large protozoa. This idea was supported by Russell and Baldwin [46] and Williams and Coleman [47], who suggested that an increase in energy levels may lead to changes in the microbial population. This finding is important because it provides a possible explanation for the observed increase in butyrate concentration as the energy levels increase during fermentation. Overall, this study highlights the relationship between energy levels and the concentration of butyrate, as well as the potential effects on microbial populations. Additional research is warranted to gain a deeper understanding of the underlying mechanisms driving these relationships and to explore how microbial populations can be enhanced to improve the efficiency of fermentation processes.

The results of the study showed that treatment with ME levels above 2.6 Mcal/kg increased TVFA, leading to the conclusion that ME levels above 2.6 Mcal/kg are sufficient to enhance TVFA synthesis in the rumen. After 48 h of fermentation, the interactions between the incubation temperatures and the energy levels were evident in the protein amount of LAB. In the treatments with ME 2.6 and 2.8 Mcal/kg, an increase in the protein amount of LAB was observed at 39 °C, but not at 41 °C. Despite controlling the energy level, a high incubation temperature still caused a reduction in the protein amounts of LAB.

5. Conclusions

The rumen environment is influenced by multiple mechanisms. In this in vitro experiment, the direct effects of different incubation temperature on microbial activity were examined. In conclusion, this study found that increased ruminal temperature elevated the NH_3-N content and reduced the TVFA and VFA concentrations due to decreased microbial protein synthesis. Varying protein and energy levels showed distinct effects. Higher protein levels raised the NH_3-N, acetate, valerate, and BCFA concentrations while lowering propionate levels, whereas increased energy levels led to lower pH and acetate levels and higher propionate, butyrate, and LAB protein levels. These findings provide insight into metabolic adaptation under different ruminal temperatures and the impacts of dietary adjustments on rumen fermentation and microbial activity.

However, given the findings of this study, adjusting the protein and energy levels may not be necessary to improve rumen fermentation and microbial protein synthesis under high-incubation-temperature conditions. It should be noted that the results of this study are based on in vitro experiments, which may not fully reflect the physiological response of the host and the complex interactions surrounding ruminant microorganisms. Therefore, further study is needed to confirm the changes in microbial protein synthesis under different temperature conditions, as well as alterations in amino acid composition resulting from shifts in the microbial composition using cattle.

Author Contributions: Conceptualization, Y.-H.J. and H.-G.L.; formal analysis, Y.-H.J.; investigation, Y.-H.J.; data curation, Y.-H.J.; writing—original draft preparation, Y.-H.J. and W.-S.K.; writing—review and editing, Y.-H.J., W.-S.K., Y.-R.K., M.-S.J., J.G.N. and H.-G.L.; supervision, H.-G.L. All authors have read and agreed to the published version of the manuscript.

Funding: This study was supported by the 'Cooperative Research Program for Agriculture Science & Technology Development (Project No. RS-2020-RD009051)' Rural Development Administration and KU Research Professor Program of Konkuk University, Republic of Korea.

Institutional Review Board Statement: All the procedures involving animals were approved by the Institutional Animal Care and Use Committee (IACUC) of Konkuk University (Approval No: KU21011).

Informed Consent Statement: Not applicable.

Data Availability Statement: The data presented in this study are available in this paper.

Conflicts of Interest: IANS Company, who was involved in the provision of data for the study, did not interfere with co-authors' access to all of the study's data, analysing and interpreting the data, preparing and publishing manuscripts independently. All authors declare no conflicts of interest.

References

1. Tajima, K.; Nonaka, I.; Higuchi, K.; Takusari, N.; Kurihara, M.; Takenaka, A.; Mitsumori, M.; Kajikawa, H.; Aminov, R.I. Influence of high temperature and humidity on rumen bacterial diversity in Holstein heifers. *Anaerobe* **2007**, *13*, 57–64. [CrossRef] [PubMed]
2. Uyeno, Y.; Sekiguchi, Y.; Tajima, K.; Takenaka, A.; Kurihara, M.; Kamagata, Y. An rRNA-based analysis for evaluating the effect of heat stress on the rumen microbial composition of Holstein heifers. *Anaerobe* **2010**, *16*, 27–33. [CrossRef] [PubMed]
3. An, Z.; Luo, G.; Abdelrahman, M.; Riaz, U.; Gao, S.; Yao, Z.; Ye, T.; Lv, H.; Zhao, J.; Chen, C. Effects of capsicum oleoresin supplementation on rumen fermentation and microbial abundance under different temperature and dietary conditions in vitro. *Front. Microbiol.* **2022**, *13*, 1005818. [CrossRef] [PubMed]
4. Bewley, J.; Einstein, M.; Grott, M.; Schutz, M. Comparison of reticular and rectal core body temperatures in lactating dairy cows. *J. Dairy Sci.* **2008**, *91*, 4661–4672. [CrossRef]
5. Kim, W.S.; Ghassemi Nejad, J.; Peng, D.Q.; Jo, Y.H.; Kim, J.; Lee, H.G. Effects of different protein levels on growth performance and stress parameters in beef calves under heat stress. *Sci. Rep.* **2022**, *12*, 8113. [CrossRef]

6. Lees, A.M.; Sejian, V.; Lees, J.; Sullivan, M.; Lisle, A.; Gaughan, J. Evaluating rumen temperature as an estimate of core body temperature in Angus feedlot cattle during summer. *Int. J. Biometeorol.* **2019**, *63*, 939–947. [CrossRef]
7. Lees, A.; Lees, J.; Lisle, A.; Sullivan, M.; Gaughan, J. Effect of heat stress on rumen temperature of three breeds of cattle. *Int. J. Biometeorol.* **2018**, *62*, 207–215. [CrossRef]
8. Clark, J.; Klusmeyer, T.; Cameron, M. Microbial protein synthesis and flows of nitrogen fractions to the duodenum of dairy cows. *J. Dairy Sci.* **1992**, *75*, 2304–2323. [CrossRef] [PubMed]
9. Bach, A.; Calsamiglia, S.; Stern, M.D. Nitrogen metabolism in the rumen. *J. Dairy Sci.* **2005**, *88*, E9–E21. [CrossRef]
10. Licitra, G.; Hernandez, T.; Van Soest, P. Standardization of procedures for nitrogen fractionation of ruminant feeds. *Anim. Feed Sci. Technol.* **1996**, *57*, 347–358. [CrossRef]
11. National Research Council; Committee on Animal Nutrition; Subcommittee on Dairy Cattle Nutrition. *Nutrient Requirements of Dairy Cattle: 2001*; National Academies Press: Cambridge, MA, USA, 2001.
12. Miller-Cushon, E.K.; Dayton, A.M.; Horvath, K.C.; Monteiro, A.P.A.; Weng, X.; Tao, S. Effects of acute and chronic heat stress on feed sorting behaviour of lactating dairy cows. *Animal* **2019**, *13*, 2044–2051. [CrossRef] [PubMed]
13. Lu, Z.; Xu, Z.; Shen, Z.; Tian, Y.; Shen, H. Dietary energy level promotes rumen microbial protein synthesis by improving the energy productivity of the ruminal microbiome. *Front. Microbiol.* **2019**, *10*, 847. [CrossRef]
14. Dewhurst, R.; Davies, D.; Merry, R. Microbial protein supply from the rumen. *Anim. Feed Sci. Technol.* **2000**, *85*, 1–21. [CrossRef]
15. Mader, T.L.; Davis, M.; Brown-Brandl, T. Environmental factors influencing heat stress in feedlot cattle. *J. Anim. Sci.* **2006**, *84*, 712–719. [CrossRef] [PubMed]
16. Kh, M. Estimation of the energetic feed value obtained from chemical analysis and in vitro gas production using rumen fluid. *Anim Res Dev* **1988**, *28*, 7–55.
17. Wilman, D.; Adesogan, A. A comparison of filter bag methods with conventional tube methods of determining the in vitro digestibility of forages. *Anim. Feed Sci. Technol.* **2000**, *84*, 33–47. [CrossRef]
18. National Academies of Sciences, Division on Earth, Life Studies, Committee on Nutrient Requirements of Beef Cattle. *Nutrient Requirements of Beef Cattle: Eighth Revised Edition*; The National Academies Press: Washington, DC, USA, 2016; p. 494.
19. Fox, D.; Tedeschi, L.; Tylutki, T.; Russell, J.; Van Amburgh, M.; Chase, L.; Pell, A.; Overton, T. The Cornell Net Carbohydrate and Protein System model for evaluating herd nutrition and nutrient excretion. *Anim. Feed Sci. Technol.* **2004**, *112*, 29–78. [CrossRef]
20. Cunniff, P.; Washington, D. Official methods of analysis of AOAC International. *J. AOAC Int.* **1997**, *80*, 127A.
21. Van Soest, P.v.; Robertson, J.B.; Lewis, B.A. Methods for dietary fiber, neutral detergent fiber, and nonstarch polysaccharides in relation to animal nutrition. *J. Dairy Sci.* **1991**, *74*, 3583–3597. [CrossRef]
22. Van Soest, P.; Robertson, J. Systems of analysis for evaluating fibrous feeds. In *Standardization of Analytical Methodology for Feeds: Proceedings...*; IDRC: Ottawa, ON, Canada, 1979.
23. Chaney, A.L.; Marbach, E.P. Modified reagents for determination of urea and ammonia. *Clin. Chem.* **1962**, *8*, 130–132. [CrossRef]
24. Makkar, H.; Sharma, O.; Dawra, R.; Negi, S. Simple determination of microbial protein in rumen liquor. *J. Dairy Sci.* **1982**, *65*, 2170–2173. [CrossRef]
25. Yadav, B.; Singh, G.; Verma, A.; Dutta, N.; Sejian, V. Impact of heat stress on rumen functions. *Vet. World* **2013**, *6*, 992. [CrossRef]
26. King, C.; Dschaak, C.; Eun, J.-S.; Fellner, V.; Young, A. Quantitative analysis of microbial fermentation under normal or high ruminal temperature in continuous cultures. *Prof. Anim. Sci.* **2011**, *27*, 319–327. [CrossRef]
27. Nonaka, I.; Takusari, N.; Tajima, K.; Suzuki, T.; Higuchi, K.; Kurihara, M. Effects of high environmental temperatures on physiological and nutritional status of prepubertal Holstein heifers. *Livest. Sci.* **2008**, *113*, 14–23. [CrossRef]
28. Attebery, J.T.; Johnson, H.D. Effects of environmental temperature, controlled feeding and fasting on rumen motility. *J. Anim. Sci.* **1969**, *29*, 734–737. [CrossRef]
29. Kadzere, C.T.; Murphy, M.R.; Silanikove, N.; Maltz, E. Heat stress in lactating dairy cows: A review. *Livest. Prod. Sci.* **2002**, *77*, 59–91. [CrossRef]
30. Makkar, H.P.; McSweeney, C.S. *Methods in Gut Microbial Ecology for Ruminants*; Springer: Berlin/Heidelberg, Germany, 2005; Volume 10.
31. Patching, J.; Rose, A. Chapter II The Effects and Control of Temperature. In *Methods in Microbiology*; Elsevier: Amsterdam, The Netherlands, 1970; Volume 2, pp. 23–38.
32. Kawai, S.; Honda, H.; Tanase, T.; Taya, M.; Iijima, S.; Kobayashi, T. Molecular cloning of Ruminococcus albus cellulase gene. *Agric. Biol. Chem.* **1987**, *51*, 59–63. [CrossRef]
33. Slyter, L.; Weaver, J. Growth factor requirements of ruminal cellulolytic bacteria isolated from microbial populations supplied diets with or without rapidly fermentable carbohydrate. *Appl. Microbiol.* **1971**, *22*, 930–932. [CrossRef]
34. Cummins, K.; Papas, A. Effect of isocarbon-4 and isocarbon-5 volatile fatty acids on microbial protein synthesis and dry matter digestibility in vitro. *J. Dairy Sci.* **1985**, *68*, 2588–2595. [CrossRef]
35. Cowley, F.; Barber, D.G.; Houlihan, A.; Poppi, D.P. Immediate and residual effects of heat stress and restricted intake on milk protein and casein composition and energy metabolism. *J. Dairy Sci.* **2015**, *98*, 2356–2368. [CrossRef]
36. Gao, S.; Guo, J.; Quan, S.; Nan, X.; Fernandez, M.S.; Baumgard, L.; Bu, D. The effects of heat stress on protein metabolism in lactating Holstein cows. *J. Dairy Sci.* **2017**, *100*, 5040–5049. [CrossRef]
37. Blümmel, M.; Makkar, H.; Becker, K. In vitro gas production: A technique revisited. *J. Anim. Physiol. Anim. Nutr.* **1997**, *77*, 24–34. [CrossRef]

38. Van der Walt, J.; Briel, B. Volatile fatty acid metabolism in sheep. 1. Average daily volatile fatty acid production in the rumen of sheep fed lucerne hay. *Onderstepoort J. Vet Res.* **1976**, *43*, 11–21. [PubMed]
39. Davidson, S.; Hopkins, B.; Diaz, D.; Bolt, S.; Brownie, C.; Fellner, V.; Whitlow, L. Effects of amounts and degradability of dietary protein on lactation, nitrogen utilization, and excretion in early lactation Holstein cows. *J. Dairy Sci.* **2003**, *86*, 1681–1689. [CrossRef]
40. Zhang, Z.; Wei, W.; Yang, S.; Huang, Z.; Li, C.; Yu, X.; Qi, R.; Liu, W.; Loor, J.J.; Wang, M. Regulation of dietary protein solubility improves ruminal nitrogen metabolism in vitro: Role of bacteria–protozoa interactions. *Nutrients* **2022**, *14*, 2972. [CrossRef]
41. Ma, T.; Tu, Y.; Zhang, N.; Deng, K.; Diao, Q. Effect of the ratio of non-fibrous carbohydrates to neutral detergent fiber and protein structure on intake, digestibility, rumen fermentation, and nitrogen metabolism in lambs. *Asian Australas J. Anim. Sci.* **2015**, *28*, 1419. [CrossRef]
42. Haddad, S.; Grant, R. Influence of nonfiber carbohydrate concentration on forage fiber digestion in vitro. *Anim. Feed Sci. Technol.* **2000**, *86*, 107–115. [CrossRef]
43. Dijkstra, J.; Forbes, J.M.; France, J. *Quantitative Aspects of Ruminant Digestion and Metabolism*; CAB International, University Press: Cambridge, UK, 2005.
44. Lanzas, C.; Sniffen, C.; Seo, S.a.; Tedeschi, L.; Fox, D. A revised CNCPS feed carbohydrate fractionation scheme for formulating rations for ruminants. *Anim. Feed Sci. Technol.* **2007**, *136*, 167–190. [CrossRef]
45. Sniffen, C.J.; O'connor, J.; Van Soest, P.J.; Fox, D.G.; Russell, J. A net carbohydrate and protein system for evaluating cattle diets: II. Carbohydrate and protein availability. *J. Anim. Sci.* **1992**, *70*, 3562–3577. [CrossRef]
46. Russell, J.B.; Baldwin, R. Comparison of maintenance energy expenditures and growth yields among several rumen bacteria grown on continuous culture. *Appl. Environ. Microbiol.* **1979**, *37*, 537–543. [CrossRef]
47. Williams, A.G.; Coleman, G.S. The rumen protozoa. In *The Rumen Microbial Ecosystem*; Springer: Berlin/Heidelberg, Germany, 1997; pp. 73–139.

animals

MDPI

Article

Effects of Active Dry Yeast Supplementation in In Vitro and In Vivo Nutrient Digestibility, Rumen Fermentation, and Bacterial Community

Haitao Liu [1,2,3], Fei Li [1,2,3], Zhiyuan Ma [1,2,3,*], Miaomiao Ma [4], Emilio Ungerfeld [5], Zhian Zhang [1,2,3], Xiuxiu Weng [1,2,3], Baocang Liu [6], Xiaoyu Deng [6] and Liqing Guo [7]

1 State Key Laboratory of Herbage Improvement and Grassland Agro-Ecosystems, Lanzhou University, Lanzhou 730020, China; liuht19@lzu.edu.cn (H.L.); lfei@lzu.edu.cn (F.L.); zhangzha20@lzu.edu.cn (Z.Z.); wengxx@lzu.edu.cn (X.W.)
2 College of Pastoral Agriculture Science and Technology, Lanzhou University, Lanzhou 730020, China
3 Key Laboratory of Grassland Livestock Industry Innovation, Ministry of Agriculture and Rural Affairs, Lanzhou 730020, China
4 Animal Husbandry Work Station of Ningxia, Yingchuan 750002, China; nannontonban@126.com
5 Centro Regional de Investigación Carillanca, Instituto de Investigaciones Agropecuarias INIA, Vilcún 4880000, Chile; emilio.ungerfeld@inia.cl
6 Aksu Tycoon Feed Co., Ltd., Aksu 843000, China; baocangliu@outlook.com (B.L.); tkxiaoyu.deng@outlook.com (X.D.)
7 Tecon Pharmaceutical Co., Ltd., Suzhou 215000, China; glqaibjl@163.com
* Correspondence: mzy@lzu.edu.cn; Tel.: +86-199-7551-9487

Simple Summary: This study evaluated the influence of four active dry yeasts (ADYs) on nutrient digestibility and rumen fermentation in lambs through in vitro and in vivo experiments. Notably, Vistacell (Vis) and Procreatin7 (Pro) showed enhanced total gas production and pH levels in vitro. Vis improved in vivo propionate molar proportion, NDF digestibility, and total VFA concentration, suggesting it is the most suitable for lamb growth. However, the study highlighted discrepancies between in vitro and in vivo outcomes, cautioning against directly translating batch culture results to live animal effects.

Abstract: This study assessed the impact of active dry yeast (ADY) on nutrient digestibility and rumen fermentation, using both in vitro and in vivo experiments with lambs. In vitro, ADYs were incubated with rumen fluid and a substrate mixture to assess gas production, pH, volatile fatty acid (VFA) profiles, and lactate concentration. In vivo, Hu lambs were randomly assigned to five dietary treatments: a control group and four groups receiving one of two dosages of either Vistacell or Procreatin7. Growth performance, nutrient digestibility, rumen fermentation parameters, and bacterial community composition were measured. Pro enhanced the propionate molar proportion while it decreased the n-butyrate molar proportion. Vis reduced the lactate concentration in vitro. In the in vivo experiment, Vis increased the propionate molar proportion and the Succinivibrionaceae_UCG-001 abundance while it decreased the n-butyrate molar proportion and the Lachnospiraceae_ND3007 abundance. Additionally, Vis showed a greater impact on improving the NDF digestibility and total VFA concentration in vivo compared to Pro. Overall, the effects of ADYs on rumen fermentation were found to vary depending on the specific ADY used, with Vis being the most suitable for lamb growth. It was observed that Vis promoted propionate fermentation and Succinivibrionaceae_UCG-001 abundance at the expense of reduced n-butyrate fermentation and Lachnospiraceae_ND3007 abundance. Importantly, differences were noted between the outcomes of the in vitro and in vivo experiments concerning the effects of ADYs on rumen fermentation, highlighting the need for caution when generalizing batch culture results to the in vivo effects of ADYs.

Keywords: active dry yeast; digestibility; fermentation; rumen

Citation: Liu, H.; Li, F.; Ma, Z.; Ma, M.; Ungerfeld, E.; Zhang, Z.; Weng, X.; Liu, B.; Deng, X.; Guo, L. Effects of Active Dry Yeast Supplementation in In Vitro and In Vivo Nutrient Digestibility, Rumen Fermentation, and Bacterial Community. *Animals* **2024**, *14*, 2916. https://doi.org/10.3390/ani14192916

Academic Editor: Lizhi Wang

Received: 13 September 2024
Revised: 7 October 2024
Accepted: 8 October 2024
Published: 9 October 2024

1. Introduction

The practice of feeding high concentrate diets to finishing feedlot lambs increases the growth rate and improves the feed efficiency but also increases the risk of metabolic disorders such as acidosis [1], which is attributed to the accumulation of VFA and/or lactate, resulting in a decreased rumen pH, negatively affecting fibrolytic microorganisms [2]. A low rumen pH and an increased rumen passage rate resulting from high concentrate feeding can also reduce rumen fiber digestion [3]. Hence, it is important to alleviate these negative effects of high concentrate diets to ensure healthy and efficient ruminant production.

The active dry yeast (ADY) *Saccharomyces cerevisae*, resulting from the brewery or baking industries, is used in the ruminant production industry as a probiotic feed additive [4]. High concentrate diets typically contain low levels of fiber, which can lead to a reduction in the fibrolytic population in the rumen [5]. Supplementing ADY to high concentrate diets of ruminants can stabilize the rumen pH and thus promote rumen fibrolytic activity [6]. Some studies have indicated that ADY can survive in the rumen and potentially interact with rumen bacteria to promote the growth of fibrolytic species [4,7]. Supplementing ADY has been proposed as a strategy to support rumen fibrolytic bacteria and enhance fiber digestion efficiency. This is of great interest in ruminants subjected to high concentrate diets. Further studies are needed to provide a more comprehensive understanding of the specific effects of ADY on rumen bacterial populations as well as the mechanisms through which ADY interacts with rumen bacteria to support fiber digestion and mitigate the negative impacts of high concentrate feeding on rumen health and function.

The present study hypothesized that the supplementation of specific ADY products in high concentrate diets fed to finishing lambs could modulate the rumen bacterial population towards a more favorable profile characterized by an increase in fibrolytic bacteria. We assessed the impact of ADY supplementation on fiber degradation, volatile fatty acid production, and the overall nutrient utilization in the rumen, aiming to improve rumen health and performance in ruminants on high concentrate diets.

2. Materials and Methods

All the animals and procedures were approved by the Animal Care and Use Committee of Lanzhou University, Lanzhou, China (approval number CPAST-2020–LI-3).

2.1. In Vitro Experiment

The in vitro fermentation experiment consisted of a control group without any ADY treatment and eight treatments with one of four commercial ADY sources at two dosages each. The four ADY sources included FB ($\geq$2.0 $\times$ 10^8 CFU/g, Angel Yeast Co., Ltd., Yichang, China), Procreatin7 ($\geq$2.0 $\times$ 10^8 CFU/g, Guangxi Danbaoli Yeast Co., Ltd., Laibin, China), Vistacell ($\geq$2.0 $\times$ 10^8 CFU/g, AB Vista, Wiltshire, UK), and YSF ($\geq$1.0 $\times$ 10^8 CFU/g, Lesaffre Industries Co., Ltd., Paris, France), with all the information of the CFU per gram provided by the manufacturers. The two doses of each product targeted 6 $\times$ 10^5 and 1.2 $\times$ 10^6 CFU/mL of fermentation fluid and were calculated considering the minimum guaranteed CFU/g informed by each manufacturer. From this point on, the four treatments that supplemented each yeast in vitro and in vivo will be referred to as FB, Pro, Vis, and YSF.

The detail of the in vitro incubation process is described in our published paper [8]. In brief, the in vitro fermentation flasks contained 0.9 g of substrate, 30 mL of rumen fluid, 60 mL of CO_2-saturated artificial saliva [9], and one ADY at one of the two dosages. The substrate incubated consisted of 25% corn silage, 15% alfalfa, 25% corn grain, 15% corn bran, 11% soybean meal, 8% cottonseed meal, and 1% limestone and contained 30.6% NDF, 24.0% starch, and 16.3% CP on a DM basis. Rumen contents were obtained from three rumen-cannulated male *Hu* lambs that were fed a commercial diet (Gansu Runmu Biotechnology Co., Ltd., Jinchang, China) containing 39.2% NDF and 13.4% CP before the morning feeding. Rumen fluids were obtained by straining rumen contents through four layers of cheesecloth and were taken to the laboratory for mixing with medium and inoculating fermentation flasks within 30 min. All the laboratory procedures involving the processing and inoculation of rumen fluid were conducted under CO_2 protection.

Each treatment, including the control, had six fermentation flasks with two flasks inoculated with each animal's rumen fluid, resulting in three biological replicates, each in turn containing two technical replicates. Before adding them to the fermentation flasks, the ADYs were suspended in ddH_2O to target a concentration of 10^8 CFU/mL according to the information of the minimum CFU per gram provided by the manufacturers. This resulted in 0.36 and 0.72 mL of ADY suspension delivered to flasks with targeted doses of 6×10^5 and 12×10^5 CUF/mL of fermentation broth, respectively. The volume of added water delivered in the ADY suspension received by the different treatments was not equalized by adding ddH_2O. The flasks were incubated at 39 °C for 24 h in a fermentation system with automatic exhaust recording exhaust gas volume (AGRS3, Beijing Boxiang Xingwang Technology Co., Ltd., Beijing, China).

After 24 h of incubation, the fermentation flasks were placed on ice to arrest the microbial activity. The bottles were opened, and the pH was measured using a portable pH meter (PHB4, Shanghai INESA & Scientific Instrument Co., Ltd., Shanghai, China). A 10-mL sample of fermentation liquid mixed with 2 mL meta-phosphoric acid (25%, g/mL) and a 5-mL sample of fermentation liquid was stored at −20 °C for subsequent analysis of VFA and lactic acid, respectively.

2.2. *Animal Trial*

Vistacell and Procreatin7 were selected for the animal trial because of their in vitro effects on pH stabilization and a decrease in lactic acid concentration. The animal trial consisted of a control treatment without any yeast supplementation and four treatments with one of Vistacell or Procreatin7 supplemented at 0.60 or 1.20 g/d, which are dosages that are recommended for non-lactating and lactating dairy cows, respectively [4]. A total of thirty healthy 4-month-old Hu lambs with a similar initial body mass (36.5 $\pm$ 3.22 kg, mean $\pm$ SD) were randomly allocated to one of the five dietary treatments. The basal diet was provided as a pelleted TMR consisting of 20% barley straw, 28% corn, 27% corn gluten feed, 14% corn germ meal, 5% cottonseed meal, 4% molasses, 1% $CaCO_3$, 0.5% NaCl, and 0.5% vitamin and trace mineral premix. The basal diet contained 31.1% NDF, 24.4% starch, and 13.7% CP (DM basis). One kilogram of premix included Fe 25 mg, Mn 40 mg, Zn 40 mg, Cu 8 mg, I 0.3 mg, Se 0.2 mg, Co 0.1 mg, vitamin A 940 IU, vitamin D 111 IU, and vitamin E 20 IU. The lambs were raised in individual pens (1.8 m $\times$ 1.25 m $\times$ 1 m) and were ad libitum fed at 8:00 and 18:00 daily to allow 15% refusals. Each ADY additive was mixed with 100 g of ground diet and was provided entirely before the morning feeding to ensure the animals ate all the provided ADY. Water was available at all times via an automatic water supply system.

Following fifteen days of feeding, the lambs experienced five days of consecutive fecal and urine collection, followed by a day of rumen fluid collection. Feces were collected in bags attached to the animals in a way to avoid urine contamination. The urine was collected in a bucket through a funnel-shaped bottom pan located underneath the pen. A 10% sample of the daily fecal output of each lamb was sampled, and samples of the 5-day fecal sampling period were pooled by animal. For the determination of total N in the urine, 20% of the daily urine was sampled and pooled according to the animal. Ten milliliters of 10%(g/mL) sulfuric acid were added to the fecal and urine samples used for N determination to avoid ammonia volatilization.

The rumen contents were collected through oral tubes before and 3 h after the morning feeding, with the earliest collected rumen fluid discarded to rule out saliva contamination. The pH of the rumen fluid was measured immediately using a portable pH meter (PHB4, Shanghai INESA & Scientific Instrument Co., Ltd., Shanghai, China). We took 9 mL of rumen fluid and added 1 mL of 25% (g/mL) meta-phosphoric acid, mixed it, centrifuged it at $2000\times g$ for 10 min, and froze the supernatant at −20 °C for subsequent analysis of the VFA concentration. Five-milliliter samples of rumen fluid were snap-frozen in liquid nitrogen and stored at −80 °C for subsequent analysis of the bacterial community composition.

2.3. Analytical

The plate counting method was used to determine the active yeast count in CFU [10]. In short, 10× serial diluted yeast suspensions were spread onto Petri dishes containing YPD solid medium (1% yeast extract, 2% peptone, 2% glucose, 2% agar, 93% ddH_2O). Then, the colonies of viable yeast cells were counted in each plate after incubation at 38 °C for 24 h.

Proximate analyses were conducted according to AOAC [11]. In brief, the CP was determined using an automatic Kieldahl apparatus (Shanghai Shengsheng Automation Analysis Instrument Co., Ltd., Shanghai, China); the NDF and ADF were measured using the Van Soest method [12] with a fiber analyzer (ANKOM A200i, ANKOM Technology, Macedon, NY, USA); the gross energy was measured using a calorimeter (C3000, IKA Laboratory Technology, Staufen, Germany).

The ammonia concentration in the rumen fluid was measured using the phenol-hypochlorite method [13]. Individual VFA concentrations were measured using a gas chromatograph (TRACE1300, Thermo Scientific, Milan, Italy). The lactic acid concentration was measured using a 3-(4,5-dimethylthiazol-2-yl)-2,5-diphenyl tetrazolium bromide (MTT) colorimetric method [14] with a commercial kit (Nanjing Jiancheng Bioengineering Institute, Nanjing, China).

Rumen fluids sampled at the two time points were pooled by animal for the analysis of the bacterial community composition. A sand-beating method [15] was performed to extract the microbial DNA in the rumen fluid. Bacterial V3-V4 amplicon sequencing was performed by Biozeron Corp., Wuhan, China, with primers of 341F (5′-CCTACGGGAGGCAGCAG-3′) and 806R (5′-GGACTACHVGGGTWTCTAAT-3′) [16].

The subsequent bioinformatic analysis based on zero-radius OTU (ZOTU) is described in detail in our previous paper [15]. In brief, ZOTU was generated by the denoise3 algorithm after quality control using usearch v5 [17]. Taxonomic annotation was performed by mothur v4 [18] against silva.nr v138 [19]. The ZOTU table was created by mapping the raw sequences to ZOTU sequences using vsearch v2 [20]. Alpha diversity and principal coordinate analysis (PcoA) were performed using an internal R script (https://github.com/sleepvet/MicrobialDiversity, accessed on 1 March 2023).

2.4. Statistical Analyses

The response variables, besides the fermentation parameters assessed in vivo, were measured only at a single time point and fitted to a mixed linear model: Response = $\mu + T_i + Animal_j + error_{ij}$, where T is a fixed effect of treatment (including control) and animal is a random effect. The response variables sampled at more than one time were fitted to a mixed linear model: Response = $\mu + T_i + \text{Sampling time}_j + T \times \text{Sampling time}_{ij} + Animal_k + error_{ijk}$, where T and sampling time are fixed effects, animal is a random effect, and sampling time was also set as a repeated measurement with an AR(1) covariance structure. When the interaction between the sampling time and treatment was significant, a reduced model was fitted for each sampling time point.

Pre-planned contrasts were used to evaluate the average of each ADY supplementation against the control treatment: Contrast (ADY vs. Con) = Control − 0.5 × ($ADY_{dose0.6}$ + $ADY_{dose1.2}$). If the contrast of a particular ADY resulted as significant ($p < 0.05$), polynomial contrasts of the dosage effect of the ADY were evaluated. Pre-planned contrasts were also used for pairwise comparison of the average of both ADY products: Contrast (ADY1 vs. ADY2) = ($ADY1_{dose0.6}$ + $ADY1_{dose1.2}$) − ($ADY2_{dose0.6}$ + $ADY2_{dose1.2}$).

All the statistics were performed in R v4.2 [21]. The linear mixed model was completed using lme module in the nlme package (https://svn.r-project.org/R-packages/trunk/nlme, accessed on 12 May 2022). The pre-planned contrast and polynomial contrast were conducted using the standardize package (https://github.com/CDEager/standardize, accessed on 12 May 2022). The relative abundance data of the bacterial community was log-transformed to meet the requirements of data distribution normalization.

3. Results

After plate culture counting, we found that the live yeast cells of FB, Pro, Vis, and YSF were 2.1×10^8, 1.5×10^8, 2.2×10^8, and 1.4×10^8 CFU/g, respectively.

3.1. In Vitro Experiment

The total gas volume increased at a high dose of Pro and Vis ($p \leq 0.04$, Table 1). Among the ADY sources, Pro produced ($p = 0.05$), and Vis tended to produce ($p = 0.10$), more total gas than YSF and FB. Pro and Vis increased or tended to increase the pH ($p \leq 0.09$). No differences among the ADYs were observed ($p \geq 0.12$). Vis increased the total VFA concentration (quadratic effect; $p = 0.01$), while other ADYs did not have any effect ($p \geq 0.65$). None of the ADYs affected the molar percentage of acetate ($p \geq 0.51$). The molar proportion of propionate was increased by Pro (quadratic effect; $p < 0.001$) and tended to increase with YSF ($p = 0.07$). A pairwise comparison of the ADY sources did not reveal any differences in the molar percentage of propionate ($p \geq 0.15$). Pro decreased ($p = 0.009$), and FB and YSF tended to decrease ($p \leq 0.10$), the molar percentage of n-butyrate. The molar percentage of n-butyrate was lower with Pro than with Vis ($p = 0.04$). The acetate to propionate molar ratio was decreased by Pro ($p = 0.002$) and tended to be decreased by YSF ($p = 0.08$). Pro had a lower acetate-to-propionate molar ratio than the other three ADYs ($p \leq 0.003$). The lactate concentration was decreased by Vis only ($p = 0.04$).

Table 1. Effects of active dry yeast (ADY) on in vitro fermentation parameters ($n = 3$).

Item	Total Gas, mL	pH	TVFA, m*M*	Molar Percentage of Individual VFA, %				Ace/Pro	Lactate, μ*M*
				Acetate	Propionate	n-Butyrate	Others [2]		
Treatment, 10^5 CFU ADY [1]/mL fermentation broth									
Control, 0	308	5.87	113	58.1	18.3	18.2	5.29	3.16	98.1
FB, 6	306	6.02	108	58.4	17.8	17.8	5.79	3.27	112
FB, 12	329	6.08	111	58.6	20.8	15.3	5.12	2.81	82.5
Pro,6	307	5.96	106	57.7	21.1	15.7	5.37	2.74	113
Pro, 12	338	6.11	126	57.7	21.8	15.1	5.12	2.64	91.8
Vis, 6	294	5.90	111	58.1	17.8	18.2	5.82	3.26	96.0
Vis, 12	335	6.08	135	58.7	20.0	15.9	5.27	2.93	73.8
YSF, 6	297	5.90	108	58.6	20.3	15.5	5.27	2.89	98.2
YSF, 12	290	5.98	112	58.3	19.1	16.6	5.77	3.04	117
SEM	10.1	0.07	6.5	0.54	0.67	0.73	0.176	0.097	9.10
p-value									
FB vs. Con	0.29	0.04 [Q]	0.65	0.52	0.20	0.10	0.44	0.24	0.95
Pro vs. Con	0.04 [Q]	0.08	0.74	0.58	<0.001 [Q]	0.009 [L]	0.85	0.002 [L]	0.73
Vis vs. Con	0.03 [Q]	0.09	0.01 [Q]	0.58	0.19	0.23	0.25	0.55	0.04 [Q]
YSF vs. Con	0.87	0.14	0.69	0.51	0.07	0.06	0.29	0.08	0.75
FB vs. Pro	0.71	0.35	0.65	0.15	0.001	0.13	0.24	<0.001	0.41
FB vs. Vis	0.82	0.26	0.05	0.89	0.47	0.58	0.64	0.47	0.05
FB vs. YSF	0.11	0.12	0.94	0.99	0.45	0.50	0.72	0.37	0.20
Pro vs. Vis	0.57	0.90	0.29	0.19	<0.001	0.04	0.11	<0.001	0.04
Pro vs. YSF	0.05	0.52	0.38	0.15	0.009	0.39	0.14	0.003	0.59
Vis vs. YSF	0.10	0.60	0.06	0.88	0.15	0.22	0.90	0.11	0.03

[1]: The four ADY sources included FB (Angel Yeast Co., Ltd., Yichang, China), Procreatin7 (Pro, Guangxi Danbaoli Yeast Co., Ltd., Laibin, China), Vistacell (Vis, AB Vista, Wiltshire, UK), and YSF (Lesaffre Industries Co., Ltd., Paris, France). The control treatment did not contain any ADY. [2]: Others included n-valerate, iso-butyrate, and iso-valerate. [L,Q]: Polynomial contrasts for dosage effects (linear, L; quadratic, Q) of individual ADYs.

3.2. In Vivo Experiment

Neither Pro nor Vis affected BM and DMI ($p \geq 0.46$, Table 2). The digestibility of DM, OM, and CP was not affected by Pro or Vis ($p \geq 0.21$). The digestibility of NDF and

ADF was not affected by Pro ($p \geq 0.40$) while Vis increased the NDF and ADF digestibility (quadratic effect; $p = 0.02$). No N balance response was affected by Pro or Vis ($p \geq 0.12$; Table 3).

Table 2. Effects of active dry yeast (ADY) on growth performance and nutrient digestibility of Hu lambs ($n = 6$).

Item	BM, kg	DMI, kg	Digestibility [2], g/kg				
			DM	OM	NDF	ADF	CP
Treatment, g ADY [1]/d							
Control, 0	38.9	1.91	652	701	369	326	782
Pro, 0.6	39.9	1.91	648	706	386	307	778
Pro, 1.2	40.0	1.77	667	703	399	375	793
Vis, 0.6	39.6	2.07	674	716	443	417	767
Vis, 1.2	40.1	1.85	662	713	417	360	795
SEM	1.76	0.074	11.4	8.9	20.8	21.1	18.4
p-value							
Pro vs. Con	0.62	0.46	0.67	0.54	0.4	0.54	0.93
Vis vs. Con	0.66	0.57	0.23	0.21	0.02 [Q]	0.02 [Q]	0.83
Pro vs. Vis	0.93	0.12	0.31	0.42	0.07	0.03	0.71

[1]: The ADY sources included Procreatin7 (Pro, Guangxi Danbaoli Yeast Co., Ltd., Laibin, China) and Vistacell (Vis, AB Vista, Wiltshire, UK). The control treatment did not contain any ADY. [2]: DM, dry matter; OM, organic matter; NDF, neutral detergent fiber; ADF, acid detergent fiber; CP, crude protein. [Q]: Polynomial contrasts for dosage effects (quadratic, Q) of individual ADYs.

Table 3. Effects of active dry yeast (ADY) on N balance of Hu lambs ($n = 6$).

Item	N Intake, g/d	Fecal N, g/d	Fecal N to N Intake Ratio, mg/g	Urinary N, g/d	Urinary N to N Intake Ratio, mg/g	N Excretion, g/d	N Excretion to N Intake Ratio, mg/g	N Retention, g/d	N Retention to N Intake Ratio, mg/g
Treatment, g ADY [1]/d									
Control, 0	42	10.1	241	8.09	192	18.2	434	23.7	565
Pro, 0.6	42	10.6	250	7.66	179	18.3	429	23.7	570
Pro, 1.2	38.7	9.24	232	8.34	218	17.6	453	21.2	548
Vis, 0.6	45.5	11.9	263	8.38	184	20.3	447	25.1	552
Vis, 1.2	40.8	9.62	231	7.58	186	17.2	417	23.6	582
SEM	1.64	1.15	20.7	0.874	19.8	1.63	29.5	1.22	29.4
p-value									
Pro vs. Con	0.46	0.84	0.92	0.39	0.33	0.69	0.82	0.56	0.82
Vis vs. Con	0.57	0.67	0.83	0.99	0.93	0.77	0.98	0.69	0.98
Pro vs. Vis	0.12	0.45	0.71	0.29	0.2	0.41	0.77	0.23	0.77

[1]: The ADY sources included Procreatin7 (Pro, Guangxi Danbaoli Yeast Co., Ltd., Laibin, China) and Vistacell (Vis, AB Vista, Wiltshire, UK). The control treatment did not contain any ADY.

There were no treatment interactions for the sampling time for any rumen fermentation variable ($p \geq 0.18$, Table 4). Pro supplementation did not affect the TVFA concentration ($p = 0.49$) while Vis supplementation tended to increase the TVFA concentration ($p = 0.08$). However, there was no difference in the effect of Pro and Vis supplementation on TVFA ($p = 0.19$). Neither Pro nor Vis supplementation affected the rumen pH ($p \geq 0.16$). Supplementation with Pro did not affect the acetate molar percentage ($p = 0.87$) while Vis supplementation tended to decrease the acetate molar percentage ($p = 0.08$), resulting in a lower acetate molar percentage than Pro ($p = 0.02$). Vis increased ($p = 0.006$), and Pro tended to increase, the propionate molar percentage ($p = 0.09$), with no differences between Pro and Vis ($p = 0.12$). Vis decreased ($p = 0.02$), and Pro tended to decrease, the n-butyrate molar percentage ($p = 0.06$), with no differences between Pro and Vis ($p = 0.48$). Neither the Pro nor Vis treatments affected the acetate-to-propionate ratio ($p = 0.50$).

Table 4. Effects of active dry yeast (ADY) on rumen fermentation parameters of Hu lambs (n = 6).

Item	TVFA, m*M*	pH	Molar Percentage of Individual VFA, %				Ac/Pr [3]
			Acetate	Propionate	n-Butyrate	Others [2]	
Treatment, g ADY [1]/d							
Control, 0	82.9	7.13	63.2	19.4	12	5.27	3.43
Pro, 0.6	91.0	7.07	63.8	22	9.01	5.03	3.06
Pro, 1.2	89.3	7.22	63.9	21.8	10.4	3.75	2.98
Vis, 0.6	93.4	7.07	61.7	23.8	9.62	5.05	2.67
Vis, 1.2	109.4	6.96	62.5	24.8	8.29	4.26	2.66
SEM	8.43	0.090	1.45	2.23	1.686	0.402	0.248
p-value							
T	0.27	0.38	0.11	0.04	0.16	0.07	0.15
Sampling time	<0.001	<0.001	<0.001	<0.001	0.008	<0.001	<0.001
T × Sampling time	0.28	0.18	0.57	0.23	0.24	0.39	0.78
Pro vs. Con	0.49	0.87	0.87	0.09	0.06	0.007 [L]	0.20
Vis vs. Con	0.08	0.32	0.08	0.006 [L]	0.02 [L]	0.04 [L]	0.01 [Q]
Pro vs. Vis	0.19	0.16	0.02	0.12	0.48	0.50	0.13

[1]: The ADY sources included Procreatin7 (Pro, Guangxi Danbaoli Yeast Co., Ltd., Laibin, China) and Vistacell (Vis, AB Vista, Wiltshire, UK). The control treatment did not contain any ADY. [2]: Others include n-valerate, iso-butyrate, and iso-valerate. [3]: Ac/Pr, acetate to propionate ratio. [L,Q]: Polynomial contrasts for dosage effects (linear, L; quadratic, Q) of individual ADYs.

The PcoA showed a separation between Vis supplementation at 1.2 g/d and the control treatment (p = 0.01, Figure 1) while Vis supplementation at 0.6 g/d did not differ from the control (p = 0.50). Both levels of Pro supplementation showed no difference from the control ($p \geq 0.23$).

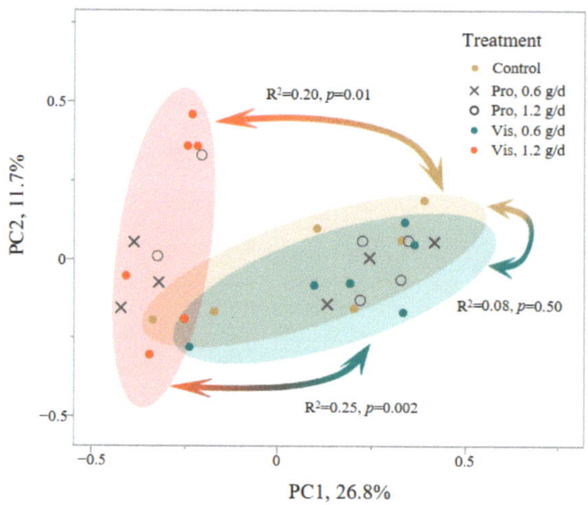

Figure 1. Principal coordinate analysis of active dry yeast (ADY) supplementation on rumen bacterial community based on Bray–Curtis dissimilarity matrix at ZOTU level. The ADY sources included Procreatin7 (Pro, Guangxi Danbaoli Yeast Co., Ltd., Laibin, China) and Vistacell (Vis, AB Vista, Wiltshire, UK). The control treatment did not contain any ADY. In particular, the three supplementation level pairs of Vis were compared using the adonis module in the vegan package.

The supplementation of Pro had no influence on the observed ZOUT and PD_whole_tree ($p \geq 0.83$, Table 5) but quadratically decreased the Pielou index (p = 0.04). The supplementation of Vis linearly decreased the Pielou index (p = 0.02). There were no differences between the Pro and Vis treatments on the alpha diversity indices of the rumen bacterial community ($p \geq 0.47$).

Table 5. Effects of active dry yeast (ADY) supplementation on rumen bacterial diversities of Hu lambs (n = 6).

Item	Observed ZOUT	PD_Whole_Tree	Pielou
Treatment, g ADY [1]/d			
Control, 0	1194	24.6	0.702
Pro, 0.6	1109	23.2	0.653
Pro, 1.2	1263	25.4	0.666
Vis, 0.6	1142	24.6	0.661
Vis, 1.2	1111	22.5	0.643
SEM	82.3	1.31	0.0168
p-value			
Pro vs. Con	0.93	0.83	0.04 ^{Q}
Vis vs. Con	0.50	0.50	0.02 ^{L}
Pro vs. Vis	0.47	0.57	0.67

[1]: The ADY sources included Procreatin7 (Pro, Guangxi Danbaoli Yeast Co., Ltd., Laibin, China) and Vistacell (Vis, AB Vista, Wiltshire, UK). The control treatment did not contain any ADY. ^{L,Q}: Polynomial contrasts for dosage effects (linear, L; quadratic, Q) of individual ADYs.

We detected a total of 150 bacterial genera using 16S rRNA gene amplicon sequencing, with the top 20 genera accounting for 89.3% of the total bacteria, making them the predominant bacteria in the rumen (Table 6). The supplementation with Pro or Vis did not affect the abundance of *Prevotella*, *Succinivibrio*, Prevotellaceae_UCG-001, *Selenomonas*, Prevotellaceae_unclassified, Rikenellaceae_RC9, *Succiniclasticum*, *Fibrobacter*, Muribaculaceae_ge, Prevotellaceae_YAB2003, *Ruminococcus*, Lachnospiraceae_unclassified, and *Oribacterium* ($p \geq 0.11$). Succinivibrionaceae_UCG-001, Selenomonadaceae_unclassified, Clostridia_UCG-014, F082_ge, Prevotellaceae_UCG-003, and *Treponema* were not influenced by Pro supplementation ($p \geq 0.14$), but Vis increased Succinivibrionaceae_UCG-001 ($p = 0.04$) and promoted a quadratic response in ($p = 0.05$). Clostridia_UCG-014 had a tendency to increase with Vis supplementation ($p = 0.10$), and the abundance of F082_ge, Prevotellaceae_UCG-003, and *Treponema* linearly decreased with an increasing Vis supplementation level ($p \leq 0.04$).

Table 6. Effects of active dry yeast (ADY) supplementation on relative abundance of top 20 rumen bacterial genera of Hu lambs (n = 6).

Genus	Treatment, g ADY [1]/d					SEM	p-Value		
	Control, 0	Pro, 0.6	Pro, 1.2	Vis, 0.6	Vis, 1.2		Pro vs. Con	Vis vs. Con	Pro vs. Vis
Prevotella	49.3	52.8	45.8	48.6	39.7	4.34	0.99	0.34	0.24
Succinivibrionaceae_UCG-001	4.19	10.8	9.62	4.56	28.6	4.75	0.30	0.04 ^{L,Q}	0.19
Succinivibrio	5.74	6.51	11.4	3.49	5.95	3.236	0.42	0.79	0.20
Prevotellaceae_UCG-001	5.68	4.53	4.17	4.93	0.95	1.707	0.53	0.19	0.40
Selenomonas	4.22	3.83	1.78	7.37	1.27	1.958	0.56	0.96	0.44
Prevotellaceae_unclassified	2.58	1.97	1.82	2.53	2.64	0.395	0.16	0.99	0.18
Rikenellaceae_RC9	2.12	1.55	1.87	1.48	1.02	0.572	0.56	0.22	0.43
Selenomonadaceae_unclassified	1.40	0.22	0.55	5.17	0.28	0.952	0.39	0.05 ^{Q}	0.01
Succiniclasticum	2.10	0.63	1.62	2.06	0.69	0.494	0.12	0.24	0.62
Fibrobacter	1.48	1.04	1.56	1.00	0.76	0.335	0.20	0.26	0.37
Muribaculaceae_ge	1.29	0.80	2.01	1.04	0.56	0.519	0.86	0.44	0.25
Prevotellaceae_YAB2003	1.01	0.83	0.99	1.76	0.64	0.527	0.88	0.77	0.59
Clostridia_UCG-014	0.73	1.05	0.83	1.09	1.34	0.235	0.46	0.10	0.26
Ruminococcus	1.02	0.82	1.08	1.09	0.61	0.332	0.85	0.67	0.77
F082_ge	1.50	0.74	1.20	0.83	0.15	0.308	0.17	0.01 ^{L}	0.10
Lachnospiraceae_unclassified	0.44	0.47	0.45	1.55	0.75	0.363	0.95	0.11	0.07
Prevotellaceae_UCG-003	1.20	0.72	0.63	0.45	0.14	0.302	0.17	0.02 ^{L}	0.21
Oribacterium	0.48	0.48	0.47	0.63	1.05	0.184	0.93	0.16	0.13
Treponema	1.04	0.65	0.42	0.41	0.25	0.275	0.14	0.04 ^{L}	0.39
Lachnospiraceae_ND3007	1.30	0.22	0.51	0.33	0.04	0.329	0.02 ^{Q}	0.01 ^{L}	0.58

[1]: The ADY sources included Procreatin7 (Pro, Guangxi Danbaoli Yeast Co., Ltd., Laibin, China) and Vistacell (Vis, AB Vista, Wiltshire, UK). The control treatment did not contain any ADY. ^{L,Q}: Polynomial contrasts for dosage effects (linear, L; quadratic, Q) of individual ADYs were run for each ADY separately plus the control.

4. Discussion

The TVFA are the main metabolic products of carbohydrate fermentation by rumen microbes [22]. Increased TVFA production reflects the extent of substrate fermentation by microbes, representing rumen fermentation activity [22]. Total gas production is also an important indicator to evaluate the activity of microbes in in vitro fermentation [23]. The results from the in vitro experiment show that only Pro and Vis enhanced the total gas production, with Vis also promoting an increase in the TVFA concentration. These findings suggest that these two additives play a role in promoting rumen fermentation. Pro and YSF increased the molar percentage of propionate to varying degrees in vitro. Similar to our results, in vitro fermentation inoculated with rumen fluid of dairy cows showed an increase in the molar percentage of propionate by ADYs [24]. In most cases, acetate is the main product of yeast fermentation, not propionate [25]. Increases in the propionate concentration as a consequence of yeast supplementation have been attributed to the substances produced by yeast metabolism promoting propionate producers in the rumen [4,24]. Propionate production incorporates metabolic hydrogen. Our results suggest that the ADY may have redirected metabolic hydrogen towards propionate production. Additionally, Vis resulted in a decreased lactate concentration. Because of these results, we selected Vis and Pro for their evaluation in vivo.

The meta-analysis by Sales [26] showed that the effects of ADY on nutrient digestibility vary substantially across studies, especially for fiber digestibility. This is consistent with our finding that Vis but not Pro improved fiber digestibility in vivo. Similar to our results, McGinn et al. [27] also found that Pro had little effect on the NDF digestibility of steers. While we verified Vis to enhance fiber digestion in finishing feedlot lambs, it had no effect on fiber digestibility in lactating dairy cows [28]. Sales [26] further noted that ADY had a lesser effect on dietary NDF digestibility under high concentrate diet conditions, which was primarily attributed to the limited action of yeast in the rumen due to the rapid rumen passage rate induced by a high concentrate diet. Even though the diets offered to lambs in our study were relatively high in concentrate content, favorable effects of Vis on fiber digestion were noted. It has been reported that the supplementation of ADY can decrease the fecal N excretion of lambs [29]. However, our results align with others that found no effect of ADY on the N balance in finishing lambs [30,31].

Vis but not Pro supplementation tended to increase the TVFA concentration in the rumen. This finding confirms that Vis has a greater impact on rumen fermentation than Pro. In vivo, as opposed to in vitro, Vis shifted rumen fermentation from n-butyrate to propionate to a greater extent than Pro. Conflicting in vitro and in vivo results have been discussed [32], underscoring the need for caution when only in vitro results are available. Additionally, Garcia Diaz et al. [33] did not observe the effect of Pro on the sheep rumen VFA profile in vivo. While there have been no reports about the effects of Vis in rumen fermentation in lambs, Vis has been reported not to affect the rumen VFA profile in lactating dairy cows [28] and finishing feedlot steers [34].

Lactate is produced in the rumen as an intermediate product of carbohydrate fermentation, which can accumulate with highly fermentable diets [35]. Studies by Lynch and Martin [36] and Lila et al. [24] found that ADY supplementation increased propionate production and reduced lactate accumulation. The authors suggested that ADY may promote lactate utilization by stimulating the growth of lactate-utilizing bacteria [24,36]. This is important because excessive lactate accumulation can lead to rumen acidosis, which can have negative impacts on animal health and productivity [35]. Malekkhahi et al. [37] related ADY to the metabolism of lactate to propionate and n-butyrate in the rumen of lactating dairy cows by *Megasphaera*. Even though the lactate concentration was decreased by Vis in our in vitro experiment, the lactate basal concentration in the control treatment was rather small, and, clearly, decreases in lactate of about 25 μM would have been unnoticeable if lactate had been metabolized to propionate and butyrate, which were at concentrations close to three orders of magnitude higher than lactate. Moreover, *Megasphaera* spp. were not detected in the major bacteria, which agrees with our presumption that lactate metabolism was quantitatively unimportant in our in vitro experiment.

The bacterial alpha diversity and beta diversity were influenced by Vis but not by Pro. This observation aligns with our findings of Vis affecting the NDF digestibility and rumen VFA profile in vivo. Pro supplementation decreased the abundance of the putative genus Selenomonadaceae_unclassified, which is presently functional information. A recent in vitro analysis indicated that the abundance of Selenomonadaceae_unclassified increased in association with a decrease in the propionate molar percentage [38]. In agreement, we observed that Pro supplementation decreased the abundance of Selenomonadaceae_unclassified and the propionate molar percentage. The relative abundances of F082_ge, Prevotellaceae_UCG-003, and Lachnospiraceae_ND3007 were decreased by Vis supplementation. F082_ge and Prevotellaceae_UCG-003 are associated with the production of propionate using the succinate pathway [39,40]. In contrast, in our study, Vis supplementation decreased the relative abundance of F082_ge and Prevotellaceae_UCG-003 but increased the molar percentage of propionate in the rumen. This indicates that other bacteria or pathways may be compensating for the decrease in the F082_ge and Prevotellaceae_UCG-003 abundance to shift the fermentation profile towards propionate. Lachnospiraceae_ND3007 produces n-butyrate from the degradation of pectins [41]. It is possible that the decreased abundance of Lachnospiraceae_ND3007 resulting from Vis supplementation contributed to account for the decrease in the molar percentage of n-butyrate in the rumen of lambs, although the basal abundance of Lachnospiraceae_ND3007 was quite low and it is, therefore, uncertain how much Lachnospiraceae_ND3007 might have contributed to n-butyrate production.

5. Conclusions

The study found that the impact of ADYs on rumen fermentation varies, with Vis yeast being optimal for lamb growth. It enhanced propionate and Succinivibrionaceae_UCG-001 while reducing n-butyrate and Lachnospiraceae_ND3007. The in vitro and in vivo results diverged, cautioning against generalizing batch culture findings.

Author Contributions: H.L.: methodology, investigation, software, validation, data curation, writing—original draft. F.L.: conceptualization, data curation, supervision, funding acquisition, project administration. Z.M.: conceptualization, methodology, investigation, software, data curation, writing—review and editing, visualization. M.M.: conceptualization, methodology, investigation, software, data curation. E.U.: data curation, software, writing—review and editing. X.W. and Z.Z.: investigation, methodology. B.L., X.D., and L.G.: conceptualization, investigation, resources. All authors have read and agreed to the published version of the manuscript.

Funding: This research was funded by the Fundamental Research Funds for the Leading Scientist Project of Qinghai Province (Grant No. 2023-NK-147).

Institutional Review Board Statement: The animal study protocol was approved by the Animal Care and Use Committee of Lanzhou University, Lanzhou, China (approval number CPAST-2020–LI-3).

Informed Consent Statement: Not applicable.

Data Availability Statement: The raw sequence reads of the bacterial amplicons are available in the Genome Sequence Archive of China National Center for Bioinformation under accession NO. PRJCA016785 (URL: https://www.cncb.ac.cn/?lang=en, accessed on 5 May 2023).

Conflicts of Interest: Authors Baocang Liu and Xiaoyu Deng were employed by the company Aksu Tycoon Feed Co., Ltd. Author Liqing Guo was employed by the company Tecon Pharmaceutical Co., Ltd. The remaining authors declare that the research was conducted in the absence of any commercial or financial relationships that could be construed as a potential conflict of interest.

References

1. Enemark, J.M.D. The monitoring, prevention and treatment of sub-acute ruminal acidosis (SARA): A review. *Vet. J.* **2008**, *176*, 32–43. [CrossRef] [PubMed]
2. Guo, T.; Wang, Z.L.; Guo, L.; Li, F.; Li, F. Effects of supplementation of nonforage fiber source in diets with different starch levels on growth performance, rumen fermentation, nutrient digestion, and microbial flora of *Hu* lambs. *Transl. Anim. Sci.* **2021**, *5*, txab65. [CrossRef] [PubMed]
3. Li, F.; Cao, Y.; Liu, N.; Yang, X.; Yao, J.; Yan, D. Subacute ruminal acidosis challenge changed in situ degradability of feedstuffs in dairy goats. *J. Dairy Sci.* **2014**, *97*, 5101–5109. [CrossRef]
4. Baker, L.M.; Kraft, J.; Karnezos, T.P.; Greenwood, S.L. The effects of dietary yeast and yeast-derived extracts on rumen microbiota and their function. *Anim. Feed Sci. Tech.* **2022**, *294*, 115476. [CrossRef]
5. Wang, R.; Wang, M.; Lin, B.; Ungerfeld, E.M.; Ma, Z.Y.; Wu, T.T.; Wen, J.N.; Zhang, X.M.; Deng, J.P.; Tan, Z.L. Associations of ruminal hydrogen and pH with fiber digestibility and microbiota composition induced by increasing starch intake in beef cattle. *Anim. Feed Sci. Tech.* **2021**, *278*, 114980. [CrossRef]
6. Amin, A.B.; Mao, S. Influence of yeast on rumen fermentation, growth performance and quality of products in ruminants: A review. *Anim. Nutr.* **2021**, *7*, 31–41. [CrossRef]
7. Ingledew, W.M.; Jones, G.A. The fate of live brevers' yeast slurry in bovine rumen fluid. *J. Inst. Brew.* **1982**, *88*, 18–20. [CrossRef]
8. Ma, Z.Y.; Zhou, J.W.; Yi, S.Y.; Wang, M.; Tan, Z.L. In vitro inoculation of fresh or frozen rumen fluid distinguishes contrasting microbial communities and fermentation induced by increasing forage to concentrate ratio. *Front. Nutr.* **2022**, *8*, 772645. [CrossRef]
9. Menke, K.H.; Raab, L.; Salewski, A.; Steingass, H.; Fritz, D.; Schneider, W. The estimation of the digestibility and metabolizable energy content of ruminant feedingstuffs from the gas production when they are incubated with rumen liquor in vitro. *J. Agric. Sci.* **1979**, *93*, 217–222. [CrossRef]
10. Cruyt, F.; Sousa, C.A.; Machado, M.D.; Soares, E.V. Improvement of the slide culture technique for the assessment of yeast viability. *J. Inst. Brew.* **2017**, *123*, 39–44. [CrossRef]
11. AOAC. *Official Methods of Analysis of AOAC International*; AOAC International: Washington, DC, USA, 2005.
12. Van Soest, P.J.; Robertson, J.B.; Lewis, B.A. Methods for dietary fiber, neutral detergent fiber, and nonstarch polysaccharides in relation to animal nutrition. *J. Dairy Sci.* **1991**, *74*, 3583–3597. [CrossRef]
13. Weatherburn, M.W. Phenol-hypochlorite reaction for determination of ammonia. *Anal. Chem.* **1967**, *39*, 971–974. [CrossRef]
14. Wang, H.; Cheng, H.; Wang, F.; Wei, D.; Wang, X. An improved 3-(4,5-dimethylthiazol-2-yl)-2,5-diphenyl tetrazolium bromide (MTT) reduction assay for evaluating the viability of Escherichia coli cells. *J. Microbiol. Meth.* **2010**, *82*, 330–333. [CrossRef] [PubMed]
15. Ma, Z.Y.; Zhang, X.M.; Wang, R.; Wang, M.; Liu, T.; Tan, Z.L. Effects of chemical and mechanical lysis on microbial DNA yield, integrity, and downstream amplicon sequencing of rumen bacteria and protozoa. *Front. Microbiol.* **2020**, *11*, 581227. [CrossRef] [PubMed]
16. Takahashi, S.; Tomita, J.; Nishioka, K.; Hisada, T.; Nishijima, M. Development of a prokaryotic universal primer for simultaneous analysis of Bacteria and Archaea using next-generation sequencing. *PLoS ONE* **2014**, *9*, e105592. [CrossRef]
17. Edgar, R.C. Search and clustering orders of magnitude faster than BLAST. *Bioinformatics* **2010**, *26*, 2460–2461. [CrossRef]
18. Schloss, P.D.; Westcott, S.L.; Ryabin, T.; Hall, J.R.; Hartmann, M.; Hollister, E.B.; Lesniewski, R.A.; Oakley, B.B.; Parks, D.H.; Robinson, C.J.; et al. Introducing mothur: Open-source, platform-independent, community-supported software for describing and comparing microbial communities. *Appl. Environ. Microb.* **2009**, *75*, 7537–7541. [CrossRef]
19. Quast, C.; Pruesse, E.; Yilmaz, P.; Gerken, J.; Schweer, T.; Yarza, P.; Peplies, J.; Glöckner, F.O. The SILVA ribosomal RNA gene database project: Improved data processing and web-based tools. *Nucleic Acids Res.* **2013**, *41*, D590–D596. [CrossRef] [PubMed]
20. Rognes, T.; Flouri, T.; Nichols, B.; Quince, C.; Mahé, F. VSEARCH: A versatile open source tool for metagenomics. *PeerJ* **2016**, *4*, e2584. [CrossRef]
21. R Core Team. R: A Language and Environment for Statistical Computing. Available online: https://www.R-project.org/ (accessed on 3 September 2020).
22. van Houtert, M.F.J. The production and metabolism of volatile fatty acids by ruminants fed roughages: A review. *Anim. Feed Sci. Tech.* **1993**, *43*, 189–225. [CrossRef]
23. Gallo, A.; Giuberti, G.; Atzori, A.S.; Masoero, F.; Gallo, A.; Giuberti, G.; Atzori, A.; Masoero, F. Short communication: *In vitro* rumen gas production and starch degradation of starch-based feeds depend on mean particle size. *J. Dairy Sci.* **2018**, *101*, 6142–6149. [CrossRef] [PubMed]
24. Lila, Z.A.; Mohammed, N.; Yasui, T.; Kurokawa, Y.; Kanda, S.; Itabashi, H. Effects of a twin strain of Saccharomyces cerevisiae live cells on mixed ruminal microorganism fermentation in vitro. *J. Anim. Sci.* **2004**, *82*, 1847–1854. [CrossRef] [PubMed]
25. Rossouw, D.; Olivares-Hernandes, R.; Nielsen, J.; Bauer, F.F. Comparative transcriptomic approach to investigate differences in wine yeast physiology and metabolism during fermentation. *Appl. Environ. Microb.* **2009**, *75*, 6600–6612. [CrossRef]
26. Sales, J. Effects of Saccharomyces cerevisiae supplementation on ruminal parameters, nutrient digestibility and growth in sheep: A meta-analysis. *Small Rumin. Res.* **2011**, *100*, 19–29. [CrossRef]
27. McGinn, S.M.; Beauchemin, K.A.; Coates, T.; Colombatto, D. Methane emissions from beef cattle: Effects of monensin, sunflower oil, enzymes, yeast, and fumaric acid. *J. Anim. Sci.* **2004**, *82*, 3346–3356. [CrossRef]

28. Ambriz-Vilchis, V.; Jessop, N.S.; Fawcett, R.H.; Webster, M.; Shaw, D.; Walker, N.; Macrae, A. Effect of yeast supplementation on performance, rumination time, and rumen pH of dairy cows in commercial farm environments. *J. Dairy Sci.* **2017**, *100*, 5449–5461. [CrossRef]
29. Tripathi, M.K.; Karim, S.A.; Chaturvedi, O.H.; Verma, D.L. Effect of different liquid cultures of live yeast strains on performance, ruminal fermentation and microbial protein synthesis in lambs. *J. Anim. Physiol. Nutr.* **2008**, *92*, 631–639. [CrossRef]
30. Kawas, J.R.; García-Castillo, R.; Fimbres-Durazo, H.; Kawas, J.; García-Castillo, R.; Fimbres-Durazo, H.; Garza-Cazares, F.; Hernández-Vidal, J.; Olivares-Sáenz, E.; Lu, C. Effects of sodium bicarbonate and yeast on nutrient intake, digestibility, and ruminal fermentation of light-weight lambs fed finishing diets. *Small Rumin. Res.* **2007**, *67*, 149–156. [CrossRef]
31. Soren, N.M.; Tripathi, M.K.; Bhatt, R.S.; Karim, S.A. Effect of yeast supplementation on the growth performance of *Malpura* lambs. *Trop. Anim. Health Pro.* **2013**, *45*, 547–554. [CrossRef]
32. Weinert-Nelson, J.R.; Ely, D.G.; Flythe, M.D.; Hamilton, T.A.; Ferrell, J.L.; Davis, B.E. *Ex vivo* fermentation of hay and corn by rumen bacteria from cattle and sheep. *Fermentation* **2023**, *9*, 929. [CrossRef]
33. Garcia Diaz, T.; Ferriani Branco, A.; Jacovaci, F.A.; Jobim, C.C.; Bolson, D.C.; Daniel, J.L.P. Inclusion of live yeast and mannan-oligosaccharides in high grain-based diets for sheep: Ruminal parameters, inflammatory response and rumen morphology. *PLoS ONE* **2018**, *13*, e193313. [CrossRef] [PubMed]
34. Williams, M.S.; AlZahal, O.; Mandell, I.B.; McBride, B.W.; Wood, K.M. The impacts of a fibrolytic enzyme additive on digestibility and performance in the grower and early finisher period, and supplemental *Saccharomyces cerevisiae* on performance and rumen health in the late finisher period for feedlot cattle. *Can. J. Anim. Sci.* **2021**, *101*, 527–547. [CrossRef]
35. Chen, L.; Shen, Y.; Wang, C.; Ding, L.; Zhao, F.; Wang, M.; Fu, J.; Wang, H. *Megasphaera elsdenii* lactate degradation pattern shifts in rumen acidosis models. *Front. Microbiol.* **2019**, *10*, 162. [CrossRef] [PubMed]
36. Lynch, H.A.; Martin, S.A. Effects of Saccharomyces cerevisiae culture and Saccharomyces cerevisiae live cells on in vitro mixed ruminal microorganism fermentation. *J. Dairy Sci.* **2002**, *85*, 2603–2608. [CrossRef]
37. Malekkhahi, M.; Tahmasbi, A.M.; Naserian, A.A.; Danesh-Mesgaran, M.; Kleen, J.L.; AlZahal, O.; Ghaffari, M.H. Effects of supplementation of active dried yeast and malate during sub-acute ruminal acidosis on rumen fermentation, microbial population, selected blood metabolites, and milk production in dairy cows. *Anim. Feed Sci. Tech.* **2016**, *213*, 29–43. [CrossRef]
38. Kim, H.; Park, T.; Kwon, I.; Seo, J. Specific inhibition of *Streptococcus bovis* by endolysin LyJH307 supplementation shifts the rumen microbiota and metabolic pathways related to carbohydrate metabolism. *J. Anim. Sci. Biotechno.* **2021**, *12*, 93. [CrossRef]
39. Jize, Z.; Zhuoga, D.; Xiaoqing, Z.; Na, T.; Jiacuo, G.; Cuicheng, L.; Bandan, P. Different feeding strategies can affect growth performance and rumen functions in *Gangba* sheep as revealed by integrated transcriptome and microbiome analyses. *Front. Microbiol.* **2022**, *13*, 908326. [CrossRef]
40. Wang, H.; He, Y.; Li, H.; Wu, F.; Qiu, Q.; Niu, W.; Gao, Z.; Su, H.; Cao, B. Rumen fermentation, intramuscular fat fatty acid profiles and related rumen bacterial populations of Holstein bulls fed diets with different energy levels. *Appl. Microbiol. Biotechnol.* **2019**, *103*, 4931–4942. [CrossRef]
41. Meehan, C.J.; Beiko, R.G. A phylogenomic view of ecological specialization in the Lachnospiraceae, a family of digestive tract-associated bacteria. *Genome Biol. Evol.* **2014**, *6*, 703–713. [CrossRef]

animals

MDPI

Article

Sargassum mcclurei Mitigating Methane Emissions and Affecting Rumen Microbial Community in In Vitro Rumen Fermentation

Shuai Li [1,2,†], Yi Sun [1,3,†], Tongjun Guo [4], Wenyou Liu [1,5], Xiong Tong [1], Zhifei Zhang [1], Jiajie Sun [2], Yufeng Yang [3], Shuli Yang [5], Dagang Li [1,*] and Li Min [1,4,*]

1 Southern Marine Science and Engineering Guangdong Laboratory (Zhuhai), Ministry of Agriculture Key Laboratory of Animal Nutrition and Feed Science in South China, Institute of Animal Science, Guangdong Academy of Agricultural Sciences, Guangzhou 510640, China; ls32649048@163.com (S.L.); sunyi@stu2022.jnu.edu.cn (Y.S.); lwy5205202023@163.com (W.L.); tongxiong@gdaas.cn (X.T.); zhangzhifei@gdaas.cn (Z.Z.)
2 Guangdong Provincial Key Laboratory of Animal Nutrition Regulation, College of Animal Science, South China Agricultural University, Guangzhou 510642, China; jiajiesun@scau.edu.cn
3 Southern Marine Science and Engineering Guangdong Laboratory (Zhuhai), College of Life Science and Technology, Jinan University, Guangzhou 510632, China; tyyf@jnu.edu.cn
4 Key Laboratory of Xinjiang feed biotechnology, Feed Research Institute, Xinjiang Academy of Animal Science, Urumqi 830000, China; guotaoxj@126.com
5 College of Life Sciences and Engineering, Foshan University, Foshan 528231, China; yangshuli77@163.com
* Correspondence: lidagang@gdaas.cn (D.L.); minli@gdaas.cn (L.M.)
† These authors contributed equally to this work.

Simple Summary: Methane (CH_4) is a by-product of ruminant microbial fermentation, leading to a 2–12% loss of feed energy. The greenhouse gas effect of methane in the atmosphere is 28 times greater than that of carbon dioxide. Recent studies have shown that supplementing feed additives can effectively reduce ruminal methane emissions. This study aimed to evaluate the effectiveness of *Sargassum mcclurei* (*S. mcclurei*) in mitigating methane emissions using different treatment methods and supplementation levels through in vitro rumen fermentation. Three supplementation levels were tested for both dried and freeze-dried treatments over 48 h of in vitro rumen fermentation. Fermentation gas production was recorded, and after fermentation, methane production, dry matter degradation (DMD), and various fermentation parameters were measured. The addition of *S. mcclurei* affected crude protein degradation (CPD) and volatile fatty acid (VFA) production. The freeze-dried *S. mcclurei* at a 2% supplementation level reduced methane emissions by 18.85%.

Abstract: Methane emissions from ruminants significantly contribute to greenhouse gases. This study explores the methane mitigation effect and mechanism of *S. mcclurei* through in vitro rumen fermentation, aiming to establish its potential as a feed additive. We investigated the effects of freeze-dried and dried *S. mcclurei* at supplementation levels of 2%, 5%, and 10% of dry matter on nutrient degradation, ruminal fermentation, methane inhibition, and microbial community structure in in vitro rumen fermentation. The freeze-dried *S. mcclurei* at 2% supplementation significantly reduced CH_4 emissions by 18.85% and enhanced crude protein degradability. However, total VFA and acetate concentrations were lower in both treatments compared to the control. The microbial shifts included a decrease in *Lachnospiraceae_NK3A20_group* and *Ruminococcus* and an increase in *Selenomonas*, *Succinivibrio*, and *Saccharofermentans*, promoting propionate production. Additionally, a significant reduction in *Methanomicrobium* was observed, indicating direct methane mitigation. Freeze-dried *S. mcclurei* at a 2% supplementation level shows potential as an effective methane mitigation strategy with minimal impact on rumen fermentation, supported by detailed insights into microbial community changes.

Keywords: rumen fermentation; seaweed; methane emissions; rumen microbiota

Citation: Li, S.; Sun, Y.; Guo, T.; Liu, W.; Tong, X.; Zhang, Z.; Sun, J.; Yang, Y.; Yang, S.; Li, D.; et al. *Sargassum mcclurei* Mitigating Methane Emissions and Affecting Rumen Microbial Community in In Vitro Rumen Fermentation. *Animals* **2024**, *14*, 2057. https://doi.org/10.3390/ani14142057

Academic Editor: Lizhi Wang

Received: 12 June 2024
Revised: 6 July 2024
Accepted: 10 July 2024
Published: 13 July 2024

1. Introduction

Greenhouse gas (GHG) emissions lead to global warming. Livestock emissions of GHGs, a significant source of the agricultural sector, account for about 15% of annual anthropogenic emissions [1]. Ruminants, particularly beef and dairy cattle, contribute most to GHG emissions from livestock [1]. Methane emissions from enteric fermentation represent about 41% of total GHG emissions from the agricultural sector [2]. It is widely recognized that the mitigation of CH_4 emissions from ruminants is crucial to address GHG emissions. Various feed additives have demonstrated the potential to mitigate ruminal CH_4 emissions by regulating pathways involved in microbial methanogenesis [3,4]. Of these, seaweed has been shown as a promising feed additive for mitigating rumen CH_4 emissions. Seaweeds are among the world's largest unexploited and renewable global biomass resources, with the total production of macroalgae reaching 32.4 million tons in 2018 across approximately 50 countries [5]. Asia, particularly China, has seaweed production contributing over 97% of the world's production with more than 18 million tons [6].

Seaweeds contain many bioactive compounds, two of which are particularly noteworthy: bromoform in red seaweed and phlorotannins in brown seaweed. Both of these compounds are known as CH_4 inhibitors [7,8]. However, the bioactive components of seaweed are affected by many factors, such as species, season, and the location of harvesting [9]. Current studies have indicated that terrestrial tannins can effectively mitigate CH_4 production [10,11]. It has been reported that phlorotannins have chemical structural similarities to terrestrial tannins [12]. The *S. mcclurei* used in our study is a brown seaweed rich in phlorotannins. Therefore, we hypothesize that phlorotannins in brown seaweed may have a CH_4-mitigating effect similar to that of terrestrial tannins. However, the effectiveness of reducing methane emissions is related to seaweed supplementation levels [7,8] and also processing methods including freeze-drying and drying [4,13]. According to reports, freeze-drying can extend the storage time of biomass in seaweed, which might play a vital role in CH_4 emission [14].

Hence, this study aims to explore the effects of the in vitro rumen fermentation of brown seaweed, *S. mcclurei*, with two treatments (freeze-drying and drying) and three supplementation levels (0, 2, 5, and 10% of DM basis) on nutrient degradation, rumen fermentation parameters, methane emission inhibition effect, and microbial community structure, thereby evaluating the use of *S. mcclurei* as a feed additive to feasibly alleviate rumen greenhouse gas emissions.

2. Materials and Methods

2.1. Preparation of Seaweed and TMR Substrate

The brown seaweed of *S. mcclurei* was harvested from Weizhou Island (Beihai, Guangxi, China). The seaweed was submerged for 1 min in freshwater and divided into two equal portions. Each portion was then dried through a spin-dryer until no water continued to flow out. One portion was frozen at −80 °C for 24 h then vacuum freeze-dried at −30 °C and 0.37 vacuum degrees for 48 h (CHRIST ALPHA2-4, Osterode, Germany). The second portion was immediately dried in an oven at 65 °C for 48 h (DHG-9075A, Shanghai, China). The samples of seaweed were milled to 1 mm and stored at −20 °C. The Total Mixed Ration (TMR) that constituted corn straw and concentrate supplementation was milled to 1 mm and stored in a silica gel desiccator.

Dry matter (DM) was determined by the achievement of constant weight at 105 °C, and organic matter (OM) was measured as the loss on combustion at 550 °C for 8 h [15]. Neutral and acid detergent fiber (NDF and ADF) content were analyzed following the methods outlined by Van Soest [16] and crude protein (CP) was analyzed by the method of Thiex [17].

2.2. In Vitro Treatment

Three healthy Holstein cows were selected as rumen fluid donors and fed twice a day at 7:00 and 17:00 with free drinking water. Rumen fluid was collected 2 h post-morning feeding, blended, and filtered separately using a four-layer cheesecloth into a pre-warmed thermos flushed with carbon dioxide (CO_2) before collection. The rumen fluid was rapidly transferred to the laboratory and mixed with rumen fluid buffer solution at a ratio of 1:2 (rumen fluid: buffer solution), as in the method of Menke [18], and maintained in water at 39 °C, ensuring continuous CO_2.

The mixed rumen fluid (75 mL) was dispensed into 100 mL incubation bottles containing 500 mg of a TMR substrate (60:40 corn straw: concentrate supplementation) and two different treatments of *S. mcclurei* (freeze-dried and dried). Four levels of supplementation (0, 2, 5, and 10% of DM basis) were set for each treatment. Six biological replicates were set for each supplementation level, and the experiment was repeated twice. Six samples were randomly selected for subsequent experimental analysis. These varying supplementation levels were designed to determine the optimal level for mitigating CH_4 emissions and to evaluate the potential adverse effects of high supplementation levels on rumen fermentation [13].

The incubation bottles were rinsed with CO_2, sealed with butyl plugs and aluminum caps, and placed into a constant temperature water bath shaker at 39 °C and 85 RPM for 48 h.

2.3. *Experimental Sample Collection and Analysis*

2.3.1. Gas Production Collection

The total gas production was collected from the incubation bottles at 2, 4, 8, 12, 24, and 48 h using a 30 mL syringe. The samples were stored in aluminum foil airbags until analysis. Concentrations of CH_4 and CO_2 were analyzed by gas chromatography with a 5A ZSF-SS column (Φ3 mm × 3 m, Support 60–80 mesh Chromosorb) and a thermal conductivity detector (SP-2060T, Beijing, China). The detector is at a constant temperature, and Argon (Ar) as the carrier gas has a flow rate of 30 mL/min.

2.3.2. Fermentation Parameter Determination

At the end of the treatment (48 h), the fermenter pH was immediately measured. Three 5 mL rumen fluid samples were placed in centrifuge tubes for analysis of VFA composition, microprotein (MCP), and ammonia nitrogen (NH_3-N) concentrations, respectively. Samples were stored at −80 °C until analysis. The concentration of VFA was analyzed using gas chromatography (Agilent 6890N, Palo alto, CA, USA) as described by [19]. The determination of NH_3-N and MCP was conducted as described by Vissers [20]. The in vitro rumen fermentation of nutrient digestibility was measured using the substrate collected in a nylon bag after incubation. The bags were washed with fresh water until clear, then over-dried at 65 °C to a constant weight. Briefly, the digestibility of DM, CP, NDF, and ADF (DMD, CPD, NDFD, and ADFD) was defined as weight loss compared to the pre-incubation weights in the incubation bottles.

2.3.3. DNA Extraction and Bacterial and Archaeal 16S rRNA Gene Sequencing and Analysis

Total genomic DNA was extracted from in vitro fermentation rumen fluid samples using the E.Z.N.A.® soil DNA Kit (Omega Bio-tek, Norcross, GA, USA) following the manufacturer's instructions. The hypervariable region V3-V4 of the bacterial 16S rRNA gene was amplified with primer pairs 338F (5′-ACTCCTACGGGAGGCAGCAG-3′) and 806R (5′-GGACTACHVGGGTWTCTAAT-3′) by a T100 Thermal Cycler PCR thermocycler (BIO-RAD, Hercules, CA, USA). The e V4-V5 region of the archaeal 16S rRNA genes was amplified by 349F (5′-GYGCASCAGKCGMGAAW-3′) and 806R (5′-GGACTACVSGGGTATCTAAT-3′). The PCR reaction mixture included 10 ng of template DNA, 0.8 μL of each primer (5 μM), 2 μL of 2.5 mM dNTPs, 0.4 μL of Fast Pfu polymerase, 4 μL of 5 × Fast Pfu buffer, and ddH2O to a final volume of 20 μL. PCR amplification cycling conditions were as follows: initial denaturation at 95 °C for 3 min, followed by 27 cycles of denaturing at 95 °C for 30 s, annealing at 55 °C for 30 s, extension at 72 °C for 45 s, single extension at 72 °C for 10 min, and ending at 4 °C. The PCR product was extracted from 2% agarose gel and purified using the PCR Clean-Up Kit (YuHua, Shanghai, China) according to the manufacturer's instructions and quantified using Qubit 4.0 (Thermo Fisher Scientific, Waltham, MA, USA).

Bioinformatic analysis of the rumen microbiota was carried out using the Majorbio Cloud platform (https://cloud.majorbio.com, accessed on 24 November 2023). Raw FASTQ files were de-multiplexed using an in-house perl script and then quality-filtered by fast version 0.19.6 and merged by FLASH version 1.2.7. Next, the primers and barcodes were removed, and chimeras were filtered to obtain valid reads. After filtration, the numbers of valid reads of bacterial and archaeal communities were 1,117,522 and 1,494,549, respectively, and the average lengths of valid reads were 419 and 375, respectively. Then, the optimized sequences were clustered into operational taxonomic units (OTUs) using UPARSE 7.1 with a 97% sequence similarity level. The most abundant sequence for each OTU was selected as a representative sequence. To minimize the effects of sequencing depth on alpha and beta diversity measure, the number of 16S rRNA gene sequences from each sample was rarefied to 20,000, which still yielded an average Good's coverage of 99.09%.

Based on the OTU information, rarefaction curves and alphas were calculated with Mothur v1.30.1. The similarity among the microbial communities in different samples was determined by principal coordinate analysis (PCoA) based on Bray–Curtis dissimilarity using the Vegan v2.5-3 package. The linear discriminant analysis (LDA) effect size (LEfSe) (http://huttenhower.sph.harvard.edu/LEfSe, accessed on 24 November 2023) was performed to identify the significantly abundant taxa (phylum to genera) of bacteria among the different groups (LDA score > 2, $p < 0.05$).

2.4. Statistical Analysis

One-factor ANOVA was used to test for significant differences in the fermenter parameter (pH, DMD, NDFD, and ADFD), the total gas produced over time, and the production of CH_4 and CO_2 during the experimental period. A one-factor ANOVA was also used to test for significant differences in VFA, MCP, and NH_3-N concentration. All ANOVA analyses were performed using IBM SPSS (27.0) software. $p < 0.05$ was considered a significant difference.

3. Results

3.1. Effects of *S. mcclurei* on the CH_4 Production and Nutrient Degradation of In Vitro Rumen Fermentation

As shown in Table 1, the total gas production (mL/g DM) was decreased ($p < 0.05$) in all treatment groups compared to control group (CON) during the incubation period. The decrease was affected by the supplementation of *S. mcclurei*, with the lowest yields observed at the 10% inclusion level for both the dried and freeze-dried treatments. The total gas production decreased by 13.96% ($p < 0.01$) and 11.61% ($p = 0.02$), respectively. Each supplementation level in the dried group had the effect of mitigating CH_4 emissions, but the difference in mitigation effect between each supplementation level was not significant. Furthermore, the 2% supplementation level in the freeze-dried group had the most effective mitigation of CH_4 emission (18.85%, $p < 0.05$), while the other two supplementation levels had no significant effect on CH_4 concentration. No significant difference in CO_2 production was observed between CON and treatment groups.

The chemical composition of the diet and *S. mcclurei* in the study is shown in Table 2. Different treatments and supplementation levels of *S. mcclurei* influenced the degradation of all analyzed nutrients in the incubation bottles (Table 3). Notably, nutrient degradation (DMD, NDFD, and ADFD) tended to be inhibited in all treatment groups when the seaweed supplementation level was 10%, compared to the CON. In the freeze-dried group, the degradation of crude protein increased with the supplementation of low levels (2% of substrate DM) of *S. mcclurei*. However, as the supplementation level increased, the degradation of crude protein was inhibited.

Table 1. The cumulative total gas and gas composition (CH_4 and CO_2) of the two treatments and different supplementation levels after in vitro rumen fermentation for 48 h.

Parameter	CON	2%	5%	10%	*p*
D					
TGP mL	142.48 ± 15.22	138.43 ± 2.13	124.77 ± 14.79	134.86 ± 4.19	0.06
CH_4 mL	12.08 ± 2.27	10.63 ± 0.78	10.11 ± 1.16	10.07 ± 0.77	0.07
CO_2 mL	89.79 ± 21.28	90.80 ± 5.76	82.28 ± 9.10	93.27 ± 3.66	0.44
TGP mL/g DM	284.62 ± 30.35 ^{A}	271.02 ± 4.17 ^{A}	237.34 ± 28.18 ^{B}	244.86 ± 7.59 ^{B}	<0.01
CH_4 mL/g DM	24.13 ± 4.45 ^{A}	20.82 ± 1.52 ^{B}	19.23 ± 2.22 ^{B}	18.29 ± 1.41 ^{B}	<0.01
CO_2 mL/g DM	179.38 ± 42.49	177.77 ± 11.30	156.52 ± 17.32	169.34 ± 6.65	0.35
F					
TGP mL	142.48 ± 15.22	136.86 ± 2.88	143.55 ± 4.30	138.51 ± 7.16	0.51
CH_4 mL	12.08 ± 2.27 ^{a}	10.00 ± 0.92 ^{b}	12.27 ± 0.88 ^{a}	11.86 ± 0.98 ^{a}	0.03
CO_2 mL	89.79 ± 21.28	90.75 ± 4.41	96.53 ± 3.71	91.18 ± 5.66	0.73
TGP mL/g DM	284.62 ± 30.35 ^{a}	267.97 ± 5.69 ^{ab}	273.08 ± 8.11 ^{ab}	251.57 ± 12.95 ^{b}	0.02
CH_4 mL/g DM	24.13 ± 4.54 ^{a}	19.58 ± 1.81 ^{b}	23.34 ± 1.67 ^{ab}	21.55 ± 1.79 ^{ab}	0.04
CO_2 mL/g DM	179.38 ± 42.49	177.70 ± 8.73	183.64 ± 7.03	165.61 ± 10.26	0.56

D, dried treatment; F, freeze-dried treatment; CON, control group; 2%, CON plus *S. mcclurei* (20 mg/g DM); 5%, CON plus *S. mcclurei* (50 mg/g DM); 10%, CON plus *S. mcclurei* (100 mg/g DM). ^{ab} Means bearing different superscripts in the same row differ significantly ($p < 0.05$). ^{AB} Means bearing different superscripts in the same row differ significantly ($p < 0.01$).

Table 2. Chemical composition of substrates used in the in vitro rumen fermentation.

Parameter (% of DM).	Corn Straw	Concentrate [1]	*S. mcclurei*
DM	26.26	93.53	NA
OM	92.85	92.81	79.17
CP	7.12	20.46	12.71
NDF	38.48	16.61	20.85
ADF	21.69	6.22	16.89
Ash	7.14	7.18	20.82

DM, dry matter; OM, organic matter; CP, crude protein; NDF, neutral detergent fiber; ADF, acid detergent fiber; NA: not available. [1] Composition of concentrate: corn 500 g/kg, DDGS (distiller's dried grains with solubles) 235 g/kg, soybean meal 220 g/kg, stone powder 10 g/kg, dicalcium phosphate 9 g/kg, multivitamin 4 g/kg, salt 10 g/kg, multi-mineral 1 g/kg, baking soda 10 g/kg, mold inhibitor 1.5 g/kg.

Table 3. Effect of the two treatments and different supplementation levels *S. mcclurei* on nutrient degradation in vitro.

Parameter	CON	2%	5%	10%	*p*
D					
DMD %	84.78 ± 2.84	84.78 ± 4.13	84.89 ± 2.21	82.61 ± 2.40	0.49
NDFD %	80.45 ± 3.66	80.85 ± 5.20	79.08 ± 3.06	78.90 ± 2.92	0.76
ADFD %	79.22 ± 3.89	78.53 ± 5.83	74.14 ± 3.78	75.44 ± 3.40	0.16
CPD %	90.83 ± 2.21	91.66 ± 2.26	91.30 ± 1.27	89.84 ± 3.27	0.38
F					
DMD %	84.78 ± 2.84	79.22 ± 2.01	83.66 ± 2.59	82.33 ± 2.56	0.33
NDFD %	80.45 ± 3.66	80.30 ± 2.57	82.16 ± 2.82	77.49 ± 3.27	0.10
ADFD %	79.22 ± 3.89	77.28 ± 2.97	75.84 ± 3.82	74.31 ± 3.73	0.14
CPD %	90.83 ± 2.21 ^{AB}	93.85 ± 2.09 ^{A}	90.85 ± 1.06 ^{AB}	87.85 ± 1.51 ^{B}	<0.01

D, dried treatment; F, freeze-dried treatment; CON, control group; DMD, degradation of dry matter; NDFD, degradation of neutral detergent fiber; ADFD, degradation of acid detergent fiber; CPD, degradation of crude protein; 2%, CON plus *S. mcclurei* (20 mg/g DM); 5%, CON plus *S. mcclurei* (50 mg/g DM); 10%, CON plus *S. mcclurei* (100 mg/g DM). ^{AB} Means bearing different superscripts in the same row differ significantly ($p < 0.01$).

3.2. Effects of S. mcclurei on the Fermentation Characteristics of In Vitro Rumen Fermentation

The effects of the seaweed on the rumen fermentation characteristics are described in Table 4. The pH and NH_3-N were not affected by any of the supplemented treatments after 48 h of incubation. However, the drying treatment did increase the concentration of MCP ($p < 0.05$), with the MCP concentration increasing linearly with the level of supplementation.

Table 4. Effect of the two treatments and different supplementation levels of *S. mcclurei* on pH, NH_3-N, MCP, and VFA profiles in vitro.

Parameter	CON	2%	5%	10%	*p*
D					
pH	6.95 ± 0.11	6.95 ± 0.11	6.94 ± 0.05	6.95 ± 0.10	0.99
NH_3-N mmol/L	14.52 ± 0.84	13.29 ± 2.10	13.92 ± 0.61	14.86 ± 0.84	0.17
MCP μg/mL	44.24 ± 8.88 ^{b}	59.16 ± 7.61 ^{a}	61.21 ± 9.43 ^{a}	62.75 ± 12.84 ^{a}	0.01
Total VFA mmol/L	79.39 ± 5.31 ^{A}	65.55 ± 5.52 ^{B}	65.72 ± 6.42 ^{B}	64.34 ± 8.34 ^{B}	<0.01
Acetate mmol/L	45.78 ± 2.43 ^{A}	37.18 ± 3.30 ^{B}	37.24 ± 3.72 ^{B}	36.35 ± 4.95 ^{B}	<0.01
Propionate mmol/L	19.36 ± 1.93	17.36 ± 1.40	17.48 ± 1.64	17.47 ± 1.96	0.17
Butyrate mmol/L	10.15 ± 0.71 ^{A}	7.85 ± 0.68 ^{B}	7.87 ± 0.74 ^{B}	7.58 ± 1.03 ^{B}	<0.01
A:P	2.37 ± 0.10 ^{A}	2.14 ± 0.09 ^{B}	2.13 ± 0.08 ^{B}	2.07 ± 0.06 ^{B}	<0.01
F					
pH	6.95 ± 0.11	6.93 ± 0.12	6.93 ± 0.09	6.98 ± 0.11	0.83
NH_3-N mmol/L	14.52 ± 0.84	14.91 ± 0.28	14.35 ± 1.02	14.33 ± 1.38	0.7
MCP μg/mL	44.24 ± 8.88	51.08 ± 15.82	57.53 ± 5.71	52.19 ± 15.02	0.32
Total VFA mmol/L	79.39 ± 5.31 ^{a}	66.43 ± 10.71 ^{b}	72.03 ± 7.56 ^{ab}	77.20 ± 2.37 ^{a}	0.02
Acetate mmol/L	45.78 ± 2.43 ^{A}	37.55 ± 6.14 ^{B}	41.48 ± 4.57 ^{AB}	44.97 ± 1.49 ^{A}	<0.01
Propionate mmol/L	19.36 ± 1.93	17.76 ± 2.78	18.04 ± 1.48	18.77 ± 0.65	0.46
Butyrate mmol/L	10.15 ± 0.71 ^{A}	7.92 ± 1.26 ^{B}	8.90 ± 1.06 ^{AB}	9.58 ± 0.36 ^{A}	<0.01
A:P	2.37 ± 0.10 ^{A}	2.11 ± 0.07 ^{B}	2.29 ± 0.10 ^{A}	2.39 ± 0.06 ^{A}	<0.01

D, dried treatment; F, freeze-dried treatment; CON, control group; 2%, CON plus *S. mcclurei* (20 mg/g DM); 5%, CON plus *S. mcclurei* (50 mg/g DM); 10%, CON plus *S. mcclurei* (100 mg/g DM), A:P, acetate/propionate ratio. ^{ab} Means bearing different superscripts in the same row differ significantly ($p < 0.05$). ^{AB} Means bearing different superscripts in the same row differ significantly ($p < 0.01$).

Supplementation with both the dried treatment ($p < 0.01$) and freeze-dried treatment ($p = 0.02$) of *S. mcclurei* significantly reduced total VFA concentrations in the in vitro rumen fermentation compared to the CON. The decrease in total VFA concentration was more pronounced in the dried group. The supplementation of two different treatments of *S. mcclurei* significantly reduced the concentrations of total VFA, acetate ($p < 0.01$), and butyrate ($p < 0.01$) in the in vitro rumen fermentation. The highest supplementation level in the dried group corresponded to the lowest total VFA, acetate, and butyrate concentrations. Conversely, the lowest total VFA, acetate ($p < 0.01$), and butyrate ($p < 0.01$) concentrations in the freeze-dried group were observed at the lowest supplementation level. The propionate concentrations in the two treatment groups showed no significant difference from those in the CON. Both the dried group ($p < 0.01$) and the 2% supplementation level in the freeze-dried group ($p < 0.01$) significantly decreased the ratio of acetate/propionate.

3.3. Changes in the Microbial Composition

3.3.1. Effects of Different Treatments on Bacterial Community at the 2% Supplementation Level

Based on the above experimental results of gas production, nutrient degradation, and fermentation characteristics, we presume that the 2% supplementation level is the best solution for seaweed addition in this trial. We further analyzed the microbial communities at the 2% supplementation level in both treatment groups and CON, and compared the effects of different treatments of *S. mcclurei* supplementation on rumen microorganisms at the same level.

As shown in Table 5, the alpha diversity of the bacterial community, observed abundance-based coverage estimator (ACE), Chao 1, Shannon, and Simpson indices did not significantly differ among different treatments ($p > 0.05$). This result shows that the supplementation of *S. mcclurei*, whether freeze-dried or dried, did not affect the diversity and richness of the bacterial community.

Table 5. Alpha diversity indices of bacteria and archaea among treatments at the 2% supplementation level in vitro.

Parameter	CON	F	D	*p*
Bacteria				
ACE	1508.08 ± 138.74	1507.12 ± 40.80	1431.51 ± 111.04	0.37
Chao 1	1496.73 ± 130.89	1482.26 ± 53.17	1415.56 ± 91.87	0.33
Shannon	5.01 ± 0.24	5.08 ± 0.19	4.91 ± 0.21	0.41
Simpson	0.04 ± 0.01	0.03 ± 0.01	0.04 ± 0.01	0.22
Archaea				
ACE	379.55 ± 139.69	267.21 ± 162.74	318.11 ± 139.8	0.43
Chao 1	378.33 ± 140.38	259.38 ± 166.58	310.96 ± 141.32	0.40
Shannon	1.21 ± 0.17	0.99 ± 0.30	1.07 ± 0.18	0.26
Simpson	0.58 ± 0.04	0.61 ± 0.05	0.63 ± 0.06	0.28

CON, control group; F, freeze-dried treatment (2% of substrate DM); D, dried treatment (2% of substrate DM); ACE, abundance-based coverage estimator.

The bacterial community beta diversity with different treatments was analyzed by principal coordinate analysis (PCoA) based on the Bray–Curtis distance, as shown in Figure 1A. ANOSIM revealed a significant difference in the bacterial community composition (R = 0.2901, p = 0.005).

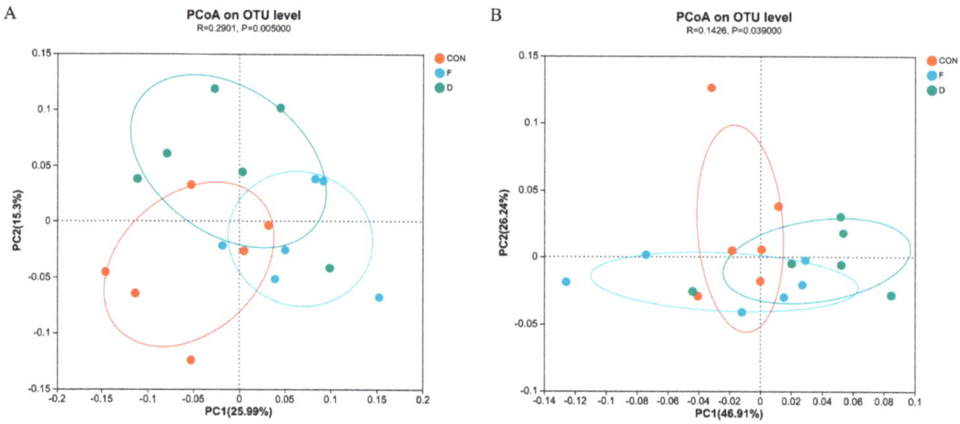

Figure 1. Principal coordinate analysis (PCoA) based on Bray–Curtis distance of the rumen bacterial (**A**) and archaeal (**B**) communities in in vitro rumen fermentation. CON, control group; F, freeze-dried treatment (2% of substrate DM); D, dried treatment (2% of substrate DM).

In this trial, at the phylum level, Bacteroidota, Firmicutes, and Proteobacteria were the dominant phyla (Figure 2A). At the genus level, 23 genera had a relative abundance of >0.1%. Twenty genera with a relative abundance of >0.1% were identified: *Ruminobacter*, *Rikenellaceae_RC9_gut_group*, *Prevotella*, *norank_f__F082*, *Selenomonas*, *Succinivibrio*, *Anaerovibri*, *norank_f__norank_o__WCHB1-41*, *Succiniclasticum*, UCG-002, *Prevotellaceae_UCG-003*, *unclassified_f__Succinivibrionaceae*, *Pseudobutyrivibrio*, *Succinivibrionaceae_UCG-002*, *Saccharofermentans*, *Lachnospiraceae_NK3A20_group*, *norank_f__Bacteroidales_RF16_group*, *Veillonellaceae_UCG-001*, probable_genus_10, *norank_f__UCG-011*, *Christensenellaceae_R-7_group*, *Acetitomaculum*, and *Oribacterium*, as shown in Figure 2B.

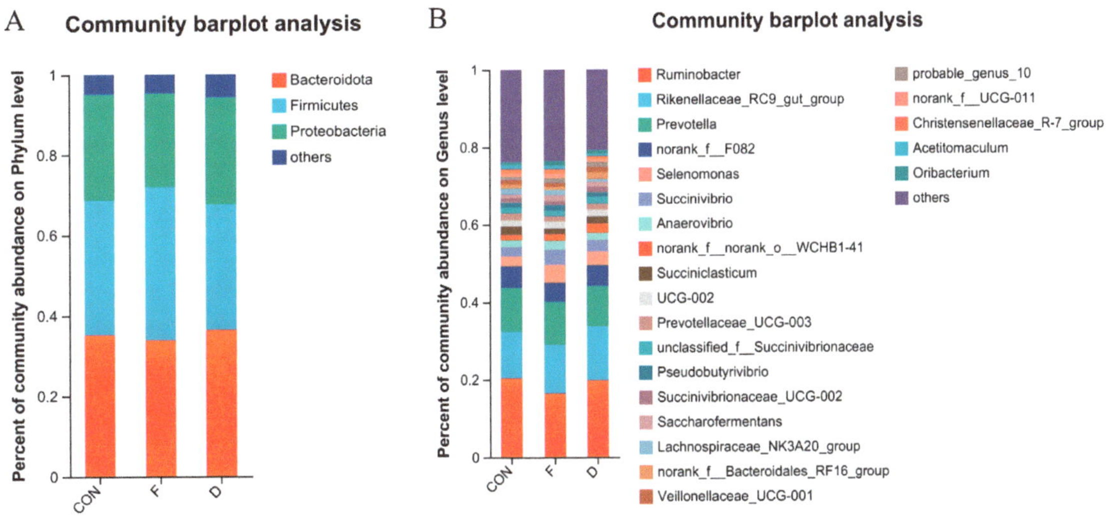

Figure 2. The rumen bacterial community composition in the CON, freeze-dried treatment, and dried treatment groups. (**A**) Phylum level; (**B**) genus level. CON, control group; F, freeze-dried treatment (2% of substrate DM); D, dried treatment (2% of substrate DM).

At the phylum level, the supplementation of *S. mcclurei* after freeze-drying treatment decreased the relative abundance of Actinobacteriota ($p < 0.05$), Fibrobacterota ($p < 0.05$), and Cyanobacteria ($p < 0.01$) in the in vitro rumen fermentation. Supplementing dried *S. mcclurei* increased the relative abundance of Fibrobacterota ($p < 0.05$) and decreased the abundance of Actinobacteriota ($p < 0.05$) and Cyanobacteria ($p < 0.01$, Figure 3A). At the genus level, the results of this experiment showed that the supplementation of *S. mcclurei* after both treatments increased the abundance of *Selenomonas* ($p < 0.05$), *Succinivibrio* ($p < 0.01$), *Saccharofermentans* ($p < 0.05$), *probable_genus_10* ($p < 0.05$), *Oribacterium* ($p < 0.05$), and *unclassified_f_Lachnospiraceae* ($p < 0.05$). Furthermore, *S. mcclurei* after drying treatment supplementation significantly increased the abundance of *norank_f__Bacteroidales_RF16_group* ($p < 0.01$), and the abundance of *Lachnospiraceae_NK3A20_group* ($p < 0.01$), *Acetitomaculum* ($p < 0.05$), and *Ruminococcus* ($p < 0.01$) were significantly decreased (Figure 3B). LEfSe analysis was performed using all microbial data to determine key bacterial groups associated with the freeze-drying and drying treatments. Figure 3C depicts the significantly different bacteria between the CON and the two treatment groups. The most significant bacterial genera in the CON were *Lachnospiraceae_NK3A20_group* and *Acetitomaculum*, and the bacterial genera with the most difference in the freeze-dried and dried groups were *Selenomonas*, *Succinivibrio*, and *RF16*, respectively.

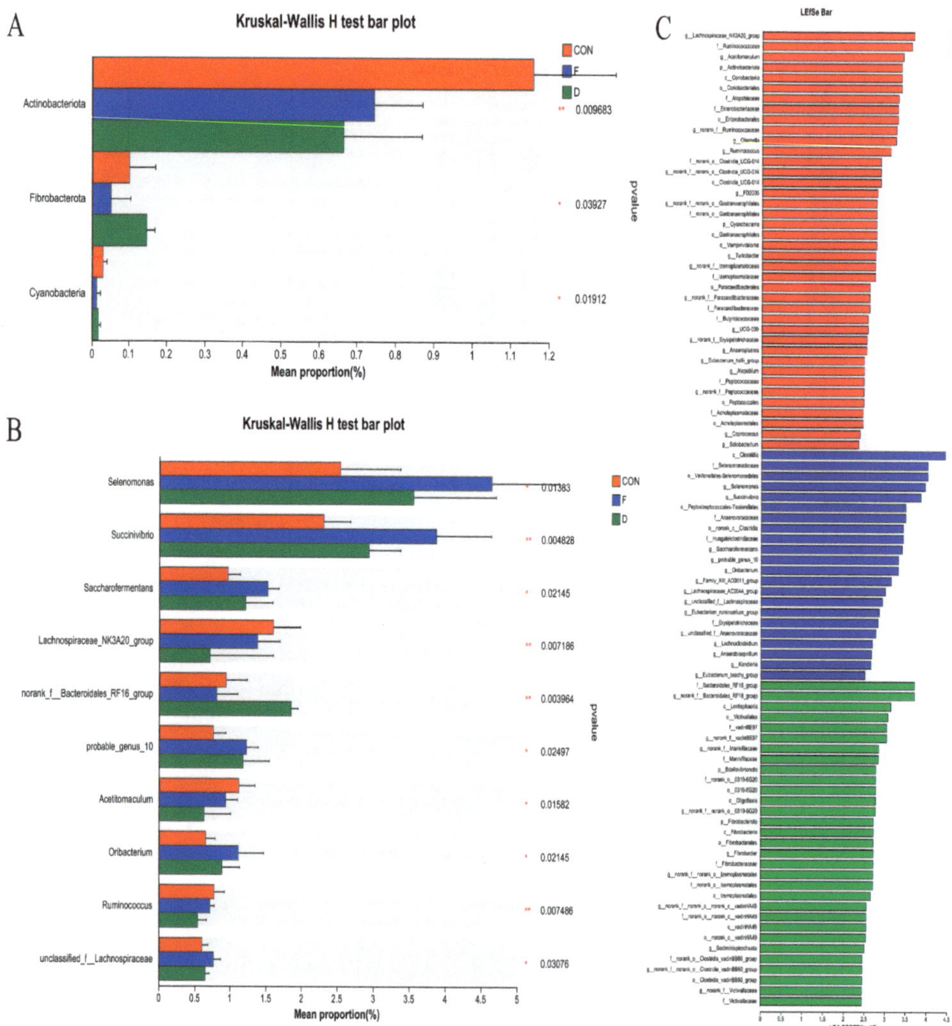

Figure 3. Principal differences in the relative abundances of bacterial phyla and histograms with LDA scores > 2 were calculated for each taxonomic unit from phylum to genus. (**A**) Phylum level; (**B**) genus level; (**C**) the LDA values for different differential species. CON, control group; F, freeze-dried treatment (2% of substrate DM); D, dried treatment (2% of substrate DM).

3.3.2. Effects of Different Treatments on the Archaeal Community at the 2% Supplementation Level

The alpha diversity of the archaeal community is detailed in Table 5, revealing no significant differences in ACE, Chao 1, Shannon, or Simpson indices among all treatment groups ($p > 0.05$). This suggests that the diversity and richness of the archaeal community were not affected by *S. mcclurei* supplementation with different treatments. The beta-diversity analysis of the archaea community is consistent with the above-mentioned bacteria and the results are depicted in Figure 1B. At the archaeal phylum level, Euryarchaeota was the most dominant phylum (Figure 4A). At the genus level, *Methanobrevibacter* was the most dominant genera (Figure 4B). The supplementation of dried or freeze-dried *S. mcclurei* had no significant difference on the archaeal community at the phylum level. At the genus level, the freeze-dried group significantly reduced the abundance of *Methanomicrobium* ($p < 0.05$), and there were no significant differences between CON and the dried group (Figure 5).

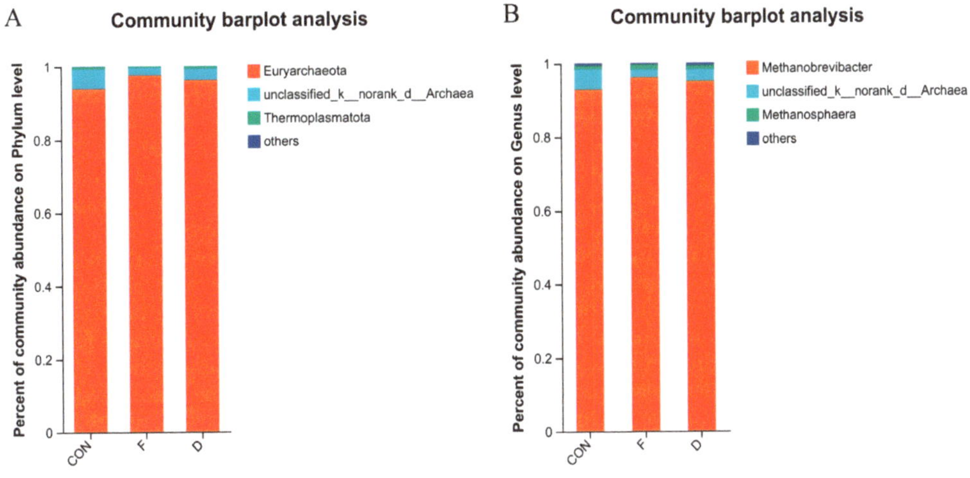

Figure 4. The rumen archaeal community composition in CON, freeze-dried treatment, and dried treatment groups. (**A**) Phylum level; (**B**) genus level. CON, control group; F, freeze-dried treatment (2% of substrate DM); D, dried treatment (2% of substrate DM).

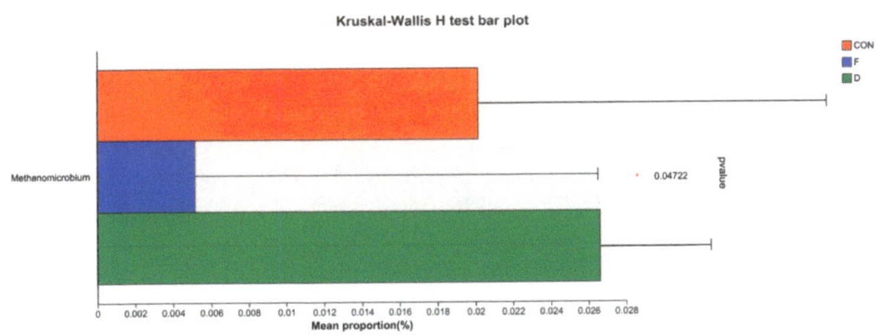

Figure 5. Principal differences in the relative abundances of the archaeal genera. CON, control group; F, freeze-dried treatment (2% of substrate DM); D, dried treatment (2% of substrate DM).

4. Discussion

The supplementation of seaweed in in vitro rumen fermentation has been reported to affect nutrient degradation, fermentation characteristics, and microbial composition [4,21,22]. In the current study, supplementation with *S. mcclurei*, a kind of brown seaweed, did not affect DMD and NDFD but lowered the CPD in the freeze-dried group when the seaweed supplementation level was 10% (Table 3). Meanwhile, a decreasing trend of ADFD was observed in the freeze-dried group as the supplementation increased, but the difference was not significant (Table 3). The results suggest that high supplementation levels also harm the degradation of ruminal fermentation nutrients. However, the 2% supplementation level has no negative impact on the degradation of nutrients, except for the degradation of crude protein. Based on the results, we speculate that a 2% *S. mcclurei* supplementation level is more suitable than a higher supplementation level in this experiment. Currently, there is limited research on the effects of *S. mcclurei* on nutrient degradation. This experiment will improve the information in this field, thereby guiding the production application.

This study explores the effects of *S. mcclurei* supplementation on the rumen fermentation parameters, providing insight into the complex dynamics of the rumen environment.

The fermentation parameters of pH, VFA, NH_3-N, and MCP reflect the stability of the internal rumen environment [23]. In this study, pH and NH_3-N concentrations showed no effect. Higher MCP concentration and lower VFA concentration were observed as a result of supplementation with *S. mcclurei* (Table 4). *S. mcclurei* contains 12% high-quality protein and when added to rumen may be utilized by rumen microorganisms to produce more MCP. The results show that, in addition to reducing VFA concentration, *S. mcclurei* supplementation has no other negative effects on rumen fermentation parameters and internal environment stability.

The dried group significantly reduced the total VFA, acetate, and butyrate concentrations, while the total VFA, acetate, and butyrate also decreased significantly when the freeze-dried group was added at the 2% level (Table 4). This observation suggests that different treatments and supplementation levels of *S. mcclurei* alter the pattern of VFA production by fermentation in the rumen. The inhibitory effect of brown seaweed on VFA concentration has also been reported. For example, Kunzel [8] observed that the supplementation of *Ascophyllum nodosum* and *Fucus vesiculosus* decreased the total VFA concentration. Beauchemin [24] reported that rumen microorganisms ferment lignocellulose through the acetate pathway, releasing H_2 and causing higher CH_4 concentrations. Simultaneously, propionic production is thought to be the major H_2 sink in the rumen when CH_4 production is inhibited [25]. Wettstein [26] also reported that decreased CH_4 concentration sometimes shifts microbial fermentation from acetate to propionate. This shows that there may be a positive correlation between acetate and CH_4, while propionate may be negatively correlated with CH_4 in rumen fermentation [27]. In this experiment, the propionate concentration was not affected by the supplementation of *S. mcclurei*, but the acetate–propionate ratio (A:P) significantly decreased. This is more conducive to the mitigation of CH_4 emissions, consistent with previous research results [13,28]. Subsequently, it is necessary to conduct further exploration of the specific relationship between VFA production and CH_4 emissions in rumen fermentation.

In the current study, different processing treatments and supplementation levels significantly affected the CH_4 concentration of in vitro rumen fermentation. It is worth noting that the freeze-dried treatment performed exceptionally well at a 2% supplementation level, decreasing CH_4 concentration by 18.85% without negatively affecting nutrient degradation (Table 1). The results demonstrate the first evidence that a low-dose (2% of substrate DM) freeze-dried treatment of *S. mcclurei* can effectively reduce CH_4 emissions compared to other brown seaweeds including *Sargassum*, *Cystoseira*, *Ascophyllum nodosum*, and *Fucus vesiculosus* [4,8,29]. As mentioned above, brown seaweed including *S. mcclurei*, inhibits CH_4 production, primarily attributed to the presence of phlorotannins, which are compounds found richly in brown seaweed. The phlorotannins vary in concentration between species and have significant differences among the different brown seaweed species [30]. Therefore, we presume that the active ingredient in *S. mcclurei*, phlorotannins, might play an important role in the process of reducing methane emissions in this experiment. Comparatively, the freeze-dried treatment is more effective in preserving phlorotannins than the dried treatment, resulting in a higher phlorotannins content in the freeze-dried group at the same supplementation level [31,32]. This observed phenomenon prompts an exploration into the potential correlation between the concentration of phlorotannins and their mitigating effect on CH_4 emission during in vitro rumen fermentation. However, there definitely exists an upper limit of the inhibitory effect on CH_4 emissions by phlorotannins. A higher supplementation level of *S. mcclurei* could not further reduce the CH_4 mitigation but might harm fermentation. This hypothesis agrees with the observed optimal performance at the 2% supplementation level in the freeze-dried treatment in our experiment, suggesting a balance between CH_4 reduction and minimal negative impact on fermentation.

CH_4 emissions from ruminants are intricately tied to the rumen microbiome, particularly focusing on methanogenic archaea that synthesize CH_4 as the end product of anaerobic fermentation [33,34]. In this study, we elucidated the effects of dried and freeze-dried treatments at the 2% supplementation level on ruminal microbial composition and functions, contributing to verifying the effect of *S. mcclurei* on methanogenesis in rumen fermentation in vitro. Our findings revealed that both dried and freeze-dried seaweed supplementation did not influence the α-diversity indices (ACE, Chao 1, Shannon, Simpson). These indices, reflecting microbial diversity and richness, indicated that bacterial diversity and richness in the rumen fermentation in vitro were not altered (Table 5).

The dominant bacteria in the rumen microbial community were identified as Bacteroidota, Firmicutes, and Proteobacteria, similar to several other studies [35–37] (Figure 2A). At the phylum level, changes in the bacterial community structure were evident (Figure 3A). The relative abundance of Actinobacteriota and Cyanobacteria significantly decreased in all treatment groups compared with CON. Conversely, the relative abundance of Fibrobacterota exhibited a trend of increase. Actinobacteriota, Fibrobacterota, and Cyanobacteria had a significant impact on starch and cellulose degradation [38,39]. These observed changes in their abundance may contribute to the decrease in total VFA concentration. To further analyze the effect of different treatments at the 2% supplementation level on in vitro rumen fermentation, we analyzed the bacterial community structure at the genus level, as shown in Figure 2B.

Intriguingly, *Selenomonas*, *Succinivibrio*, and *Saccharofermentans*' relative abundance significantly increased in the freeze-dried group (Figure 3B). Xue [33] reported that *Selenomonas* is positively correlated with several genera of the family *Succinivibrionaceae*, including *Succinivibrio*. *Selenomonas* can ferment starch to produce lactate, acetate, and propionate, while *Succinivibrio* and *Saccharofermentans* are involved in utilizing fermentation products to produce succinate, lactate, acetate, and formate and transfer H_2 away from methanogenesis [40]. The higher relative abundance of these genera in the freeze-dried group was associated with significant decreases in CH_4 concentration compared to the CON. These genera were positively correlated with propionate concentration; however, there was no increase in propionate concentration in this study. We hypothesized that the relatively low abundance of *Selenomonas* (2.53–4.63%), *Succinivibrio* (2.30–3.87%), and *Saccharofermentans* (0.96–1.52%) might not have a significant effect on propionate production. The production of acetate bacteria such as *Lachnospiraceae_NK3A20_group*, *Ruminococcus*, and *Acetitomaculum* [41,42] exhibited a significant decrease in relative abundance in both treatment groups, with a more pronounced effect in the dried group (Figure 3B). These changes agree with a lower acetate concentration in both treatment groups compared to CON, with the dried group evidencing the lowest acetate concentration. The reduction in the relative abundance of these bacteria contributed to the observation of a decrease in acetate production. *Probable_genus_10*, for which the relative abundance significantly increased in both treatment groups, was also shown to promote propionate production [36]. Overall, *S. mcclurei* supplementation in both treatments changes the rumen fermentation pattern by altering the relative abundances of bacteria, which allows more H_2 to be transferred to the pathway of propionate production, thereby mitigating CH_4 emission. Compared with the two treatments, the freeze-dried treatment demonstrated a more pronounced effect. It produced more propionate (2.07%), more effectively mitigating the negative effect on acetate production (0.81%), and demonstrated superior efficacy in inhibiting CH_4 emissions. This understanding of the intricate relationships between microbial communities, rumen fermentation, and CH_4 production highlights the potential of freeze-dried *S. mcclurei* as a promising supplement for mitigating CH_4 emissions in ruminant systems.

Archaea are widely present in the rumen and can utilize H_2 to maintain the fermentation environment of rumen microorganisms and the production of CH_4 [33]. In our trial, the phylum Euryarchaeota was the most dominant (Figure 4A), aligning with the findings of Liu [43]. Euryarchaeota is recognized as a classic methanogen in the rumen. Thaumarchaeota, identified in our study, are considered among the most abundant archaea globally. Both archaea species are thought to have an impact on climate change [44]. At the genus level (Figure 4B), *Methanobrevibacter*, *unclassified_k__norank_d__Archaea*, and *Methanosphaera* were identified. *Methanobrevibacter*, the predominant rumen methanogen, produces CH_4 from CO_2 via the hydrogenotrophic pathway and correlates positively with

CH_4 production [45]. Conversely, *Methanosphaera*, known for consuming methanol to produce CH_4 through the methylotrophic pathway, has been associated with negative correlations with CH_4 production [46]. Remarkably, the only notable change observed was in the relative abundance of *Methanomicrobium*, which significantly decreased in the freeze-dried group compared to CON (Figure 5). *Methanobrevibacter*, belonging to the hydrogenotrophic methanogens within the phylum Euryarchaeota, also exhibits positive correlations with CH_4 production, deriving CH_4 through the reduction of CO_2 [45].

5. Conclusions

This study investigated the effect of supplementing freeze-dried and dried *S. mcclurei* on in vitro rumen fermentation, microbiota, and CH_4 production using the same basal total mixed ration. The freeze-dried treatment at the 2% addition level increased the degradation rate of crude protein, had a positive effect on the rumen fermentation in vitro, and reduced CH_4 emissions by 18.85%. The supplementation influenced the microbial composition by increasing the relative abundance of *Selenomonas*, *Succinivibrio*, and Saccharofermentans while decreasing the abundance of *Methanomicrobium*. The increases in the relative abundance of these bacteria change the ruminal fermentation pattern, allowing more hydrogen to be transferred to the propionate production pathway and competitively inhibiting CH_4 production. The decreases in the relative abundance of *Methanomicrobium* directly reduce CH_4 production in in vitro rumen fermentation. The findings of this study offer valuable insights into the potential of freeze-dried *S. mcclurei* supplementation as a strategy for CH_4 reduction in ruminants and pave the way for practical application.

Author Contributions: Conceptualization, L.M. and D.L.; methodology, S.L.; software, S.L.; validation, S.L. and Y.S.; formal analysis, S.L.; investigation, S.L., Y.S. and W.L.; resources, S.L. and W.L.; data curation, S.L.; writing—original draft preparation, S.L.; writing—review and editing, T.G., X.T., Z.Z., J.S., Y.Y., S.Y. and L.M.; visualization, S.L. and Y.S.; supervision, X.T., Z.Z., J.S., L.M. and D.L.; project administration, L.M. and D.L.; funding acquisition, T.G., Y.Y., S.Y, L.M. and D.L. All authors have read and agreed to the published version of the manuscript.

Funding: This study was supported by the Southern Marine Science and Engineering Guangdong Laboratory (Zhuhai) (SML2021SP203), the Key Laboratory of Xinjiang Feed Biotechnology, Guangzhou Science and Technology Planning Project (2023A04J0793), the Guangdong Modern Agro-industry Technology Research System (2023KJ114), the Guangdong Science and Technology Planning Project (KTP20210043), and Research Project of Department of Education of Guangdong Province (2022ZDZX4041).

Institutional Review Board Statement: The study protocol was approved by the Experimental Animal Ethics Committee of the Institute of Animal Science, Guangdong Academy of Agricultural Sciences (project number 2024004).

Informed Consent Statement: Not applicable.

Data Availability Statement: All data supporting the present study are reported in this study. Sequence data presented in this study are openly available in the NCBI database.

Conflicts of Interest: The authors declare no conflicts of interest.

References

1. Gerber, P.J.; Steinfeld, H.; Henderson, B.; Mottet, A.; Opio, C.; Dijkman, J.; Falcucci, A.; Tempio, G. *Tackling Climate Change through Livestock a Global Assessment of Emissions and Mitigation Opportunities*; CABI: New York, NY, USA, 2013.
2. Dillon, J.A.; Stackhouse-Lawson, K.R.; Thoma, G.J.; Gunter, S.A.; Rotz, C.A.; Kebreab, E.; Riley, D.G.; Tedeschi, L.O.; Villalba, J.; Mitloehner, F.; et al. Current state of enteric methane and the carbon footprint of beef and dairy cattle in the United States. *Anim. Front.* **2021**, *11*, 57–68. [CrossRef]
3. Hristov, A.N.; Melgar, A.; Wasson, D.; Arndt, C. Symposium review: Effective nutritional strategies to mitigate enteric methane in dairy cattle. *J. Dairy Sci.* **2022**, *105*, 8543–8557. [CrossRef] [PubMed]
4. Machado, L.; Magnusson, M.; Paul, N.A.; de Nys, R.; Tomkins, N. Effects of marine and freshwater macroalgae on in vitro total gas and methane production. *PLoS ONE* **2014**, *9*, e85289. [CrossRef] [PubMed]

5. Blikra, M.J.; Altintzoglou, T.; Lvdal, T.; Rogns, G.; Skipnes, D.; Skra, T.; Sivertsvik, M.; Fernández, E.N. Seaweed products for the future: Using current tools to develop a sustainable food industry. *Trends Food Sci. Technol.* **2021**, *118*, 765–776. [CrossRef]
6. Chopin, T.; Tacon, A.G.J. Importance of Seaweeds and Extractive Species in Global Aquaculture Production. *Rev. Fish. Sci. Aquac.* **2020**, *29*, 139–148. [CrossRef]
7. Roque, B.M.; Brooke, C.G.; Ladau, J.; Polley, T.; Marsh, L.J.; Najafi, N.; Pandey, P.; Singh, L.; Kinley, R.; Salwen, J.K.; et al. Effect of the macroalgae Asparagopsis taxiformis on methane production and rumen microbiome assemblage. *Anim. Microbiome* **2019**, *1*, 3. [CrossRef]
8. Kunzel, S.; Yergaliyev, T.; Wild, K.J.; Philippi, H.; Petursdottir, A.H.; Gunnlaugsdottir, H.; Reynolds, C.K.; Humphries, D.J.; Camarinha-Silva, A.; Rodehutscord, M. Methane Reduction Potential of Brown Seaweeds and Their Influence on Nutrient Degradation and Microbiota Composition in a Rumen Simulation Technique. *Front. Microbiol.* **2022**, *13*, 889618. [CrossRef] [PubMed]
9. Britton, D.; Schmid, M.; Revill, A.T.; Virtue, P.; Nichols, P.D.; Hurd, C.L.; Mundy, C.N. Seasonal and site-specific variation in the nutritional quality of temperate seaweed assemblages: Implications for grazing invertebrates and the commercial exploitation of seaweeds. *J. Appl. Phycol.* **2021**, *33*, 603–616. [CrossRef]
10. Belanche, A.; Jones, E.; Parveen, I.; Newbold, C.J. A Metagenomics Approach to Evaluate the Impact of Dietary Supplementation with Ascophyllum nodosum or Laminaria digitata on Rumen Function in Rusitec Fermenters. *Front. Microbiol.* **2016**, *7*, 299. [CrossRef]
11. Wang, Y.; Alexander, T.W.; Mcallister, T.A. In vitro effects of phlorotannins from Ascophyllum nodosum (brown seaweed) on rumen bacterial populations and fermentation. *J. Sci. Food Agric.* **2010**, *89*, 2252–2260. [CrossRef]
12. Choi, Y.; Lee, S.J.; Kim, H.S.; Eom, J.S.; Jo, S.U.; Guan, L.L.; Seo, J.; Kim, H.; Lee, S.S.; Lee, S.S. Effects of seaweed extracts on in vitro rumen fermentation characteristics, methane production, and microbial abundance. *Sci. Rep.* **2021**, *11*, 24092. [CrossRef] [PubMed]
13. Kinley, R.D.; De Nys, R.; Vucko, M.J.; Machado, L.; Tomkins, N.W. The red macroalgae Asparagopsis taxiformis is a potent natural antimethanogenic that reduces methane production during in vitro fermentation with rumen fluid. *Anim. Prod. Sci.* **2016**, *56*, 282. [CrossRef]
14. Ratti, C. Hot air and freeze-drying of high-value foods: A review. *J. Food Eng.* **2001**, *49*, 311–319. [CrossRef]
15. Goering, H.K.; Van Soest, P.J. Forage fiber analysis. USDA agricultural research service. In *Handbook Number 379*; US Department of Agriculture Superintendent of Documents, US Government Printing Office: Washington, DC, USA, 1970.
16. Van Soest, P.; Robertson, J.; Lewis, B. Symposium: Carbohydrate methodology, metabolism, and nutritional implications in dairy cattle. *J. Dairy Sci.* **1991**, *74*, 3583–3597. [CrossRef] [PubMed]
17. Thiex, N. Evaluation of analytical methods for the determination of moisture, crude protein, crude fat, and crude fiber in distillers dried grains with solubles. *J. AOAC Int.* **2009**, *92*, 61–73. [CrossRef]
18. Menke. Estimation of the energetic feed value obtained from chemical analysis and in vitro gas production using rumen fluid. *Anim. Res. Dev.* **1988**, *28*, 7–55.
19. Wischer, G.; Boguhn, J.; Steingaß, H.; Schollenberger, M.; Hartung, K.; Rodehutscord, M. Effect of monensin on in vitro fermentation of silages and microbial protein synthesis. *Arch. Anim. Nutr.* **2013**, *67*, 219–234. [CrossRef] [PubMed]
20. Vissers, A.M.; Pellikaan, W.F.; Bouwhuis, A.; Vincken, J.P.; Gruppen, H.; Hendriks, W.H. *Laminaria digitata* phlorotannins decrease protein degradation and methanogenesis during in vitro ruminal fermentation. *J. Sci. Food Agric.* **2018**, *98*, 3644–3650. [CrossRef] [PubMed]
21. Belanche, A.; Ramos-Morales, E.; Newbold, C.J. In vitro screening of natural feed additives from crustaceans, diatoms, seaweeds and plant extracts to manipulate rumen fermentation. *J. Sci. Food Agric.* **2016**, *96*, 3069–3078. [CrossRef]
22. Maia, M.R.; Fonseca, A.J.; Oliveira, H.M.; Mendonça, C.; Cabrita, A.R. The Potential Role of Seaweeds in the Natural Manipulation of Rumen Fermentation and Methane Production. *Sci. Rep.* **2016**, *6*, 32321. [CrossRef]
23. Zhang, F.; Wang, Y.; Wang, H.; Nan, X.; Guo, Y.; Xiong, B. Calcium propionate supplementation has Minor effects on major ruminal bacterial community composition of early lactation dairy cows. *Front. Microbiol.* **2022**, *13*, 847488. [CrossRef]
24. Beauchemin, K.A.; McGinn, S.M. Methane emissions from feedlot cattle fed barley or corn diets1. *J. Anim. Sci.* **2005**, *83*, 653–661. [CrossRef]
25. Ungerfeld, E.M. Shifts in metabolic hydrogen sinks in the methanogenesis-inhibited ruminal fermentation: A meta-analysis. *Front. Microbiol.* **2015**, *6*, 37. [CrossRef] [PubMed]
26. Wettstein, H.R.; Machmüller, A.; Kreuzer, M. Effects of raw and modified canola lecithins compared to canola oil, canola seed and soy lecithin on ruminal fermentation measured with rumen simulation technique. *Anim. Feed Sci. Technol.* **2000**, *85*, 153–169. [CrossRef]
27. Ungerfeld, E.M. Metabolic hydrogen flows in rumen fermentation: Principles and possibilities of interventions. *Front. Microbiol.* **2020**, *11*, 589. [CrossRef]
28. Machado, L.; Magnusson, M.; Paul, N.A.; Kinley, R.; de Nys, R.; Tomkins, N. Identification of bioactives from the red seaweed Asparagopsis taxiformis that promote antimethanogenic activity in vitro. *J. Appl. Phycol.* **2016**, *28*, 3117–3126. [CrossRef]
29. Thorsteinsson, M.; Weisbjerg, M.R.; Lund, P.; Bruhn, A.; Hellwing, A.L.F.; Nielsen, M.O. Effects of dietary inclusion of 3 Nordic brown macroalgae on enteric methane emission and productivity of dairy cows. *J. Dairy Sci.* **2023**, *106*, 6921–6937. [CrossRef]

30. Lopes, G.; Sousa, C.; Silva, L.R.; Pinto, E.; Andrade, P.B.; Bernardo, J.; Mouga, T.; Valentão, P. Can Phlorotannins Purified Extracts Constitute a Novel Pharmacological Alternative for Microbial Infections with Associated Inflammatory Conditions. *PLoS ONE* **2012**, *7*, e31145. [CrossRef] [PubMed]
31. Vucko, M.J.; Magnusson, M.; Kinley, R.D.; Villart, C.; de Nys, R. The effects of processing on the in vitro antimethanogenic capacity and concentration of secondary metabolites of Asparagopsis taxiformis. *J. Appl. Phycol.* **2017**, *29*, 1577–1586. [CrossRef]
32. Magnusson, M.; Mata, L.; De Nys, R.; Paul, N.A. Biomass, lipid and fatty acid production in large-scale cultures of the marine macroalga Derbesia tenuissima (Chlorophyta). *Mar. Biotechnol.* **2014**, *16*, 456–464. [CrossRef]
33. Xue, M.-Y.; Sun, H.-Z.; Wu, X.-H.; Liu, J.-X.; Guan, L.L. Multi-omics reveals that the rumen microbiome and its metabolome together with the host metabolome contribute to individualized dairy cow performance. *Microbiome* **2020**, *8*, 64. [CrossRef]
34. Thauer, R.K.; Kaster, A.-K.; Seedorf, H.; Buckel, W.; Hedderich, R. Methanogenic archaea: Ecologically relevant differences in energy conservation. *Nat. Rev. Microbiol.* **2008**, *6*, 579–591. [CrossRef]
35. Won, M.-Y.; Oyama, L.B.; Courtney, S.J.; Creevey, C.J.; Huws, S.A. Can rumen bacteria communicate to each other? *Microbiome* **2020**, *8*, 23. [CrossRef]
36. Liang, J.; Fang, W.; Chang, J.; Zhang, G.; Ma, W.; Nabi, M.; Zubair, M.; Zhang, R.; Chen, L.; Huang, J.; et al. Long-term rumen microorganism fermentation of corn stover in vitro for volatile fatty acid production. *Bioresour. Technol.* **2022**, *358*, 127447. [CrossRef]
37. Plaizier, J.C.; Danscher, A.-M.; Azevedo, P.A.; Derakhshani, H.; Andersen, P.H.; Khafipour, E. A Grain-Based SARA Challenge Affects the Composition of Epimural and Mucosa-Associated Bacterial Communities throughout the Digestive Tract of Dairy Cows. *Animals* **2021**, *11*, 1658. [CrossRef]
38. Li, J.; Zhao, S.; Meng, Z.; Gao, Y.; Miao, J.; Mao, S.; Jin, W. Effects of Fumarate and Nitroglycerin on In Vitro Rumen Fermentation, Methane and Hydrogen Production, and on Microbiota. *Biology* **2023**, *12*, 1011. [CrossRef]
39. Gharechahi, J.; Vahidi, M.F.; Bahram, M.; Han, J.-L.; Ding, X.-Z.; Salekdeh, G.H. Metagenomic analysis reveals a dynamic microbiome with diversified adaptive functions to utilize high lignocellulosic forages in the cattle rumen. *ISME J.* **2021**, *15*, 1108–1120. [CrossRef]
40. Mizrahi, I.; Wallace, R.J.; Moraïs, S. The rumen microbiome: Balancing food security and environmental impacts. *Nat. Rev. Microbiol.* **2021**, *19*, 553–566. [CrossRef]
41. Waters, J.L.; Ley, R.E. The human gut bacteria Christensenellaceae are widespread, heritable, and associated with health. *BMC Biol.* **2019**, *17*, 83. [CrossRef]
42. Cheng, Z.; Meng, Z.; Tan, D.; Datsomor, O.; Zhan, K.; Lin, M.; Zhao, G. Effects of supplementation of sodium acetate on rumen fermentation and microbiota in postpartum dairy cows. *Front. Microbiol.* **2022**, *13*, 1053503. [CrossRef]
43. Liu, Z.; Wang, K.; Nan, X.; Cai, M.; Yang, L.; Xiong, B.; Zhao, Y. Synergistic Effects of 3-Nitrooxypropanol with Fumarate in the Regulation of Propionate Formation and Methanogenesis in Dairy Cows In Vitro. *Appl. Environ. Microbiol.* **2022**, *88*, e01908-21. [CrossRef] [PubMed]
44. Baker, B.J.; De Anda, V.; Seitz, K.W.; Dombrowski, N.; Santoro, A.E.; Lloyd, K.G. Author Correction: Diversity, ecology and evolution of Archaea. *Nat. Microbiol.* **2020**, *5*, 976. [CrossRef] [PubMed]
45. Jia, P.; Dong, L.-F.; Tu, Y.; Diao, Q.-Y. Bacillus subtilis and Macleaya cordata extract regulate the rumen microbiota associated with enteric methane emission in dairy cows. *Microbiome* **2023**, *11*, 229. [CrossRef]
46. Fricke, W.F.; Seedorf, H.; Henne, A.; Krüer, M.; Liesegang, H.; Hedderich, R.; Gottschalk, G.; Thauer, R.K. The Genome Sequence of Methanosphaera stadtmanae Reveals Why This Human Intestinal Archaeon Is Restricted to Methanol and H2 for Methane Formation and ATP Synthesis. *J. Bacteriol.* **2006**, *188*, 642–658. [CrossRef]

animals

MDPI

Article

Effect of *Brosimum alicastrum* Foliage on Intake, Kinetics of Fermentation and Passage and Microbial N Supply in Sheep Fed *Megathyrsus maximus* Hay

Vicente Valdivia-Salgado [1], Ever del Jesús Flores-Santiago [2,*], Luis Ramírez-Avilés [3], José Candelario Segura-Correa [3], Jesús Miguel Calzada-Marín [3] and Juan Carlos Ku-Vera [3]

1 Livestock Consultant, Estelí C.P. 31000, Nicaragua; viznat1414@yahoo.com.mx

2 Southeast University Regional Unit, Autonomous University Chapingo, Carr. Teapa-Vicente Guerrero km 7.5, Teapa C.P. 86807, Tabasco, Mexico

3 Faculty of Veterinary Medicine and Animal Science, University of Yucatan, Carr. Merida-Xmatkuil km 15.5, Merida C.P. 97100, Yucatan, Mexico; luis.ramirez@correo.uady.mx (L.R.-A.); jose.segura@correo.uady.mx (J.C.S.-C.); calzadamarinmiguel@hotmail.com (J.M.C.-M.); kvera@correo.uady.mx (J.C.K.-V.)

* Correspondence: ejfloress@chapingo.mx

Simple Summary: The foliage of the tree *Brosimum alicastrum* is widely employed by small farmers in Latin America to supplement cattle, sheep and goats during the dry season. Foliage is manually cut, and the branches are chopped and fed directly, being highly palatable to ruminant species. The foliage remains green throughout the year, and the tree can be defoliated twice per year. Cattle can also directly browse the foliage from the tree, but it is mostly used in cut and carry systems and transported in carts to farms. The experiment hereby described showed that the crude protein fraction of the foliage of *B. alicastrum* is highly fermentable in the rumen, supplying nitrogen for growth of the microbial population. As the percent of foliage of *B. alicastrum* in the ration was increased, dry matter intake by sheep was concomitantly increased, as was the rate of the passage of digesta through the rumen and the supply of microbial nitrogen to the small intestine. The foliage of *B. alicastrum* is a valuable feedstuff for ruminants in the tropics and can be used as a cheap source of protein during the critical periods of the year to improve weight gain and milk yield in a sustainable way.

Abstract: An experiment was carried out to assess the effect of the incorporation of sun-dried foliage of *Brosimum alicastrum* into rations based on hay of *Megathyrsus maximus* on intake, rumen fermentation, kinetics of passage, microbial nitrogen supply to the small intestine, apparent digestibility in Pelibuey hair sheep. Four rations were randomly allotted to four rumen-cannulated lambs (BW = 37.4 $\pm$ 4.9 kg) using a 4 $\times$ 4 Latin square design to assess the effect of increasing levels (0, 15, 30 and 45% DM basis) of foliage of *Brosimum alicastrum* on a basal ration of *M. maximus*. Organic matter intake and water consumption increased linearly ($p < 0.01$) with increasing levels of *B. alicastrum* in the ration. The rate and potential extent of rumen fermentation of OM and CP of *B. alicastrum* were 10.6%/h and 86.6% and 11.4%/h and 95.2%, respectively, but no effect ($p > 0.05$) was found on the potential rumen degradation of DM (40.2%) or on the rate of degradation of DM (0.033%/h) of *M. maximus*, although a positive effect was found in the rumen degradation rate of NDF ($p < 0.05$). VFA and ammonia concentration in the rumen and the rate of passage of solids and liquids through the rumen (k_1) increased linearly ($p < 0.01$) with increasing levels of *B. alicastrum*. Rumen pH was not affected by the incorporation of *B. alicastrum* ($p > 0.05$). Microbial nitrogen supply to the small intestine ($p < 0.001$), apparent digestibility of dry matter ($p < 0.01$) and NDF ($p < 0.05$) of the rations were also significantly increased as a result of the incorporation of *B. alicastrum* foliage. Results from this experiment suggest that the foliage of *Brosimum alicastrum* can be readily incorporated at around 30% of the ration of dry matter in hair sheep with beneficial effects on feed intake, rate of passage and microbial N supply to the lower tract.

check for updates

Citation: Valdivia-Salgado, V.; Flores-Santiago, E.d.J.; Ramírez-Avilés, L.; Segura-Correa, J.C.; Calzada-Marín, J.M.; Ku-Vera, J.C. Effect of *Brosimum alicastrum* Foliage on Intake, Kinetics of Fermentation and Passage and Microbial N Supply in Sheep Fed *Megathyrsus maximus* Hay. *Animals* **2024**, *14*, 1144. https://doi.org/10.3390/ani14081144

Academic Editors: Lizhi Wang and Manuel Fondevila

Received: 12 December 2023
Revised: 21 March 2024
Accepted: 3 April 2024
Published: 9 April 2024

Keywords: *Brosimum alicastrum*; hair sheep; rumen fermentation; voluntary intake

1. Introduction

Brosimum alicastrum Swartz (Spanish common name: ramon) is a tropical tree that has been used since ancient times by the Maya population of Central America for various purposes, including food, shade and forage [1–3]. The foliage of *Brosimum alicastrum* is widely utilized by small-scale dairy and cattle farmers in Latin America as a source of forage; its yield may reach 2 tons of DM/ha/year [4]. It is one of the few trees that remain green during the six-month dry season. However, relatively little is known regarding its nutritive value for ruminants. Medina-Figueras [5] was the first to describe the apparent digestibility of DM and CP of the foliage and seeds of *B. alicastrum* by cattle. Yerena et al. [6] determined the apparent DM digestibility (67%) of *B. alicastrum* foliage by cattle. Recently, Sarmiento-Franco et al. [7] proposed that the seeds of the *Brosimum alicastrum* tree may represent a good source of nutrients for different animal species given their high starch content (71.2%), and Soberanis-Poot et al. [8] advanced that *Brosimum alicastrum* may be considered a rational option for animal production systems in view of the need for adaptation to climate change.

The present work was carried out to assess the effect of incorporating increasing levels of sun-dried *Brosimum alicastrum* foliage into a poor-quality grass (*Megathyrsus maximus*) ration on voluntary intake, rumen digestion, kinetics of passage throughout the gastrointestinal tract, microbial N supply to the small intestine and apparent digestibility by Pelibuey hair sheep.

2. Materials and Methods

2.1. Animals

Animal care was conducted under the supervision and approval of the Academic Committee of the Faculty of Veterinary Medicine and Animal Science, University of Yucatan, according to the regulations established for the handling and welfare of experimental animals. Four entire male Pelibuey sheep (initial average BW = 37.4 ± 4.9 kg; 9 months old) fitted with rumen cannulas (7.5 cm diameter; Bar Diamond, Parma, ID, USA) were used. During the experimental period, sheep were housed indoors in metabolic crates (60 × 150 cm) to allow for the complete separation and recovery of feces and urine. They were treated against internal and external parasites and given an intramuscular injection of vitamins A, D and E before the start of the experiment. Two Zebu bulls weighing 427 ± 32 kg BW (3 years old) fitted with rumen cannulas (10 cm internal diameter, Bar Diamond Parma, ID, USA) were used to measure rumen degradation of *Brosimum alicastrum* foliage. Cattle were chosen to assess kinetics of fermentation of *B. alicastrum*, due to their larger capacity to hold incubation bags in the rumen compared to sheep.

2.2. Experimental Rations and Design

Foliage of *Brosimum alicastrum* was harvested from mature trees and sun-dried on concrete floor for two days. *Brosimum alicastrum* and *Megathyrsus maximus* were ground twice in a grinder with a screen 0.5 cm diameter to avoid selection by sheep and mixed in the following ratios: 0:100; 15:85, 30:70 and 45:55 on a DM basis. In all rations, 9.4% cane molasses (DM basis) was included. Levels of incorporation of *B. alicastrum* were allocated on the premise that, under practical conditions, farmers would not incorporate foliage of *B. alicastrum* as more than a third of the ration. Rations were not corrected for equal N content to test the direct effect of *B. alicastrum* level on response variables. The chemical compositions of *B. alicastrum* foliage, *Megathyrsus maximus* hay and the rations are shown on Table 1. The experiment was designed as a 4 × 4 Latin square with four periods of 20 d (10 d for adaptation and 10 d for measurements). Bulls used to evaluate rumen degradation of *Brosimum alicastrum*, were fed hay of *Megathyrsus maximus*.

Table 1. Chemical composition (% DM) of *Brosimum alicastrum* and *Megathyrsus maximus* and of the experimental rations containing increasing levels of *B. alicastrum* foliage fed to Pelibuey sheep.

Item	Forage			Level of *B. alicastrum* (% DM)			
	Brosimum alicastrum	*Megathyrsus maximus*	*Molasses*	0	15	30	45
DM *	89.3	92.1	78.1	90.4	90.2	90.3	89.9
Organic matter	90.4	89.4	85.2	89.0	89.2	89.3	89.4
Crude protein	15.7	5.6	4.1	5.1	6.8	8.2	9.6
Neutral detergent fiber	37.5	80.0	—	72.5	66.7	60.9	55.2
Acid detergent fiber	28.5	48.4	—	43.9	41.2	38.5	35.8
Hemicellulose	9.0	31.6	—	28.6	25.5	22.5	19.4
Lignin	12.0	17.8	—	15.4	15.3	14.5	13.8

* As percentage of wet weight; DM, Dry Matter.

2.3. Voluntary Feed Intake and Digestibility

Experimental rations (Table 1) were offered ad libitum every morning at 08:00 and a 10% refusal, relative to the amount offered, was allowed. Water was freely available. Feed intake was measured as the difference between the amount offered and that refused the following day. In vivo apparent digestibility of the feed components was measured by total collection of feces at 08:00 h for 7 days. Daily feces collection was made during the last 7 days of each experimental period. Feces were weighed, bulked, mixed, and a 10% aliquot was taken and stored in plastic bags at −20 °C until analysis.

2.4. Rumen Fermentation

Kinetics of rumen fermentation of DM, OM, NDF and CP of *B. alicastrum* and *M. maximus* were measured by means of the nylon bag technique [9]. Nylon bag dimensions were 5 × 7 cm for sheep and 10 × 20 cm for bulls, with pore size of 53 μm (Bar Diamond, Parma, ID, USA). A total of 3 grams of sample were placed into each bag and incubated in the rumen for 6, 9, 12, 24, 48, 72 and 96 h in sheep, and 5 g of sample were placed into each bag and incubated in the rumen at the same incubation times in bulls. Samples (nylon bags) were introduced by triplicate into the rumen. After incubation, bags were washed in a washing machine until the water was clear and dried in a forced-air oven at 60 °C for 72 h. Rumen disappearance data were fitted into the model $p = a + b\,(1 - e^{-ct})$ [10] to estimate degradation constants a, b and c, where p is percent degradation at time t; a is the zero time intercept; b represents the asymptote to the equation; c is the fractional degradation rate of b and t is time of incubation. Effective degradability was estimated, employing the equation $a + bc/(c + k)$, where k is the determined outflow rate from the rumen (per hour) [10]. The washing loss at zero time was estimated following the procedures described by López et al. [11].

2.5. Purine Derivative Excretion

Microbial N supply to the small intestine was estimated by the purine derivative excretion technique [12]. Daily urinary output was collected in about 50 mL of 1 M H_2SO_4 to prevent ammonia-N loss (final urine pH < 3), and every 24 h, urine collection was diluted with tap water (5:1) (to prevent precipitation of uric acid during storage), filtered through glass wool and sampled. Urine samples were stored at –20 °C before analysis of purine derivatives and total N.

The amount of microbial purines absorbed was estimated from equation below [12].

$$Y = 0.84X + (0.15W^{0.75}\, e^{-0.25x})$$

where Y is the total (mmol/d) urinary excretion of purine derivatives; X is the exogenous absorbed purines (mmol/d); 0.84 is the proportion of purine derivatives excreted in the urine; 0.15 is the endogenous purine derivative excretion (mmol/d); $W^{0.75}$ is the metabolic BW and 0.25 is the rate constant for the replacement of de novo synthesis of endogenous purines by exogenous purines. Daily supply of microbial N to the small intestine was estimated following the model described by Chen et al. [13], namely:

Microbial N supply (g/d) = 70X/(0.83 × 0.116 × 1000) = 0.727X, where 70 is the N content of purines (mg/mmol), 0.83 is the digestibility of microbial purines and 0.116 is the proportion of microbial N as purine N.

2.6. Rumen pH, Volatile Fatty Acid and Ammonia Concentrations

Ruminal liquor was withdrawn from the rumen of each sheep using a vacuum pump at 0, 3, 6 and 9 h post-prandial for pH, VFA and ammonia determination. Rumen pH was measured in rumen liquor after withdrawal with a portable potentiometer (Cole Parmer, Vernon Hills, IL, USA) previously calibrated with buffers of pH 4 and 7. Immediately after pH determination, a sample (20 mL) of rumen liquor was filtered through four layers of cheesecloth and then treated with 20 mL of an HCl solution (0.2 M) and kept at 4 °C until analyzed for ammonia-N concentration. Concentration of ammonia (NH_3-N) in rumen liquor was determined with a specific ion electrode (Corning Ammonia Electrode Cat. No. 476130) as suggested by Galyean and Chabot [14]. A sample (4 mL) of rumen liquor was taken after filtering, 1 mL of metaphosphoric acid solution was added, and it was kept at 4 °C until analyzed for VFA molar proportions by gas chromatography (Hewlett-Packard 5890; equipped with a flame ionization detector; the column was HP—FFAP 30 m × 0.53 mm).

2.7. Rate of Passage of Liquids and Solids

Kinetics of liquid flow was estimated by administering polyethylenglycol (PEG, molecular weight 4000; Merck, Darmstadt, Germany) into the rumen. Twelve grams of PEG were dissolved in distilled water (50 mL) and infused through the rumen cannula; samples of rumen liquor were then taken at 0, 2, 4, 8, 12, 18, 24, 36 and 48 h after dosing. PEG was assayed by the turbidimetric technique [15]. Regression techniques (marker concentration vs. time) were employed to estimate the volume, outflow rate and turnover rate of liquids from the rumen of each sheep. The rate of passage of solid digesta was estimated by the Cr-mordanting technique [16]. Forty grams of Cr-mordanted fiber were introduced through the rumen cannula, and fecal samples were taken directly from the rectum at 0, 6, 8, 10, 12, 16, 24, 36, 48, 72, 96 and 120 h after dosing to describe the excretion curve of the marker. The two-compartment model [17] was used to estimate kinetics of passage through the gastrointestinal tract:

$$Y = A\, e^{k1(t-TT)} - A\, e^{k2(t-TT)}$$

where k_1 is the rate of passage through the rumen; k_2 is the rate of passage through the caecum and proximal colon and TT is the transit time through the gastrointestinal tract.

2.8. Chemical Analysis

Ration samples were collected daily during each of the four experimental periods and pooled across days within period for assay of DM, OM and N [18]. NDF and ADF were determined as described by Goering and Van Soest [19]. Urinary allantoin was determined by the Rimini–Schryver reaction [20] and uric acid by the technique described by Guerci [21], following the method of Caraway. Xanthine and hypoxanthine were enzymatically converted to uric acid by treating the urine with xanthine oxidase (EC 1.2.3.2) and added to the amount of uric acid [22]. Metabolizable energy concentration of the rations was estimated from the digestible organic matter in dry matter (DOMD %) based on the equation described by MAFF [23]: ME = 0.15 × DOMD %.

2.9. Statistical Analysis

Data were analyzed with a 4 × 4 (period x animal) Latin square design using the GLM procedure of SAS [24]. The statistical model was:

$$Y_{ijk} = \mu + \rho_i + \mathbf{A}_j + \mathbf{T}_k + \varepsilon_{ijk}$$

where Y_{ijk} = Analysed variable; μ = General mean; ρ_i = Period (i = 1, 2,... k); A_j = Animal (j = 1, 2, ... k); T_k = Treatment (k = 1, 2, 3, 4); ε_{ijk} = Standard error.

Polynomial orthogonal contrasts were employed to analyze the linear, quadratic and cubic components of the response to the increasing levels of *Brosimum alicastrum* foliage in the rations. Data on *Brosimum alicastrum* rumen degradation were not statistically analyzed; just the means were taken.

3. Results

B. alicastrum foliage was higher in CP content, similar in DM and OM content and lower in NDF, ADF, hemicellulose and lignin content than *Megathyrsus maximus* hay (Table 1). Dry matter and OM contents of the rations were similar. The crude protein content in the rations increased, while the cell wall components decreased with graded levels of *B. alicastrum* in the rations (Table 1).

Organic matter and CP of *B. alicastrum* were rapidly and extensively degraded in the rumen (Table 2). The dry matter, OM, NDF and ME intakes increased linearly ($p < 0.01$) as the level of *B. alicastrum* in the ration was increased (Table 3). The incorporation of *B. alicastrum* did not significantly affect the potential degradation of DM or OM in the rumen of the basal ration of low-quality *Megathyrsus maximus* hay (Table 4) but increased the degradation rate of NDF ($p < 0.05$). the effective degradability of DM, OM and NDF were linearly decreased ($p < 0.01$) as the level of incorporation of *B. alicastrum* into the ration was increased. The rate of passage of solid digesta through the rumen (k_1) and through the caecum and proximal colon (k_2) increased linearly ($p < 0.05$), whereas the transit time (TT) through the gastrointestinal tract diminished ($p < 0.05$) as the level of incorporation of *B. alicastrum* into the ration was augmented (Table 5). The incorporation of *B. alicastrum* foliage also led to a significant ($p < 0.01$) linear increase in the turnover rate and in the outflow rate of liquids from the rumen (Table 5). No significant differences between treatments were detected for rumen volume.

Table 2. Rumen degradation (%) of DM, OM, CP and NDF of *Brosimum alicastrum* foliage in two Zebu bulls (Values are means of feed component degradation from bags incubated in the rumen in triplicate in each bull).

Item	DM	OM	CP	NDF
A	21.2	20.5	33.8	ND
B	65.7	66.1	61.4	ND
a	27.3	23.4	35.5	0.23
b	59.5	63.2	59.7	71.9
PD	86.8	86.6	95.2	72.1
c, h^{-1}	0.11	0.11	0.12	0.09

DM, Dry Matter; OM, Organic Matter; CP, Crude Protein; NDF, Neutral Detergent Fiber; n = 4; ND = Not Determined; $p = a + b\,(1 - e^{-ct})$ [10], where PD = $(a + b)$ is potential degradability; a is the intercept of the curve; b is the asymptote of equation; c is the fractional degradation rate of b and t is time of incubation; A is zero time washing loss; B is the insoluble but degradable fraction = $(a + b) - (A)$.

The concentration of ammonia-N in the rumen was significantly ($p < 0.05$) increased by the incorporation of *B. alicastrum* foliage into the ration (Table 6). The rumen ammonia-N concentration decreased gradually postprandially across treatments (Table 6). The rumen pH was kept within normal limits for cellulolysis (6.4–6.9). There were no significant effects ($p > 0.05$) of supplementation on the VFA concentration at different times of sampling; for this reason, data were not presented. However, a significant effect was found in the VFA average concentration ($p < 0.05$), but molar proportions were not altered ($p > 0.05$).

Table 3. Voluntary intake of rations containing increasing levels of *Brosimum alicastrum* foliage by Pelibuey sheep.

Item	Level of *B. alicastrum* (% DM)				SEM	Response [a]
	0	15	30	45		
DM intake g/d						
Megathyrsus	463	653	701	654	68.1	L **
Brosimum	0.00	115	301	535	26.7	L **
Molasses	48.0	79.7	104	123	-	-
Total (g/day)	511	848	1106	1313	106	L **
DM g/kg$^{0.75}$	35.1	54.7	70.2	84.4	5.80	L ***
OM g/d	464	758	1032	1191	80.0	L ***
OM g/kg$^{0.75}$	31.9	48.8	65.7	76.5	4.1	L **
NDF g/d	364	552	675	771	55.0	L **
NDF g/kg$^{0.75}$	25.1	35.6	42.9	49.7	2.9	L ***
Water consumption	2.58	2.71	4.98	4.96	0.65	L *

[a] L = Linear response; * $p < 0.05$; ** $p < 0.01$; *** $p < 0.001$. DM, Dry Matter; OM, Organic Matter; NDF, Neutral Detergent Fiber; ME, Metabolic Energy; MJ, Megajoules; BW, Body Weight; SEM, Standard Error of the Mean.

Table 4. Rumen degradation constants (%) of DM and OM of *Megathyrsus maximus* incubated in Pelibuey sheep fed increasing levels of *Brosimum alicastrum* foliage.

Item	Level of *B. alicastrum* (% DM)				SEM	Response [a]
	0	15	30	45		
			Dry matter			
A	10.9					ND
a	5.09	4.36	4.66	1.69		ND
b	37.2	36.7	36.9	38.3	0.60	NS
PD	42.3	41.1	41.6	40.0	0.60	NS
c h^{-1}	0.029	0.035	0.036	0.044	0.40	NS
Effective degradability	29.8	26.9	26.7	24.6	0.70	L **
			Organic matter			
a	2.31	3.1	1.52	0.76		ND
b	39.2	37.5	38.3	37.8	0.65	NS
PD	41.5	40.6	39.9	38.6	0.80	NS
c, h^{-1}	0.029	0.034	0.035	0.039	0.40	NS
Effective degradability	27.9	24.8	24.2	22.5	0.80	L **
			NDF			
a	4.10	3.60	3.80	2.10	NS	0.13
b	32.5	33.2	32.7	32.3	NS	1.4
PD	36.6	36.8	36.5	34.3	NS	1.3
c, h^{-1}	0.032	0.033	0.032	0.048	L *	0.003
Effective degradability	23.9	20.9	20.3	17.5	L **	1

[a] NS, Not Significant; ND, Not Determined; NDF, Neutral Detergent Fiber; DM, Dry Matter; OM, Organic Matter; SEM, Standard Error of the Mean; L, linear; * $p < 0.05$. ** $p < 0.01$. $p = a + b\,(1 - e^{-ct})$ [10], where PD = $(a + b)$ is potential degradability; a is the intercept of the curve; b is the asymptote of equation; c is the fractional degradation rate of b and t is time of incubation; A is zero time washing loss; B is the insoluble but degradable fraction = $(a + b) - (A)$.

Urinary excretion of allantoin, uric acid, xanthine and hypoxanthine (Table 7) increased linearly as the level of *B. alicastrum* in the ration was increased. Consequently, the supply of microbial N to the small intestine increased linearly ($p < 0.001$) as the level of *B. alicastrum* in the ration was increased (Table 7). Nitrogen intake, N excretion in feces and urine, as well as the N retention, increased linearly ($p < 0.001$) with the incorporation of *B. alicastrum* foliage. The apparent in vivo digestibility of DM, OM and NDF increased linearly with graded levels of *B. alicastrum* in the ration (Table 8).

Table 5. Kinetics of solid digesta and liquids through the gastrointestinal tract of Pelibuey sheep fed increasing levels of *Brosimum alicastrum* foliage.

Item	Level of *B. alicastrum* (% DM)					
	0	15	30	45	SEM	Response [a]
Solid digesta						
k_1 (per h)	0.0147	0.0275	0.028	0.0412	0.003	L ***
k_2 (per h)	0.0385	0.0725	0.063	0.076	0.009	L *
TT (hours)	3.84	3.54	2.99	2.63	0.34	L *
Liquids						
Rumen volume (L)	10.9	9.08	10.5	12.6	1.08	NS
Outflow rate (L/h)	0.98	1.03	1.52	2.51	0.25	L **
Turnover rate (times/d)	2.15	2.73	3.49	4.78	0.50	L **

[a] L, Linear response; NS, Not Significant; DM, Dry Matter; SEM, Standard Error of the Mean; * $p < 0.05$; ** $p < 0.01$; *** $p < 0.001$. k_1, Rate of passage of solids through the rumen (%/h); k_2, Rate of passage of solids through the caecum and proximal colon (%/h); TT, Transit time in h through the gastrointestinal tract.

Table 6. Concentration of ammonia-N (mg/100 mL rumen liquor) and VFAs in the rumen of Pelibuey sheep fed increasing levels of *Brosimum alicastrum* foliage.

Time after Feeding (h)	Level of *B. alicastrum* (% DM) 0	15	30	45	SEM	Response [a]
pH	6.69	6.55	6.52	6.61	0.07	NS
N-NH_3 (mg/100 mL)						
0	8.6	11.7	13.5	18.5	1.9	L *
3	7.56	11.7	12.7	17.6	1	L ***
6	5.43	9.82	11.4	15.8	1.2	L **
9	4.76	8.32	11.5	16.6	2	L **
Mean	6.58	10.38	11.27	17.12		
VFAs (mmol/L)	99.7	116.6	118.9	133.7	7.3	L *
VFAs mol/100 mmol)						
Acetic acid	74.8	73.1	73.4	72.2	1.4	NS
Propionic acid	16.4	17.7	16.6	17.3	0.5	NS
Butyric acid	7	7.4	7.9	8.6	0.6	NS
Others [b]	1.7	1.8	1.9	1.8	0.31	NS

[a] L, Linear response; * $p < 0.05$; ** $p < 0.01$; *** $p < 0.001$; [b] sum of isobutyrate, valerate, isovalerate. VFA, Volatile Fat Acids; N-NH^3, Ammoniacal Nitrogen; NS, Not Significant; DM, Dry Matter; SEM, Standard Error of the Mean.

Table 7. Urinary excretion of purine derivatives and microbial N supply to the small intestine and nitrogen balance in Pelibuey sheep fed increasing levels of *Brosimum alicastrum* foliage.

Item	Level of *B. alicastrum* (% DM)				SEM	Response [a]
	0	15	30	45		
Allantoin (mmol/d)	2.29	4.25	6.42	7.24	0.76	L **
Uric acid (mmol/d)	1.14	1.72	2.66	3.65	0.26	L ***
Xanthine + Hypoxanthine (mmol/d)	0.2	0.26	0.41	0.45	0.07	L *
Total (mmol/d)	3.63	6.23	9.5	11.34	0.91	L ***
MNSSI (g/d)	2.2	4.92	7.93	9.72	0.89	L ***
N intake (g/d)	4.7	10.1	16	21.1	0.9	L **
N excretion (g/d)						
Fecal	3.3	5.1	6.5	9.2	0.6	L **
Urinary	3.1	4.3	5.9	7.3	0.55	L **
Total (g/d)	6.4	9.4	12.4	16.5	0.57	L **
N retention (g/d)	-1.7	0.6	3.5	4.6	0.7	L ***

[a] L, Linear response; SEM, Standard Error of the Mean; * $p < 0.05$; ** $p < 0.01$; *** $p < 0.001$; MNSSI, Microbial N supply to the small intestine.

Table 8. Apparent digestibility (%) of DM, OM and NDF in Pelibuey sheep fed increasing levels of foliage of *Brosimum alicastrum* tree.

Item	Level of *B. alicastrum* (% DM)					
	0	15	30	45	SEM	Response [a]
DM	35.7	45.1	46.9	49.6	1.7	L **
OM	40.5	49.2	49.4	52	3.5	NS
NDF	33.2	44.4	45.9	47.9	3.1	L *

[a] L, linear response; NS, Not Significant; SEM, Standard Error of the Mean; DM, Dry Matter; OM, Organic Matter; NDF, Neutral Detergent Fiber; * $p < 0.05$; ** $p < 0.01$.

4. Discussion

The positive effect of the incorporation of foliage of *B. alicastrum* into the ration on DM and OM intake could be explained despite the increase in the intake of degradable N in the rumen [25–27]. It is generally recognized that the concentration of NDF in forages is the main factor limiting dry matter intake [28,29]. In this study, the incorporation of foliage of *B. alicastrum* not only reduced the NDF content in the ration, but it also increased its rumen degradation rate.

Foliage of *Brosiumum alicastrum* increased the total DM and DM intake of the basal ration by (98%) and (52%), respectively, at the level of 30% incorporation; these values are higher than those obtained with other foliages, such as *Sesbania sesban*, *Chamaecytisus palmensis*, *Acacia agustiniana* and *Leucaena leucocephala* [30–32]. Another possible explanation for the high DM intake induced by *B. alicastrum* could be related to the activity and content of phenolic compounds; this forage had 2.4% of extractable phenolics, with a protein precipitating capacity (PPC) of 16 µg/mg of bovine albumin serum, values much lower compared with the 3.5% and PPC of 146 µg/mg in *Acacia pennatula* of extractable phenolics and PPC, respectively. The content of phenolic compounds and their activity may have had an influence on N utilization and OM intake [26].

The high rumen fermentation of OM, CP and NDF of *B. alicastrum* were similar to those found for *Sesbania sesban* by Kamatali et al. [33]. Hovell et al. [34] have shown that the high ruminal degradation of forages is generally associated with a low NDF content, as found in foliage of *Brosimum alicastrum* (37.5%). In addition, the high rate and extent of degradation of *B. alicastrum* in the rumen (Table 2) could had a positive effect on feed intake. Rumen degradation of forage has been positively associated with feed intake [25,35,36]. The particulate passage rate was increased with the inclusion of *B. alicastrum* in the ration, and this effect has been associated with intake [30,37,38]. The increase in the NDF and OM degradation rate of *Megathyrsus maximus*, as well as the high rumen degradation rate of *B. alicastrum*, could explain the positive effect on *Megathyrsus maximus* intake. Species with high ruminal degradation promote less substitutive effects [30]. On the other hand, in other species, there is little effect on intake of the basal ration, or the effect is substitutive [30–32]. The increase in OM intake of *Megathyrsus maximus* could have considerable practical application in the tropics, where it is important to obtain the highest efficiency of utilization of low-quality forage.

Metabolizable energy intake was 0.4, 0.93, 1.4 and 1.7 times the maintenance requirements for hair sheep compared with the ME requirement reported by Kawas and Huston [39] in hair sheep. This increase in ME intake may be associated to the higher DOM intake.

Although potential DM degradability was not affected by the incorporation of foliage of *B. alicastrum* into the ration, the degradation rate of DM, OM and NDF were increased. Trends for an increase in DM degradation rate when tree fodders are used as supplements with low-quality forages have been reported [40,41], probably associated with a high ammonia-N concentration in rumen liquor [30,40]. Bonsi et al. [31] reported the highest degradation rate, with high levels of ammonia-N in the rumen. Even when the levels of ammonia-N in all treatments were higher than those recommended by Satter and Slyter [42] and Boniface et al. [43], for maximum growth in vitro and maximum degradation in the rumen, respectively, the ration without the incorporation of *B. alicastrum* (control), with a high concentration of ammonia-N, showed a lower degradation rate. The control ration without the incorporation of *B. alicastrum* foliage was most likely deficient in rumen degradable protein, as evidenced by the low supply of microbial N to the small intestine. This result suggests that the amount of nitrogen ingested and the N:ME ratio may have influenced rumen degradation. An important role may have been played by the N:ME ratio in the degradation rate; estimations carried out following the method used by Umunna et al. [30], with N and OM slowly degradable in the rumen, gave values of 1.45 and 1.54 g of N per MJ ME, which is closer to the 1.34 g value given by ARC [44]. The reduction observed in the effective degradation of DM, OM and NDF of *Megathyrsus maximus* may be explained by the increase in the passage rate through the rumen, which leave less time available for microbial attack, since effective degradability decreases with an increase in the rate of passage from the rumen [45].

The incorporation of foliage of *B. alicastrum* into the ration augmented the ammonia-N concentration in the rumen liquor, and this may have increased the particulate passage rate [46]. On the other hand, microbial attack, chewing and rumination are responsible for the reduction in the particulate size of forage, as it is known that only particles with the critical size can pass through the rumen and into the omasum [47], as well as those particles with higher specific gravity [48]—chewing and rumination promote changes in the physical structure of forage, increasing their specific gravity [28] due to the increase in their capacity for hydration [49]. The increase observed in the ammonia-N concentration in the rumen with the incorporation of *Brosimum alicastrum* was similarly recorded in Pelibuey sheep by Alayón et al. [25], when they incorporated foliage of *Gliricidia sepium* in a basal ration of *Cynodon nlemfuensis*. A linear increase in the ammonia-N concentration was reported by Bonsi et al. [40], using *Sesbania sesban* as a supplement in sheep. In the experiment hereby described, all treatments showed values for ammonia-N higher than those suggested by Satter and Slyter [42] for maximum microbial growth under in vitro conditions and those reported by Boniface et al. [43] for maximum fermentation of OM in tropical grasses. The increase in rumen ammonia-N concentration may be due to the sgreater N intake and to the higher rumen degradation of N in *Brosimum alicastrum*.

A large volume of liquid in the rumen increases the dilution rate and rumen outflow in sheep [50]. The results of the present experiment showed that the dilution rate had a positive relationship with water consumption ($r^2 = 0.77$). Furthermore, the outflow rate and turnover rate were positively related to DM intake ($r^2 = 0.75$ and $r^2 = 0.83$, respectively). Rations that promote high DM intake induce a high outflow rate compared to those that result in a low DM intake [31,40]. Bonsi et al. [40] found a significant effect on the kinetics of liquids in sheep supplemented with *Sesbania sesban*, which has a similar rumen degradation profile compared to foliage of *B. alicastrum*.

The incorporation of foliage of *B. alicastrum* into the ration increased the microbial nitrogen supply (MNS; g/day). Bonsi et al. [40] and Karda and Dryden [51] also reported a greater MNS (g/day) and efficiency in small ruminants supplemented with multipurpose trees. Likewise, Valdivia et al. [52] reported a higher MNS in cows in the silvopastoral system with *Leucaena leucocephala* compared with those grazing *Megathyrsus maximus* grass only. The microbial nitrogen supply to the small intestine showed a positive relation to the particulate passage rate ($r^2 = 0.87$) and outflow rate ($r^2 = 0.82$). Microbial protein synthesis has been related to the particulate passage rate [53], as well as the outflow rate [50,54]. Umunna et al. [55] related improvements in MNS to an increase in OM intake; in this study, MNS was related positively correlated with OM intake ($r^2 = 0.89$). The increase in MNS (g/day) may be due to a better N:energy relationship in the rumen, promoted by feeding foliage of *B. alicastrum*. The rations' soluble N:ME in the OM, estimated following the method suggested by Umunna et al. [30] were: 1.74, 1.95, 2.10 g and 2.2 g N per MJ ME for 0%, 15%, 30% and 45% levels of *B. alicastrum*. The proportions of slowly degradable N to slowly fermentable ME (g N/MJ) were: 0.68, 1.12, 1.45 and 1.54 for 0%, 15%, 30% and 45% of *B. alicastrum*, respectively. Compared with 1.34. g N per MJ ME reported by ARC [44], this suggest that ME was deficient in relation to fermentable N, in the control and 15% *B. alicastrum* rations. However, the rations slowly degradable N:OM with 0% and 15% of *B. alicastrum* had good levels of energy but not enough N; this may affect MNS. In diets with 30% and 45% of *B. alicastrum*, there was a good balance of N:ME; the small differences in the efficiency of MNS with these rations may be explained by this. Although ammonia-N concentration increased with incorporation of *B. alicastrum* in the ration and the values were higher than those suggested by Satter and Slyter [42] for maximal microbial growth, it was poorly related to MNS (r^2 = 0.30). The values of the efficiency of MNS found in the rations supplemented with 30% and 45% *B. alicastrum* were similar to those found by Bonsi et al. [40] with *Sesbania sesban* and higher than values reported by Masama et al. [56] and Hindrichsen et al. [57] with *Leucaena leucocephala*. The differences with leucaena could be due to a lower rumen degradation and slower degradation rate of N in Leucaena. This may be influenced by phenolic compounds—low fermentation of N affects microbial nitrogen synthesis [55]. *Brosiumum alicastrum* has extractable phenolics with low PPC; for this reason, probably, *B. alicastrum* increased MNS and the efficiency even when it had a lower N content than that in the foliage of others tree species.

The better N balance in the rations incorporating foliage of *Brosimum alicastrum* may be due to the increased N intake and improvement in the efficiency of MNS. A better efficiency of MNS, along with an increase in N retention in the current study, are in agreement with the findings of Bonsi et al. [31], Masama et al. [56] and Karda and Dryden [51], which were drawn from the use of tree foliage as supplements in tropical rations. The nitrogen intake increased by 3% in the ration with 45% *Brosimum alicastrum* compared with the ration with 30%. In the same proportion, there was an increase in N retention, probably due to the good balance of N:ME, which improved the efficiency of MNS. This agrees with the results of this experiment, where N retention has been improved using multipurpose trees as a supplement in ruminant rations [31,41,57].

The linear increase in the VFA concentrations in the rumen liquor with the incorporation of foliage of *B. alicastrum* into the ration may be associated with the high degradation rate of *B. alicastrum* and the increase in the degradation rate of *M. maximus*. Concentrations of VFAs found in this study were higher than the values reported by Hindrichsen et al. [57] and Muinga et al. [41] when they used foliage of *Leucaena leucocephala* as a supplement. These differences might be related to the high rumen degradation of *B. alicastrum* compared with Leucaena and to the increase in the degradation rate of the DM, OM and NDF of *M. maximus*. The absence of changes in the rumen fermentation pattern with supplementation with *B. alicastrum* in this experiment are consistent with reports by Hindrichsen et al. [57] and Muinga et al. [41].

The linear increase in the outflow and turnover rates recorded with the incorporation of *B. alicastrum* into the ration could possibly be explained by the increase in water consumption (Table 2). This increase in water consumption was probably ws associated with changes in osmotic pressure of the rumen liquid. Volatile fatty acids [50] and N-NH_3 [46] contribute to increased osmotic pressure in the rumen. At high rumen osmotic pressure, ruminants increase water consumption [58]. In this experiment, the VFA and N-NH_3 concentrations in the rumen liquid were increased, as well as water consumption. High quantities of liquid in the rumen increase the dilution rate and rumen outflow in sheep [50], and these results show that the dilution rate had a positive relationship with water consumption (r^2 = 0.77). Furthermore, the outflow rate and turnover rate were positively related to DM intake r^2 = 0.75; r^2 = 0.83, respectively. Bonsi et al. [40] found a significant effect on the kinetics of liquid in sheep supplemented with forage of *Sesbania sesban*, which has a similar rumen degradation to *B. alicastrum*; however, differences in rumen volume were not observed.

The improvement in the apparent digestibility of DM, NDF and OM as foliage of *B. alicastrum* was increased in the ration may be caused by a reduction in NDF content of the whole ration, as NDF has a negative effect on apparent digestibility [35]. Effective rumen degradability of the rations could have been increased by the incorporation of *Brosimum alicastrum*, considering the proportion of each ingredient in the ration and their effective rumen degradation. Assuming that the molasses was fermented completely in the rumen and taking into consideration the values for k_1, the effective rumen degradability was 36.4%, 39.21%, 44.38% and 45.75% for 0%, 15%, 30% and 45% of *B. alicastrum*. This may help to explain why digestibility tended to increase; the small difference among the rations with 30% and 45% *Brosimum alicastrum* was closer to the estimated values of effective rumen degradation. The high rumen degradation of foliage of *Brosimum alicastrum* (Table 2) could be contributing to better apparent digestibility of the ration. Bonsi et al. [40] found a greater effect in digestibility with *Sesbania sesban* than with *Leucaena leucocephala*, attributing this effect to the higher rumen degradation of sesbania. Furthermore, the increase in the rumen degradation rate of NDF, DM and OM of *M. maximus* (Table 4) could have influenced apparent digestibility. Orskov et al. [9] observed a positive relation between apparent digestibility and rumen degradation rate.

5. Conclusions

Results from this experiment demonstrate that the foliage of *Brosimum alicastrum* is a good source of nutrients for Pelibuey hair sheep, increasing dry matter intake, rumen ammonia-N concentration, outflow from the rumen, efficiency of microbial nitrogen synthesis and microbial nitrogen supply to the small intestine when incorporated into low-quality basal hay rations of tropical grass and may therefore be considered as a valuable feed resource for the development of sustainable systems of sheep production in the tropical areas of Latin America.

Author Contributions: Conceptualization, V.V.-S., E.d.J.F.-S. and J.C.K.-V.; methodology, V.V.-S., E.d.J.F.-S. and J.C.K.-V.; validation, E.d.J.F.-S., J.C.K.-V. and V.V.-S.; formal analysis, V.V.-S. and J.C.S.-C.; investigation, V.V.-S.; resources, J.C.K.-V.; data curation, V.V.-S. and J.C.K.-V.; writing—original draft preparation, V.V.-S., E.d.J.F.-S., J.C.K.-V., J.M.C.-M., L.R.-A. and J.C.S.-C. writing—review and editing, E.d.J.F.-S., J.M.C.-M., V.V.-S. and J.C.K.-V.; visualization, E.d.J.F.-S., J.C.K.-V. and V.V.-S.; supervision, J.C.K.-V.; project administration, V.V.-S. and J.C.K.-V.; funding acquisition, V.V.-S. and J.C.K.-V. All authors have read and agreed to the published version of the manuscript.

Funding: This research received no external funding.

Institutional Review Board Statement: The animal protocol used in this study was approved by the University of Yucatan Institutional Animal Care and Use Committee guidelines (UADY), protocol no. CB-CCBA-D-2021-001.

Informed Consent Statement: Not applicable.

Data Availability Statement: The data presented in this study are available on request from the corresponding author. The data are not publicly available due to the institutional policies of the University of Yucatan.

Acknowledgments: The first author thanks Secretaria de Relaciones Exteriores (SRE) of Mexico for supporting his MSc scholarship at the Faculty of Veterinary Medicine and Animal Science of the University of Yucatan, Mexico.

Conflicts of Interest: The authors declare no conflicts of interest.

References

1. National Research Council (NRC). *Underexploited Tropical Plants with Promising Economic Value*; The National Academies Press: Washington, DC, USA, 1975.
2. Ayala, A.; Bird, M. *Brosimum alicastrum*—The riches of Ramon trees in Mexico. *Agrofor. Today* **1997**, *9*, 26–27.
3. Morell-Hart, S.; Dussol, L.; Fedick, S.L. Agriculture in the ancient Maya lowlands (Part 1): Paleoethnobotanical residues and new perspectives on plant management. *J. Archaeol. Res.* **2022**, *31*, 561–615. [CrossRef]
4. Ayala, A.; Sandoval, S.M. Establecimiento y producción temprana de forraje de Ramón (*Brosimum alicastrum* Swartz) en plantaciones a altas densidades en el norte de Yucatán, México. *Agrofor. Am.* **1995**, *7*, 10–16. Available online: https://www.fao.org/3/x6305s/x6305s.pdf (accessed on 1 October 2023).
5. Medina-Figueras, A. Coeficientes de Digestibilidad del Ramón. Bachelor's Thesis, Universidad Nacional Autónoma de México, Mexico City, Mexico, 1949.

6. Yerena, F.; Ferreiro, H.M.; Elliot, R.; Godoy, R.; Preston, T.R. Digestibilidad de Ramón (*Brosimum alicastrum*), *Leucaena leucocephala*, pasto Buffel (*Cenchrus ciliaris*) y pulpa y bagazo de Henequén (*Agave fourcroydes*). *Prod. Anim. Trop.* **1978**, *3*, 70–73.
7. Sarmiento-Franco, L.; Montfort-Grajales, S.; Sandoval-Castro, C. La semilla del árbol Ramón (*Brosimum alicastrum* Swartz): Alternativa alimentaria energética para animales de producción y seres humanos. *Bioagrociencias* **2022**, *15*, 19–27. [CrossRef]
8. Soberanis-Poot, M.; Torres-Lugo, R.; Solorio-Sánchez, F.J. *Brosimum alicastrum:* Componente clave en la producción animal y adaptación al cambio climático. *Rev. Árbol Ramón Óox Proy. Larqué* **2022**, *1*, 12–14.
9. Orskov, E.R.; Hovell, F.D.D.B.; Mould, F. Uso de la técnica de la bolsa de nylon para la evaluación de los alimentos. *Prod. Anim. Trop.* **1980**, *5*, 213–233.
10. Orskov, E.R.; McDonald, I. The estimation of protein degradability in the rumen from incubation measurements weighted according to rate of passage. *J. Agric. Sci.* **1979**, *92*, 499–503. [CrossRef]
11. López, S.; Hovell, F.D.; McLeod, A. Osmotic pressure, water kinetics and volatile fatty acid absorption in the rumen of sheep sustained by intragastric infusions. *Br. J. Nutr* **1994**, *71*, 153–168. [CrossRef]
12. Chen, X.B.; Hovell, F.D.D.; Orskov, E.R.; Brown, D.S. Excretion of purine derivatives by ruminants: Effect of exogenous nucleic acid supply on purine derivative excretion by sheep. *Br. J. Nutr.* **1990**, *63*, 131–142. [CrossRef]
13. Chen, X.B.; Chen, Y.K.; Franklin, M.F.; Orskov, E.R.; Shand, W.J. The effect of feed intake and body weight on purine derivative excretion and microbial protein supply in sheep. *J. Anim. Sci.* **1992**, *70*, 1534–1542. [CrossRef]
14. Galyean, M.L.; Chabot, R.C. Effects of sodium bentonite, buffer salts, cement kiln dust, and clinoptilolite on rumen characteristics of beef steers fed a high roughage diet. *J. Anim. Sci.* **1981**, *52*, 1197–1204. [CrossRef]
15. Malawer, S.J.; Powell, D.W. An improved turbidimetric analysis of polyethylene glycol using an emulsifier. *Gastroenterology* **1967**, *53*, 250–256. [CrossRef]
16. Uden, P.; Colucci, P.E.; Van-Soest, P.J. Investigation of chromium, cerium and cobalts as markers in digesta. Rate of passage studies. *J. Sci. Food Agric.* **1980**, *31*, 625–632. [CrossRef] [PubMed]
17. Grovum, W.L.; Williams, V.J. Rate of passage of digesta in sheep. 4. Passage of marker through the alimentary tract and the biological relevance of rate-constants derived from the changes in concentration of marker in faeces. *Br. J. Nutr.* **1973**, *30*, 313–329. [CrossRef] [PubMed]
18. AOAC. *Official Methods of Analysis*, 15th ed.; Association of Official Analytical Chemist: Washington, DC, USA, 1990.
19. Goering, H.K.; Van Soest, P.J. *Forage Fiber Analyses: Apparatus, Reagent, Procedures and some Applications*; Handbook No. 379; US Department of Agriculture: Washington, DC, USA, 1970; 20p.
20. Young, E.G.; Conway, C.F. On the estimation of allantoin by the Rimini-Schryver reaction. *J. Biol. Chem.* **1942**, *142*, 839–853. [CrossRef]
21. Guerci, A.A. *Laboratorio: Métodos de Análisis Clínicos y su Interpretación*; Librería El Ateneo: Buenos Aires, Argentina, 1975; 333p.
22. Fujihara, T.; Orskov, E.R.; Reeds, P.J.; Kyle, D.J. The effects of protein infusion on urinary excretion of purine derivatives in ruminants nourished by intragastric nutrition. *J. Agric. Sci. Camb.* **1987**, *109*, 7–12. [CrossRef]
23. Ministry of Agriculture, Fisheries and Food, MAFF. *Energy Allowances and Feeding Systems for Ruminants*; Technical Bulletin 33; Her Majesty's Stationery Service: London, UK, 1975; 79p. Available online: https://wellcomecollection.org/works/ey8cqebf (accessed on 1 November 2023).
24. SAS Institute. *SAS/STAT 9.4 User's Guide*; SAS Institute: Cary, NC, USA, 2012.
25. Alayón, J.A.; Ramírez-Avilés, L.; Ku-Vera, J.C. Intake, rumen digestion, digestibility and microbial nitrogen supply in sheep fed Cynodon nlemfuensis supplemented with Gliricidia sepium. *Agrofor. Syst.* **1998**, *41*, 115–126.
26. Nherera, F.V.; Ndlovu, L.R.; Dzowela, B.H. Utilization of *Leucaena diversifolia*, *Leucaena esculenta*, *Leucaena pallida* and Calliandra calothyrsusas nitrogen supplements for growing gotas fed Maite stover. *Anim. Feed Sci. Technol.* **1998**, *74*, 15–28. [CrossRef]
27. Alawa, J.P.; Fishwick, G.; Parkings, J.J.; Hemingway, R.G. A note on the effects of dietary protein degradability in the rumen on the voluntary intake and digestibility of barley straw by lactating beef cows. *Anim. Sci.* **1987**, *44*, 446–499. [CrossRef]
28. Van Soest, P.J. *Nutritional Ecology of the Ruminant*, 2nd ed.; Cornell University Press: Ithaca, NY, USA, 1994; Available online: http://www.jstor.org/stable/10.7591/j.ctv5rf668 (accessed on 1 November 2023).
29. Lippke, H. Regulation of voluntary intake of ryegrass and sorghum forages in cattle by indigestible neutral detergent fiber. *Anim. Sci.* **1986**, *63*, 1459–1468. [CrossRef]
30. Umunna, N.N.; Nsahlai, I.V.; Osuji, P.O. Degradability of forage protein supplements and their effects on the kinetics of digestion and passage rate. *Small. Rumin. Res.* **1995**, *17*, 145–152. [CrossRef]
31. Bonsi, M.L.K.; Tuah, A.K.; Osuji, P.O.; Nsahlai, V.I.; Umunna, N.N. The effect of protein supplement source or supply pattern on the intake, digestibility, rumen kinetics, nitrogen utilization and growth of Ethiopian Menz sheep fed teff straw. *Anim. Feed Sci. Technol.* **1996**, *64*, 11–25. [CrossRef]
32. Hove, L.; Topps, J.H.; Sibanda, S.; Ndlovu, L.R. Nutrient intake and utilization by goats fed dried leaves of the shrub legumes *Acacia angustissima*, *Calliandra calothyrsus* an *Leucaena leucocephala* as supplements to native pasture hay. *Anim. Feed Sci. Technol.* **2001**, *91*, 95–106. [CrossRef]
33. Kamatali, P.; Teller, E.; Vanbelle, M.; Collignon, G.; Foulon, M. *In situ* degradability of organic matter, crude protein and cell wall of various tree forages. *Anim. Prod.* **1992**, *55*, 29–34. [CrossRef]
34. Hovell, F.D.D.; Ngambi, J.W.W.; Barber, W.P.; Kyle, D.J. The voluntary intake of hay by sheep in relation to its degradability in the rumen as measured in nylon bags. *Anim. Prod.* **1986**, *42*, 111–118.

35. Carro, M.D.; Lopez, S.; Gonzalez, J.S.; Ovejero, F.J. The use of the rumen degradation characteristics of hay as predictors of its voluntary intake by sheep. *Anim. Prod.* **1991**, *52*, 133–139. [CrossRef]
36. Nandra, K.S.; Hendry, A.; Dobos, R.C. A study of voluntary intake and digestibility of roughage in relation to their degradation characteristics and retention time in the rumen. *Anim. Feed Sci. Technol.* **1993**, *43*, 227–237. [CrossRef]
37. Shem, M.N.; Orskov, E.R.; Kimambo, A.E. Prediction of voluntary intake, digestible dry-matter intake and growth rate of cattle from the degradation characteristics of tropical foods. *Anim. Sci.* **1995**, *60*, 65–74. [CrossRef]
38. Rees, M.C.; Minson, D.J. Intake, digestibility and rumen characteristics of sheep given grass fertilized with phosphorus. *Aust. J. Agric. Res.* **1982**, *33*, 629–636. [CrossRef]
39. Kawas, J.R.; Huston, J.E. Nutrient requirements of hair sheep in tropical and subtropical regions. In *Hair Sheep Production in Tropical and Subtropical Regions*; Shelton, M., Figueiredo, E.A.P., Eds.; University of California: Davis, CA, USA, 1990; pp. 37–58.
40. Bonsi, M.L.K.; Osuji, P.O.; Nsahlai, I.V.; Tuah, A.K. Grade levels of *Sesbania sesban* and *Leucaena leucocephala* as supplements of teff straw given to Ethiopian Menz sheep. *Anim. Prod.* **1994**, *59*, 235–244. [CrossRef]
41. Muinga, R.W.; Topps, J.H.; Rooke, J.A.; Thorpe, W. Effect of supplementation with *Leucaena leucocephala* and maize brand on voluntary foof intake, digestibility, live weigh and milk yield of *Bos indicus* x *Bos taurus* dairy cows and rumen fermentation in steers offered *Pennisetum purpureum ad libitum* in the semi-humid tropics. *Anim. Sci.* **1995**, *60*, 13–23. [CrossRef]
42. Satter, L.D.; Slyter, L.L. Effect of ammonia concentration on ruminal microbial protein production in vitro. *Br. J. Nutr.* **1974**, *32*, 199–208. [CrossRef] [PubMed]
43. Boniface, A.N.; Murray, R.M.; Hogan, J.P. Optimum levels of ammonia in the rumen liquor of cattle fed tropical pasture hay. *Proc. Aus. Soc. Anim. Prod.* **1986**, *16*, 151–154.
44. ARC—Agricultural Research Council. *The Nutrients Requirement of Ruminant Livestock*; Technical Review by an Agricultural Research Council Working Party; Common Wealth Agricultural Bureau: Farnham Royal, UK, 1990; pp. 12–13.
45. Mgheni, D.M.; Kimambo, A.E.; Sundstol, A.E.; Madsen, J. Influence of urea treatment or supplementation on degradation, intake and growth performance of goats fed rice straws diets. *Anim. Feed Sci. Technol.* **1993**, *40*, 209–220. [CrossRef]
46. Oosting, S.J.; Waanders, A. The effects of rumen ammonia nitrogen concentration on intake and digestion of wheat straw by goats. *Anim. Feed Sci. Technol.* **1993**, *43*, 31–40. [CrossRef]
47. Welch, J.G. Rumination, particle size and passage from the rumen. *J. Anim. Sci.* **1982**, *54*, 885–894. [CrossRef]
48. Welch, J.G. Physical parameters of fiber affecting passage from the rumen. *J. Dairy Sci.* **1986**, *69*, 2750–2754. [CrossRef]
49. Hooper, A.P.; Welch, J.G. Funtional specific gravity of ground hay samples in ionic solutions. *J. Dairy. Sci.* **1985**, *68*, 848–856. [CrossRef]
50. Harrison, D.G.; Beever, D.E.; Thomson, D.J.; Osbourn, D.F. Manipulation of rumen fermentation in sheep by increasing the rate of flow of water from the rumen. *J. Agric. Sci. Camb.* **1975**, *85*, 93–101. [CrossRef]
51. Karda, L.W.; Dryden, G. McL. Effect of leucaena supplementation level, and provision of urea on utilization by sheep of Rhodes grass hay. *Aust. J. Exp. Agric.* **2001**, *41*, 155–160. [CrossRef]
52. Valdivia, V.; Sandoval-Castro, C.A.; Otaduy, K.; Ku-Vera, J.C. Microbial-N supply and milk yield cows in a silvopastoral system with and without access to the forage tree and energy supplementation during the dry season. *Proc. Brit. Soc. Anim. Sci.* **2007**, 131. [CrossRef]
53. Djouvinov, D.S.; Todorov, N.A. Influence of dry matter intake and passage rate on microbial protein synthesis in the rumen of sheep and its estimation by canulation and non-invasive method. *Anim. Feed Sci. Technol.* **1994**, *48*, 289–304. [CrossRef]
54. Dewhurst, R.J.; Webster, A.J.F. A note on the effect of plane nutrition on fractional outflow rates from the rumen and urinary allantoin excretion by wether sheep. *Anim. Prod.* **1992**, *54*, 445–448. [CrossRef]
55. Umunna, N.N.; Osuji, P.O.; Nsahlai, I.V.; Khalili, H.; Mohamed-Saleem, M.A. Effect of supplementing oat hay with lablab, sesbania, tagasaste or wheat middlings on voluntary intake, N utilization and weight gain of Ethiopian Menz sheep. *Small Rumin. Res.* **1995**, *18*, 113–120. [CrossRef]
56. Masama, E.; Topps, J.H.; Ngongoni, N.T.; Maasdorp, B.V. Effects of supplementation with foliage from the tree legumes *Acacia angustissima*, *Cajanus cajan*, *Calliandra calothyrsus* and *Leucaena leucocephala* on feed intake, digestibility and nitrogen metabolism of sheep gives maize stover ad libitum. *Anim. Feed Sci. Technol.* **1997**, *69*, 233–240. [CrossRef]
57. Hindrichsen, I.K.; Osuji, P.O.; Odenyo, A.A.; Madsen, J.; Hevelplund, T. Effects of supplementation of basal diet of maize stover with different amounts of Leucaena diversifolia on intake, digestibility, nitrogen metabolism and rumen parameters in sheep. *Anim. Feed Sci. Technol.* **2002**, *98*, 131–142. [CrossRef]
58. López, S.; France, J.; Dhanoa, M.S. A correction for particulate matter loss when applying the polyester-bag method. *Br. J. Nutr.* **1994**, *71*, 135–137. [CrossRef]

MDPI AG
Grosspeteranlage 5
4052 Basel
Switzerland
Tel.: +41 61 683 77 34

Animals Editorial Office
E-mail: animals@mdpi.com
www.mdpi.com/journal/animals

www.ingramcontent.com/pod-product-compliance
Lightning Source LLC
Chambersburg PA
CBHW041345210926
51850CB00013B/2994
* 9 7 8 3 7 2 5 8 3 9 1 4 8 *